# Third Edition

# SOLID STATE CHEMISTRY

## An Introduction

# Third Edition

# SOLID STATE CHEMISTRY

## An Introduction

## Lesley E. Smart
## Elaine A. Moore

Taylor & Francis
Taylor & Francis Group

Boca Raton  London  New York  Singapore

A CRC title, part of the Taylor & Francis imprint, a member of the
Taylor & Francis Group, the academic division of T&F Informa plc.

Published in 2005 by
CRC Press
Taylor & Francis Group
6000 Broken Sound Parkway NW, Suite 300
Boca Raton, FL 33487-2742

QM LIBRARY
(MILE END)

International Standard Book Number-10: 0-7487-7516-1 (Hardcover)
International Standard Book Number-13: 978-0-7487-7516-3 (Hardcover)
Library of Congress Card Number 2004058533

---

### Library of Congress Cataloging-in-Publication Data

---

Smart, Lesley.
   Solid state chemistry : an introduction / Lesley Smart and Elaine Moore.--3rd ed.
      p. cm.
   Includes bibliographical references and index.
   ISBN 0-7487-7516-1 (alk. paper)
   1. Solid state chemistry. I. Moore, Elaine (Elaine A.) II. Title.

QD478.S53 2005
541'.0421--dc22                                          2004058533

---

Taylor & Francis Group
is the Academic Division of T&F Informa plc.

**Visit the Taylor & Francis Web site at
http://www.taylorandfrancis.com**

**and the CRC Press Web site at
http://www.crcpress.com**

*Dedicated to*

---

*Graham, Sam, Rosemary, and Laura*

# Preface to the Third Edition

Solid state and materials chemistry is a rapidly moving field, and the aim of this edition has been to bring the text as up to date as possible with new developments. A few changes of emphasis have been made along the way.

Single crystal X-ray diffraction has now been reduced in Chapter 2 to make way for a wider range of the physical techniques used to characterize solids, and the number of synthetic techniques has been expanded in Chapter 3. Chapter 5 now contains a section on fuel cells and electrochromic materials. In Chapter 6, the section on low-dimensional solids has been replaced with sections on conducting organic polymers, organic superconductors, and fullerenes. Chapter 7 now covers mesoporous solids and ALPOs, and Chapter 8 includes a section on photonics. Giant magnetoresistance (GMR) and colossal magnetoresistance (CMR) have been added to Chapter 9, and $p$-wave (triplet) superconductors to Chapter 10. Chapter 11 is new, and looks at the solid state chemical aspects of nanoscience.

We thank our readers for the positive feedback on first two editions and for the helpful advice which has led to this latest version.

As ever, we thank our friends in the Chemistry Department at the OU, who have been such a pleasure to work with over the years, and have made enterprises such as this possible.

# Preface to the Second Edition

We were very pleased to be asked to prepare a second edition of this book. When we tried to decide on the changes (apart from updating) to be made, the advice from our editor was "if it ain't broke, don't fix it." However, the results of a survey of our users requested about five new subjects but with the provisos that nothing was taken out, that the book didn't get much longer, and, above all, that it didn't increase in price! Therefore, what you see here is an attempt to do the impossible, and we hope that we have satisfied some, if not all, of the requests.

The main changes from the first edition are two new chapters: Chapter 2 on X-ray diffraction and Chapter 3 on preparative methods. A short discussion of symmetry elements has been included in Chapter 1. Other additions include an introduction to ALPOs and to clay minerals in Chapter 7 and to ferroelectrics in Chapter 9. We decided that there simply was not enough room to cover the Phase Rule properly and for that we refer you to the excellent standard physical chemistry texts, such as Atkins. We hope that the book now covers most of the basic undergraduate teaching material on solid state chemistry.

We are indebted to Professor Tony Cheetham for kindling our interest in this subject with his lectures at Oxford University and the beautifully illustrated articles that he and his collaborators have published over the years. Our thanks are also due to Dr. Paul Raithby for commenting on part of the manuscript.

As always, we thank our colleagues at the Open University for all their support and especially the members of the lunch club, who not only keep us sane, but also keep us laughing. Finally, thanks go to our families for putting up with us and particularly to our children for coping admirably with two increasingly distracted academic mothers — our book is dedicated to them.

**Lesley E. Smart and Elaine A. Moore**
Open University, Walton Hall, Milton Keynes

# Preface to the First Edition

The idea for this book originated with our involvement in an Open University inorganic chemistry course (S343: Inorganic Chemistry). When the Course Team met to decide the contents of this course, we felt that solid state chemistry had become an interesting and important area that must be included. It was also apparent that this area was playing a larger role in the undergraduate syllabus at many universities, due to the exciting new developments in the field.

Despite the growing importance of solid state chemistry, however, we found that there were few textbooks that tackled solid state theory from a chemist's rather than a physicist's viewpoint. Of those that did most, if not all, were aimed at final year undergraduates and postgraduates. We felt there was a need for a book written from a chemist's viewpoint that was accessible to undergraduates earlier in their degree programme. This book is an attempt to provide such a text.

Because a book of this size could not cover all topics in solid state chemistry, we have chosen to concentrate on structures and bonding in solids, and on the interplay between crystal and electronic structure in determining their properties. Examples of solid state devices are used throughout the book to show how the choice of a particular solid for a particular device is determined by the properties of that solid.

Chapter 1 is an introduction to crystal structures and the ionic model. It introduces many of the crystal structures that appear in later chapters and discusses the concepts of ionic radii and lattice energies. Ideas such as close-packed structures and tetrahedral and octahedral holes are covered here; these are used later to explain a number of solid state properties.

Chapter 2 introduces the band theory of solids. The main approach is via the tight binding model, seen as an extension of the molecular orbital theory familiar to chemists. Physicists more often develop the band model via the free electron theory, which is included here for completeness. This chapter also discusses electronic conductivity in solids and in particular properties and applications of semiconductors.

Chapter 3 discusses solids that are not perfect. The types of defect that occur and the way they are organized in solids forms the main subject matter. Defects lead to interesting and exploitable properties and several examples of this appear in this chapter, including photography and solid state batteries.

The remaining chapters each deal with a property or a special class of solid. Chapter 4 covers low-dimensional solids, the properties of which are not isotropic. Chapter 5 deals with zeolites, an interesting class of compounds used extensively in industry (as catalysts, for example), the properties of which strongly reflect their structure. Chapter 6 deals with optical properties and Chapter 7 with magnetic properties of solids. Finally, Chapter 8 explores the exciting field of superconductors, particularly the relatively recently discovered high temperature superconductors.

The approach adopted is deliberately nonmathematical, and assumes only the chemical ideas that a first-year undergraduate would have. For example, differential calculus is used on only one or two pages and non-familiarity with this would not hamper an understanding of the rest of the book; topics such as ligand field theory are not assumed.

As this book originated with an Open University text, it is only right that we should acknowledge the help and support of our colleagues on the Course Team, in particular Dr. David Johnson and Dr. Kiki Warr. We are also grateful to Dr. Joan Mason who read and commented on much of the script, and to the anonymous reviewer to whom Chapman & Hall sent the original manuscript and who provided very thorough and useful comments.

The authors have been sustained through the inevitable drudgery of writing by an enthusiasm for this fascinating subject. We hope that some of this transmits itself to the student.

**Lesley E. Smart and Elaine A. Moore**
OU, Walton Hall, Milton Keynes

# About the Authors

**Lesley E. Smart** studied chemistry at Southampton University. After completing a Ph.D. in Raman spectroscopy, also at Southampton, she moved to a lectureship at the Royal University of Malta. After returning to the United Kingdom, she took an SRC Fellowship to Bristol University to work on X-ray crystallography for 3 years. Since 1977, she has worked at the Open University as a lecturer, and then senior lecturer (2000), in inorganic chemistry. At the Open University, she has been involved in the production of undergraduate courses in inorganic and physical chemistry. Most recently, she was the coordinating editor of *The Molecular World* course, which has been copublished with the RSC as a series of eight books. She was also an author on two of these, *The Third Dimension* and *Separation, Purification and Identification*.

Her research interests are in the characterization of the solid state, and she has over 40 publications in single-crystal Raman studies, X-ray crystallography, zintl phases, pigments, and heterogeneous catalysis.

*Solid State Chemistry* was first produced in 1992. Since then, it has been translated into French, German, Spanish, and Japanese.

**Elaine A. Moore** studied chemistry as an undergraduate at Oxford University and then stayed on to complete a D.Phil. in theoretical chemistry with Peter Atkins. After a 2-year, postdoctoral position at Southampton, she joined the Open University in 1975 as course assistant, becoming a lecturer in Chemistry in 1977 and Senior lecturer in 1998. She has produced OU teaching texts in chemistry for courses at levels 1, 2, and 3 and has written texts in astronomy at level 2. The text *Molecular Modelling and Bonding*, which forms part of the OU Level 2 Chemistry Course, was copublished by the Royal Society of Chemistry as part of *The Molecular World* series. She oversaw the introduction of multimedia into chemistry courses and designed multimedia material for levels 1 and 2. She is coauthor, with Dr. Rob Janes of the Open University, of *Metal-Ligand Bonding*, which is part of a level 3 Course in Inorganic Chemistry and copublished with the Royal Society of Chemistry.

Her research interests are in theoretical chemistry applied to solid state systems and to NMR spectroscopy. She is author or coauthor on over 40 papers in scientific journals. She was coauthor of an article in *Chemical Reviews* on nitrogen NMR spectroscopy of metal nitrosyl complexes.

# BASIC SI UNITS

| Physical quantity (and symbol) | Name of SI unit | Symbol for unit |
|---|---|---|
| Length (l) | Metre | m |
| Mass (m) | Kilogram | kg |
| Time (t) | Second | s |
| Electric current (I) | Ampere | A |
| Thermodynamic temperature (T) | Kelvin | K |
| Amount of substance (n) | Mole | mol |
| Luminous intensity ($I_v$) | Candela | cd |

# DERIVED SI UNITS

| Physical quantity (and symbol) | Name of SI unit | Symbol for SI derived unit and definition of unit |
|---|---|---|
| Frequency ($v$) | Hertz | $Hz\ (= s^{-1})$ |
| Energy ($U$), enthalpy ($H$) | Joule | $J\ (= kg\ m^2\ s^{-2})$ |
| Force | Newton | $N\ (= kg\ m\ s^{-2} = J\ m^{-1})$ |
| Power | Watt | $W\ (= kg\ m^2\ s^{-3} = J\ s^{-1})$ |
| Pressure ($p$) | Pascal | $Pa\ (= kg\ m^{-1}\ s^{-2} = N\ m^{-2} = J\ m^{-3})$ |
| Electric charge ($Q$) | Coulomb | $C\ (= A\ s)$ |
| Electric potential difference ($V$) | Volt | $V\ (= kg\ m^2\ s^{-3}\ A^{-1} = J\ A^{-1}\ s^{-1})$ |
| Capacitance ($c$) | Farad | $F\ (= A^2\ s^4\ kg^{-1}\ m^{-2} = A\ s\ V^{-1} = A^2\ s^2\ J^{-1})$ |
| Resistance ($R$) | Ohm | $\Omega\ (= V\ A^{-1})$ |
| Conductance ($G$) | Siemen | $S\ (= A\ V^{-1})$ |
| Magnetic flux density ($B$) | Tesla | $T\ (= V\ s\ m^{-2} = J\ C^{-1}\ s\ m^{-2})$ |

# SI PREFIXES

| $10^{-18}$ | $10^{-15}$ | $10^{-12}$ | $10^{-9}$ | $10^{-6}$ | $10^{-3}$ | $10^{-2}$ | $10^{-1}$ | $10^{3}$ | $10^{6}$ | $10^{9}$ | $10^{12}$ | $10^{15}$ | $10^{18}$ |
|---|---|---|---|---|---|---|---|---|---|---|---|---|---|
| alto | femto | pico | nano | micro | milli | centi | deci | kilo | mega | giga | tera | peta | exa |
| a | f | p | n | μ | m | c | d | k | M | G | T | P | E |

# FUNDAMENTAL CONSTANTS

| Constant | Symbol | Value |
|---|---|---|
| Speed of light in a vacuum | $c$ | $2.997925 \times 10^8$ m s$^{-1}$ |
| Charge of a proton | $e$ | $1.602189 \times 10^{-19}$ C |
| Charge of an electron | $-e$ | |
| Avogadro constant | $N_A$ | $6.022045 \times 10^{23}$ mol$^{-1}$ |
| Boltzmann constant | $k$ | $1.380662 \times 10^{-23}$ J K$^{-1}$ |
| Gas constant | $R = N_A k$ | $8.31441$ J K$^{-1}$ mol$^{-1}$ |
| Faraday constant | $F = N_A e$ | $9.648456 \times 10^4$ C mol$^{-1}$ |
| Planck constant | $h$ | $6.626176 \times 10^{-34}$ J s |
| | $\hbar = \dfrac{h}{2\pi}$ | $1.05457 \times 10^{-34}$ J s |
| Vacuum permittivity | $\varepsilon_0$ | $8.854 \times 10^{-12}$ F m$^{-1}$ |
| Vacuum permeability | $\mu_0$ | $4\pi \times 10^{-7}$ J s$^2$ C$^{-2}$ m$^{-1}$ |
| Bohr magneton | $\mu_B$ | $9.27402 \times 10^{-24}$ J T$^{-1}$ |
| Electron $g$ value | $g_e$ | $2.00232$ |

# MISCELLANEOUS PHYSICAL QUANTITIES

| Name of physical quantity | Symbol | SI unit |
|---|---|---|
| Enthalpy | $H$ | J |
| Entropy | $S$ | J K$^{-1}$ |
| Gibbs function | $G$ | J |
| Standard change of molar enthalpy | $\Delta H_m^{\ominus}$ | J mol$^{-1}$ |
| Standard of molar entropy | $\Delta S_m^{\ominus}$ | J K$^{-1}$ mol$^{-1}$ |
| Standard change of molar Gibbs functionz | $\Delta G_m^{\ominus}$ | J mol$^{-1}$ |
| Wave number | $\sigma \left( = \dfrac{1}{\lambda} \right)$ | cm$^{-1}$ |
| Atomic number | $Z$ | Dimensionless |
| Conductivity | $\sigma$ | S m$^{-1}$ |
| Molar bond dissociation energy | $D_m$ | J mol$^{-1}$ |
| Molar mass | $M \left( = \dfrac{m}{n} \right)$ | kg mol$^{-1}$ |

# THE GREEK ALPHABET

| | | | | | | |
|---|---|---|---|---|---|---|
| alpha | A | α | nu | N | ν |
| beta | B | β | xi | Ξ | ξ |
| gamma | Γ | γ | omicron | O | o |
| delta | Δ | δ | pi | Π | π |
| epsilon | E | ε | rho | P | ρ |
| zeta | Z | ζ | sigma | Σ | σ |
| eta | H | η | tau | T | τ |
| theta | Θ | θ | upsilon | Y | υ |
| iota | I | ι | phi | Φ | φ |
| kappa | K | κ | chi | X | χ |
| lambda | Λ | λ | psi | Ψ | ψ |
| mu | M | μ | omega | Ω | ω |

# PERIODIC CLASSIFICATION OF THE ELEMENTS

| Group | I(1) | II(2) | (3) | (4) | (5) | (6) | (7) | (8) | (9) | (10) | (11) | (12) | III(13) | IV(14) | V(15) | VI(16) | VII(17) | 0(18) |
|---|---|---|---|---|---|---|---|---|---|---|---|---|---|---|---|---|---|---|
| 1st Period | 1 H | | | | | | | | | | | | | | | | | 2 He |
| 2nd Period | 3 Li | 4 Be | | | | | | | | | | | 5 B | 6 C | 7 N | 8 O | 9 F | 10 Ne |
| 3rd Period | 11 Na | 12 Mg | | | | | | | | | | | 13 Al | 14 Si | 15 P | 16 S | 17 Cl | 18 Ar |
| 4th Period | 19 K | 20 Ca | 21 Sc | 22 Ti | 23 V | 24 Cr | 25 Mn | 26 Fe | 27 Co | 28 Ni | 29 Cu | 30 Zn | 31 Ga | 32 Ge | 33 As | 34 Se | 35 Br | 36 Kr |
| 5th Period | 37 Rb | 38 Sr | 39 Y | 40 Zr | 41 Nb | 42 Mo | 43 Tc | 44 Ru | 45 Rh | 46 Pd | 47 Ag | 48 Cd | 49 In | 50 Sn | 51 Sb | 52 Te | 53 I | 54 Xe |
| 6th Period | 55 Cs | 56 Ba | 71 Lu | 72 Hf | 73 Ta | 74 W | 75 Re | 76 Os | 77 Ir | 78 Pt | 79 Au | 80 Hg | 81 Tl | 82 Pb | 83 Bi | 84 Po | 85 At | 86 Rn |
| 7th Period | 87 Fr | 88 Ra | 103 Lr | 104 Rf | 105 Db | 106 Sg | 107 Bh | 108 Hs | 109 Mt | 110 Ds | 111 Rg | 112 Uub | 113 Uut | 114 Uuq | 115 Uup | | | |

transition elements

typical elements

lanthanides

| 57 La | 58 Ce | 59 Pr | 60 Nd | 61 Pm | 62 Sm | 63 Eu | 64 Gd | 65 Tb | 66 Dy | 67 Ho | 68 Er | 69 Tm | 70 Yb |
|---|---|---|---|---|---|---|---|---|---|---|---|---|---|

actinides

| 89 Ac | 90 Th | 91 Pa | 92 U | 93 Np | 94 Pu | 95 Am | 96 Cm | 97 Bk | 98 Cf | 99 Es | 100 Fm | 101 Md | 102 No |
|---|---|---|---|---|---|---|---|---|---|---|---|---|---|

# Table of Contents

## Third Edition

# SOLID STATE CHEMISTRY

## An Introduction

# 1 An Introduction to Crystal Structures

In the last decade of the twentieth century, research into solid state chemistry expanded very rapidly, fuelled partly by the dramatic discovery of 'high temperature' ceramic oxide superconductors in 1986, and by the search for new and better materials. We have seen immense strides in the development and understanding of nano-technology, micro- and meso-porous solids, fuel cells, and the giant magnetoresistance effect, to mention but a few areas. It would be impossible to cover all of the recent developments in detail in a text such as this, but we will endeavour to give you a flavour of the excitement that some of the research has engendered, and perhaps more importantly the background with which to understand these developments and those which are yet to come.

All substances, except helium, if cooled sufficiently form a solid phase; the vast majority form one or more **crystalline** phases, where the atoms, molecules, or ions pack together to form a regular repeating array. This book is concerned mostly with the structures of metals, ionic solids, and extended covalent structures; structures which do not contain discrete molecules as such, but which comprise extended arrays of atoms or ions. We look at the structure and bonding in these solids, how the properties of a solid depend on its structure, and how the properties can be modified by changes to the structure.

## 1.1 INTRODUCTION

To understand the solid state, we need to have some insight into the structure of simple crystals and the forces that hold them together, so it is here that we start this book. Crystal structures are usually determined by the technique of **X-ray crystallography.** This technique relies on the fact that the distances between atoms in crystals are of the same order of magnitude as the wavelength of X-rays (of the order of 1 Å or 100 pm): a crystal thus acts as a three-dimensional diffraction grating to a beam of X-rays. The resulting diffraction pattern can be interpreted to give the internal positions of the atoms in the crystal very precisely, thus defining interatomic distances and angles. (Some of the principles underlying this technique are discussed in Chapter 2, where we review the physical methods available for characterizing solids.) Most of the structures discussed in this book will have been determined in this way.

The structures of many inorganic crystal structures can be discussed in terms of the simple packing of spheres, so we will consider this first, before moving on to the more formal classification of crystals.

## 1.2  CLOSE-PACKING

Think for the moment of an atom as a small hard sphere. Figure 1.1 shows two
possible arrangements for a layer of such identical atoms. On squeezing the square
layer in Figure 1.1(a), the spheres would move to the positions in Figure 1.1(b) so
that the layer takes up less space. The layer in Figure 1.1(b) (layer A) is called **close-
packed**.

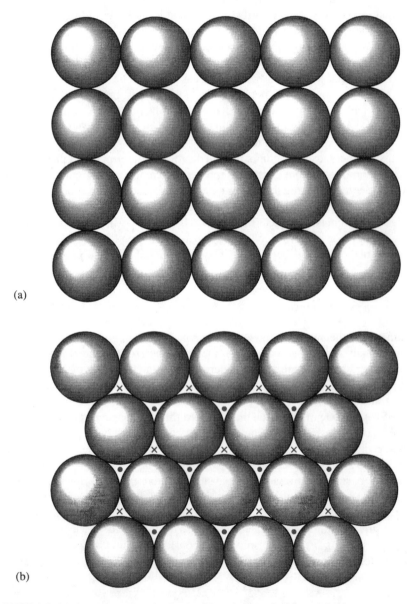

(a)

(b)

**FIGURE 1.1**  (a) A square array of spheres; (b) a close-packed layer of spheres.

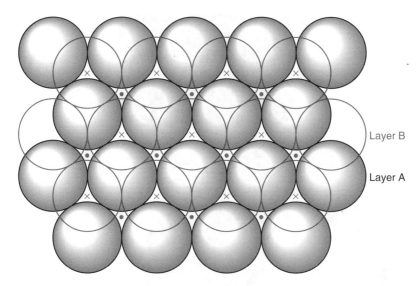

**FIGURE 1.2** Two layers of close-packed spheres.

To build up a close-packed structure in three-dimensions we must now add a second layer (layer B). The spheres of the second layer sit in half of the hollows of the first layer: these have been marked with dots and crosses. The layer B in Figure 1.2 sits over the hollows marked with a cross (although it makes no difference which type we chose). When we add a third layer, there are two possible positions where it can go. First, it could go directly over layer A, in the unmarked hollows: if we then repeated this stacking sequence we would build up the layers ABABABA . . . and so on. This is known as **hexagonal close-packing (*hcp*)** (Figure 1.3(a)). In this structure, the hollows marked with a dot are never occupied by spheres, leaving very small channels through the layers (Figure 1.3(b)).

Second, the third layer could be positioned over those hollows marked with a dot. This third layer, which we could label C, would not be directly over either A or B, and the stacking sequence when repeated would be ABCABCAB . . . and so on. This is known as **cubic close-packing (*ccp*)** (Figure 1.4). (The names *hexagonal* and *cubic* for these structures arise from the resulting symmetry of the structure — this will be discussed more fully later on.)

Close-packing represents the most efficient use of space when packing identical spheres — the spheres occupy 74% of the volume: the **packing efficiency** is said to be 74%. Each sphere in the structure is surrounded by *twelve* equidistant neighbours — six in the same layer, three in the layer above and three in the layer below: the **coordination number** of an atom in a close-packed structure is thus 12.

Another important feature of close-packed structures is the shape and number of the small amounts of space trapped in between the spheres. Two different types of space are contained within a close-packed structure: the first we will consider is called an **octahedral hole**. Figure 1.5(a) shows two close-packed layers again but now with the octahedral holes shaded. Six spheres surround each of these holes: three in layer A and three in layer B. The centres of these spheres lay at the corners

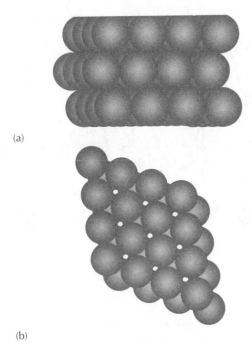

(a)

(b)

**FIGURE 1.3** (a) Three *hcp* layers showing the ABAB . . . stacking sequence; (b) three *hcp* layers showing the narrow channels through the layers.

of an octahedron, hence the name (Figure 1.5(b)). If $n$ spheres are in the array, then there are also $n$ octahedral holes.

Similarly, Figure 1.6(a) shows two close-packed layers, now with the second type of space, **tetrahedral holes**, shaded. Four spheres surround each of these holes with centres at the corners of a tetrahedron (Figure 1.6(b)). If $n$ spheres are in the array, then there are $2n$ tetrahedral holes.

The octahedral holes in a close-packed structure are much bigger than the tetrahedral holes — they are surrounded by six atoms instead of four. It is a matter of simple geometry to calculate that the radius of a sphere that will just fit in an

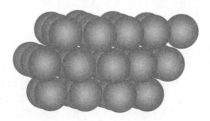

**FIGURE 1.4** Three *ccp* layers.

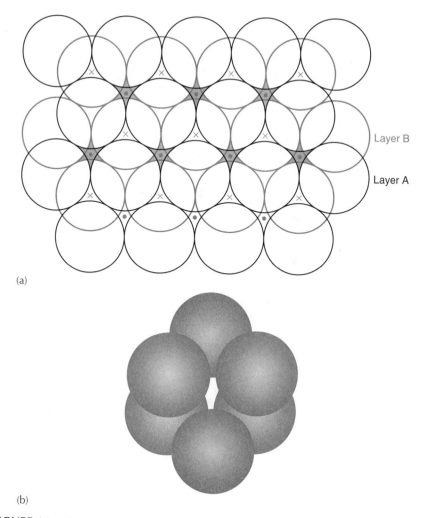

(a)

(b)

**FIGURE 1.5** (a) Two layers of close-packed spheres with the enclosed octahedral holes shaded; (b) a computer representation of an octahedral hole.

octahedral hole in a close-packed array of spheres of radius $r$ is $0.414r$. For a tetrahedral hole, the radius is $0.225r$ (Figure 1.7).

Of course, innumerable stacking sequences are possible when repeating close-packed layers; however, the hexagonal close-packed and cubic close-packed are those of maximum simplicity and are most commonly encountered in the crystal structures of the noble gases and of the metallic elements. Only two other stacking sequences are found in perfect crystals of the elements: an ABAC repeat in La, Pr, Nd, and Am, and a nine-layer repeat ABACACBCB in Sm.

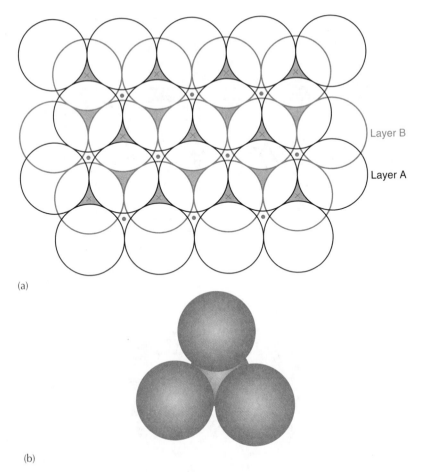

(a)

(b)

**FIGURE 1.6** (a) Two layers of close-packed spheres with the tetrahedral holes shaded; (b) a computer representation of a tetrahedral hole.

## 1.3 BODY-CENTRED AND PRIMITIVE STRUCTURES

Some metals do not adopt a close-packed structure but have a slightly less efficient packing method: this is the **body-centred cubic structure *(bcc)***, shown in Figure 1.8. (Unlike the previous diagrams, the positions of the atoms are now represented here — and in subsequent diagrams — by small spheres which do not touch: this is merely a device to open up the structure and allow it to be seen more clearly— the whole question of atom and ion size is discussed in Section 1.6.4.) In this structure an atom in the middle of a cube is surrounded by *eight identical and equidistant* atoms at the corners of the cube — the coordination number has dropped from twelve to eight and the packing efficiency is now 68%, compared with 74% for close-packing.

The simplest of the cubic structures is the **primitive cubic structure**. This is built by placing square layers like the one shown in Figure 1.1(a), directly on top of one another. Figure 1.9(a) illustrates this, and you can see in Figure 1.9(b) that each atom sits at the corner of a cube. The coordination number of an atom in this structure is six.

In the image, labels read "Layer B" and "Layer A" on the right side.

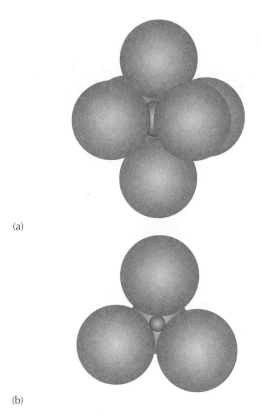

(a)

(b)

**FIGURE 1.7** (a) A sphere of radius $0.414r$ fitting into an octahedral hole; (b) a sphere of radius $0.225r$ fitting into a tetrahedral hole.

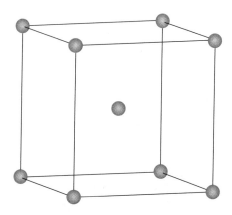

**FIGURE 1.8** Body-centred cubic array.

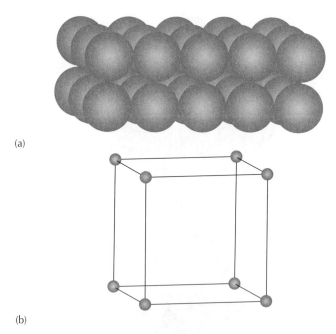

(a)

(b)

**FIGURE 1.9** (a) Two layers of a primitive cubic array; (b) a cube of atoms from this array.

The majority of metals have one of the three basic structures: *hcp*, *ccp*, or *bcc*. Polonium alone adopts the primitive structure. The distribution of the packing types among the most stable forms of the metals at 298 K is shown in Figure 1.10. As we noted earlier, a very few metals have a mixed *hcp/ccp* structure of a more complex type. The structures of the actinides tend to be rather complex and are not included.

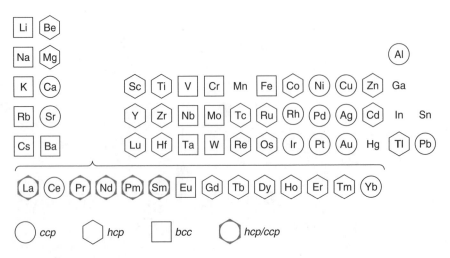

**FIGURE 1.10** Occurrence of packing types among the metals.

## 1.4 SYMMETRY

Before we take the discussion of crystalline structures any further, we will look at the symmetry displayed by structures. The concept of symmetry is an extremely useful one when it comes to describing the shapes of both individual molecules and regular repeating structures, as it provides a way of describing similar features in different structures so that they become unifying features. The symmetry of objects in everyday life is something that we tend to take for granted and recognize easily without having to think about it. Take some simple examples illustrated in Figure 1.11. If you imagine a mirror dividing the spoon in half along the plane indicated, then you can see that

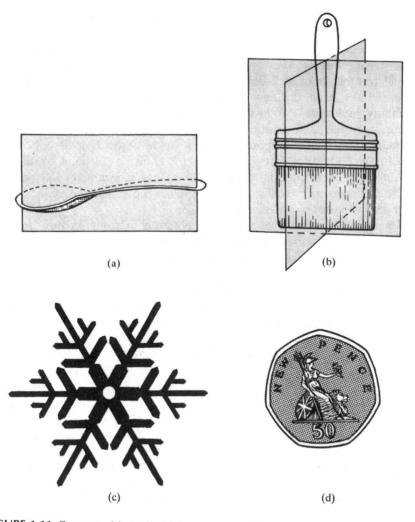

(a)　　　　　(b)

(c)　　　　　(d)

FIGURE 1.11 Common objects displaying symmetry: (a) a spoon, (b) a paintbrush, (c) a snowflake, and (d) a 50p coin.

one-half of the spoon is a mirror image or reflection of the other. Similarly, with the paintbrush, only now two mirror planes at right angles divide it.

Objects can also possess rotational symmetry. In Figure 1.11(c) imagine an axle passing through the centre of the snowflake; in the same way as a wheel rotates about an axle, if the snowflake is rotated through $\frac{1}{6}$ of a revolution, then the new position is indistinguishable from the old. Similarly, in Figure 1.11(d), rotating the 50p coin by $\frac{1}{7}$ of a revolution brings us to the same position as we started (ignoring the pattern on the surface). The symmetry possessed by a single object that describes the repetition of identical parts of the object is known as its **point symmetry**.

Actions such as rotating a molecule are called **symmetry operations**, and the rotational axes and mirror planes possessed by objects are examples of **symmetry elements**.

Two forms of symmetry notation are commonly used. As chemists, you will come across both. The **Schoenflies** notation is useful for describing the point symmetry of individual molecules and is used by spectroscopists. The **Hermann–Mauguin** notation can be used to describe the point symmetry of individual molecules but in addition can also describe the relationship of different molecules to one another in space — their so-called **space-symmetry** — and so is the form most commonly met in crystallography and the solid state. We give here the Schoenflies notation in parentheses after the Hermann–Mauguin notation.

### 1.4.1 Axes of Symmetry

As discussed previously for the snowflake and the 50p coin, molecules and crystals can also possess rotational symmetry. Figure 1.12 illustrates this for several molecules.

In Figure 1.12(a) the rotational axis is shown as a vertical line through the O atom in $OF_2$; rotation about this line by $180°$ in the direction of the arrow, produces an identical looking molecule. The line about which the molecule rotates is called an **axis of symmetry**, and in this case, it is a twofold axis because we have to perform the operation twice to return the molecule to its starting position.

Axes of symmetry are denoted by the symbol $n$ ($C_n$), where $n$ is the order of the axis. Therefore, the rotational axis of the $OF_2$ molecule is 2 ($C_2$).

The $BF_3$ molecule in Figure 1.12(b) possesses a threefold axis of symmetry, 3 ($C_3$), because each $\frac{1}{3}$ of a revolution leaves the molecule looking the same, and three turns brings the molecule back to its starting position. In the same way, the $XeF_4$ molecule in (c) has a fourfold axis, 4 ($C_4$), and four quarter turns are necessary to bring it back to the beginning. All linear molecules have an $\infty$ ($C_\infty$) axis, which is illustrated for the $BeF_2$ molecule in (d); however small a fraction of a circle it is rotated through, it always looks identical. The smallest rotation possible is $1/\infty$, and so the axis is an infinite-order axis of symmetry.

### 1.4.2 Planes of Symmetry

Mirror planes occur in isolated molecules and in crystals, such that everything on one side of the plane is a mirror image of the other. In a structure, such a mirror

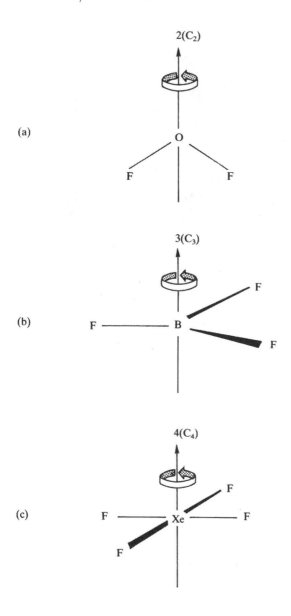

**FIGURE 1.12** Axes of symmetry in molecules: (a) twofold axis in $OF_2$, (b) threefold axis in $BF_3$, (c) fourfold axis in $XeF_4$, and (d) $\infty$–fold axis in $BeF_2$.

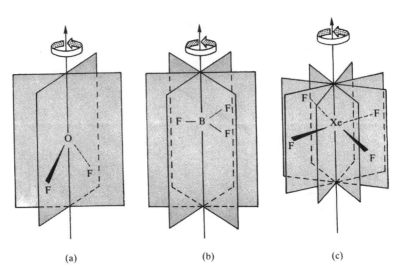

**FIGURE 1.13** Planes of symmetry in molecules: (a) planes of symmetry in $OF_2$, (b) planes of symmetry in $BF_3$, and (c) planes of symmetry in $XeF_4$.

plane is known as a **plane of symmetry** and is given the symbol **m** ($\sigma$). Molecules may possess one or more planes of symmetry, and the diagrams in Figure 1.13 illustrate some examples. The planar $OF_2$ molecule has two planes of symmetry (Figure 1.13(a)), one is the plane of the molecule, and the other is at right angles to this. For all planar molecules, the plane of the molecule is a plane of symmetry. The diagrams for $BF_3$ and $XeF_4$ (also planar molecules) only show the planes of symmetry which are perpendicular to the plane of the molecule.

### 1.4.3 INVERSION

The third symmetry operation that we show in this section is called **inversion** through a centre of symmetry and is given the symbol $\overline{1}$ (i). In this operation you have to imagine a line drawn from any atom in the molecule, through the centre of symmetry and then continued for the same distance the other side; if for every atom, this meets with an identical atom on the other side, then the molecule has a centre of symmetry. Of the molecules in Figure 1.12, $XeF_4$ and $BeF_2$ both have a centre of symmetry, and $BF_3$ and $OF_2$ do not.

### 1.4.4 INVERSION AXES AND IMPROPER SYMMETRY AXES

The final symmetry element is described differently by the two systems, although both descriptions use a combination of the symmetry elements described previously. The Hermann–Mauguin **inversion axis** is a combination of rotation and inversion and is given the symbol $\overline{n}$. The symmetry element consists of a rotation by $1/n$ of a revolution about the axis, followed by inversion through the centre of symmetry. An example of an $\overline{4}$ inversion axis is shown in Figure 1.14 for a tetrahedral molecule such as $CF_4$. The molecule is shown inside a cube as this makes it easier to see the

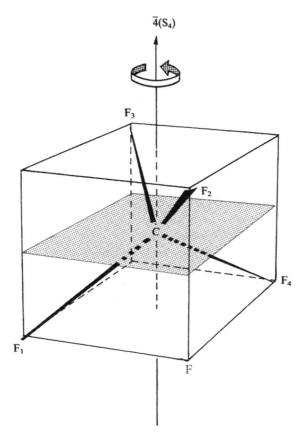

**FIGURE 1.14** The $\overline{4}$ ($S_4$) inversion (improper) axis of symmetry in the tetrahedral $CF_4$ molecule.

symmetry elements. Rotation about the axis through 90° takes $F_1$ to the position shown as a dotted F; inversion through the centre then takes this atom to the $F_3$ position.

The equivalent symmetry element in the Schoenflies notation is the **improper axis of symmetry, $S_n$**, which is a combination of rotation and reflection. The symmetry element consists of a rotation by $1/n$ of a revolution about the axis, followed by reflection through a plane at right angles to the axis. Figure 1.14 thus presents an $S_4$ axis, where the $F_1$ rotates to the dotted position and then reflects to $F_2$. The equivalent inversion axes and improper symmetry axes for the two systems are shown in Table 1.1.

### 1.4.5 SYMMETRY IN CRYSTALS

The discussion so far has only shown the symmetry elements that belong to individual molecules. However, in the solid state, we are interested in regular arrays of

**TABLE 1.1**
**Equivalent symmetry elements in the Schoenflies and Hermann-Mauguin Systems**

| Schoenflies | Hermann–Mauguin |
|---|---|
| $S_1 \equiv m$ | $\bar{2} \equiv m$ |
| $S_2 \equiv i$ | $\bar{1} \equiv i$ |
| $S_3$ | $\bar{6}$ |
| $S_4$ | $\bar{4}$ |
| $S_6$ | $\bar{3}$ |

atoms, ions, and molecules, and they too are related by these same symmetry elements. Figure 1.15 gives examples (not real) of how molecules could be arranged in a crystal. In (a), two $OF_2$ molecules are related to one another by a plane of symmetry; in (b), three $OF_2$ molecules are related to one another by a threefold axis of symmetry; in (c), two $OF_2$ molecules are related by a centre of inversion. Notice that in both (b) and (c), the molecules are related in space by a symmetry element that they themselves do not possess, this is said to be their **site symmetry**.

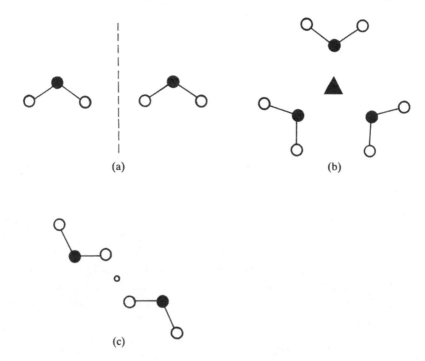

**FIGURE 1.15** Symmetry in solids: (a) two $OF_2$ molecules related by a plane of symmetry, (b) three $OF_2$ molecules related by a threefold axis of symmetry, and (c) two $OF_2$ molecules related by a centre of inversion.

## 1.5 LATTICES AND UNIT CELLS

Crystals are regular shaped solid particles with flat shiny faces. It was first noted by Robert Hooke in 1664 that the regularity of their external appearance is a reflection of a high degree of internal order. Crystals of the same substance, however, vary in shape considerably. Steno observed in 1671 that this is not because their internal structure varies but because some faces develop more than others do. The angle between similar faces on different crystals of the same substance is always identical. The constancy of the interfacial angles reflects the internal order within the crystals. Each crystal is derived from a basic 'building block' that continuously repeats, in all directions, in a perfectly regular way. This building block is known as the **unit cell**.

To talk about and compare the many thousands of crystal structures that are known, there has to be a way of defining and categorizing the structures. This is achieved by defining the shape and symmetry of each unit cell as well as its size and the positions of the atoms within it.

### 1.5.1 LATTICES

The simplest regular array is a line of evenly spaced objects, such as those depicted by the commas in Figure 1.16(a). There is a dot at the same place in each object: if we now remove the objects leaving the dots, we have a line of equally spaced dots, spacing $a$, (Figure 1.16(b)). The line of dots is called the **lattice**, and each **lattice point** (dot) must have *identical surroundings*. This is the only example of a one-dimensional lattice and it can vary only in the spacing $a$. Five two-dimensional lattices are possible, and examples of these can be seen every day in wallpapers and tiling.

### 1.5.2 ONE- AND TWO-DIMENSIONAL UNIT CELLS

The unit cell for the one-dimensional lattice in Figure 1.16(a) lies between the two vertical lines. If we took this unit cell and repeated it over again, we would reproduce the original array. Notice that it does not matter where in the structure we place the

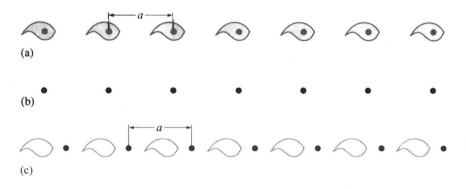

**FIGURE 1.16** A one-dimensional lattice (a,b) and the choice of unit cells (c).

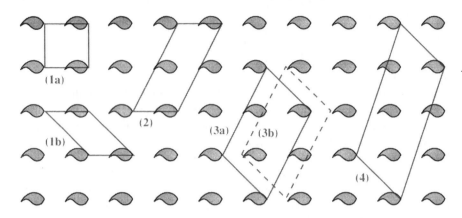

**FIGURE 1.17** Choice of unit cell in a square two-dimensional lattice.

lattice points as long as they each have identical surroundings. In Figure 1.16(c), we have moved the lattice points and the unit cell, but repeating this unit cell will still give the same array — we have simply moved the origin of the unit cell. There is never one unique unit cell that is 'correct.' Many can always be chosen, and the choice depends both on convenience and convention. This is equally true in two and three dimensions.

The unit cells for the two-dimensional lattices are parallelograms with their corners at equivalent positions in the array (i.e., the corners of a unit cell are lattice points). In Figure 1.17, we show a square array with several different unit cells depicted. All of these, if repeated, would reproduce the array: it is conventional to choose the smallest cell that fully represents the symmetry of the structure. Both unit cells (1a) and (1b) are the same size but clearly (1a) shows that it is a square array, and this would be the conventional choice. Figure 1.18 demonstrates the same principles but for a centred rectangular array, where (a) would be the conventional choice because it includes information on the centring; the smaller unit cell (b) loses this information. It is always possible to define a non-centred oblique unit cell, but doing so may lose information about the symmetry of the lattice.

Unit cells, such as (1a) and (1b) in Figure 1.17 and (b) in Figure 1.18, have a lattice point at each corner. However, they each contain one lattice point because four adjacent unit cells share each lattice point. They are known as **primitive unit cells** and are given the symbol **P**. The unit cell marked (a) in Figure 1.18 contains

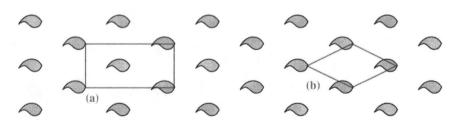

**FIGURE 1.18** Choice of unit cell in a centred-rectangular lattice.

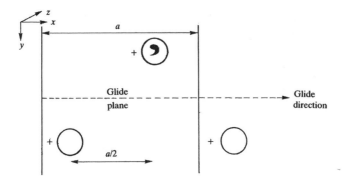

**FIGURE 1.19** An *a* glide perpendicular to *b*.

two lattice points — one from the shared four corners and one totally enclosed within the cell. This cell is said to be **centred** and is given the symbol **C**.

### 1.5.3 TRANSLATIONAL SYMMETRY ELEMENTS

Section 1.4 introduced the idea of symmetry, both in individual molecules and for extended arrays of molecules, such as are found in crystals. Before going on to discuss three-dimensional lattices and unit cells, it is important to introduce two more symmetry elements; these elements involve translation and are only found in the solid state.

The **glide plane** combines translation with reflection. Figure 1.19 is an example of this symmetry element. The diagram shows part of a repeating three-dimensional structure projected on to the plane of the page; the circle represents a molecule or ion in the structure and there is distance *a* between identical positions in the structure. The + sign next to the circle indicates that the molecule lies above the plane of the page in the *z* direction. The plane of symmetry is in the *xz* plane perpendicular to the paper, and is indicated by the dashed line. The symmetry element consists of reflection through this plane of symmetry, followed by translation. In this case, the translation can be either in the *x* or in the *z* direction (or along a diagonal), and the translation distance is half of the repeat distance in that direction. In the example illustrated, the translation takes place in the *x* direction. The repeat distance between identical molecules is *a*, and so the translation is by *a*/2, and the symmetry element is called an *a* glide. You will notice two things about the molecule generated by this symmetry element: first, it still has a + sign against it, because the reflection in the plane leaves the *z* coordinate the same and second, it now has a comma on it. Some molecules when they are reflected through a plane of symmetry are enantiomorphic, which means that they are not superimposable on their mirror image: the presence of the comma indicates that this molecule could be an enantiomorph.

The **screw axis** combines translation with rotation. Screw axes have the general symbol $n_i$ where *n* is the rotational order of the axis (i.e., twofold, threefold, etc.), and the translation distance is given by the ratio *i/n*. Figure 1.20 illustrates a $2_1$ screw axis. In this example, the screw axis lies along *z* and so the translation must be in

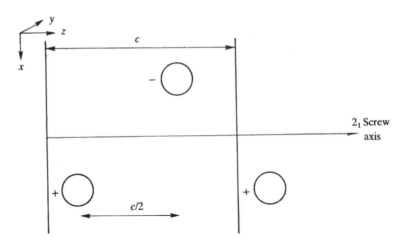

**FIGURE 1.20** A $2_1$ screw axis along $z$.

the $z$ direction, by $c/2$, where $c$ is the repeat distance in the $z$ direction. Notice that in this case the molecule starts above the plane of the paper (indicated by the + sign) but the effect of a twofold rotation is to take it below the plane of the paper (– sign). Figure 1.21 probably illustrates this more clearly, and shows the different effects that rotational and screw axes of the same order have on a repeating structure. Rotational and screw axes produce objects that are superimposable on the original. All other symmetry elements — glide plane, mirror plane, inversion centre, and inversion axis — produce a mirror image of the original.

### 1.5.4 THREE-DIMENSIONAL UNIT CELLS

The unit cell of a three-dimensional lattice is a parallelepiped defined by three distances $a$, $b$, and $c$, and three angles $\alpha$, $\beta$, and $\gamma$, as shown in Figure 1.22. Because the unit cells are the basic building blocks of the crystals, they must be space-filling (i.e., they must pack together to fill all space). All the possible unit cell shapes that can fulfill this criterion are illustrated in Figure 1.23 and their specifications are listed in Table 1.2. These are known as the **seven crystal systems** or **classes**. These unit cell shapes are determined by minimum symmetry requirements which are also detailed in Table 1.2.

The three-dimensional unit cell includes *four* different types (see Figure 1.24):

1. The **primitive** unit cell — symbol **P** — has a lattice point at each corner.
2. The **body-centred** unit cell — symbol **I** — has a lattice point at each corner and one at the centre of the cell.
3. The **face-centred** unit cell — symbol **F** — has a lattice point at each corner and one in the centre of each face.
4. The **face-centred** unit cell — symbol **A, B,** or **C** — has a lattice point at each corner, and one in the centres of one pair of opposite faces (e.g., an A-centred cell has lattice points in the centres of the $bc$ faces).

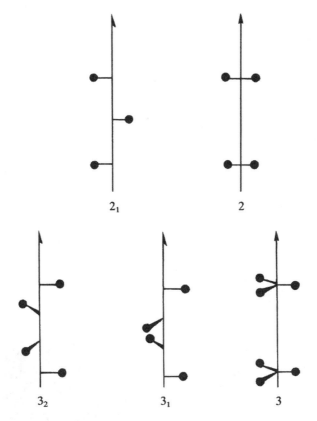

FIGURE 1.21 Comparison of the effects of twofold and threefold rotation axes and screw axes.

When these four types of lattice are combined with the 7 possible unit cell shapes, 14 permissible **Bravais lattices** (Table 1.3) are produced. (It is not possible to combine some of the shapes and lattice types and retain the symmetry requirements listed in Table 1.2. For instance, it is not possible to have an A-centred, cubic, unit cell; if only two of the six faces are centred, the unit cell necessarily loses its cubic symmetry.)

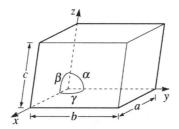

FIGURE 1.22 Definition of axes, unit cell dimensions, and angles for a general unit cell.

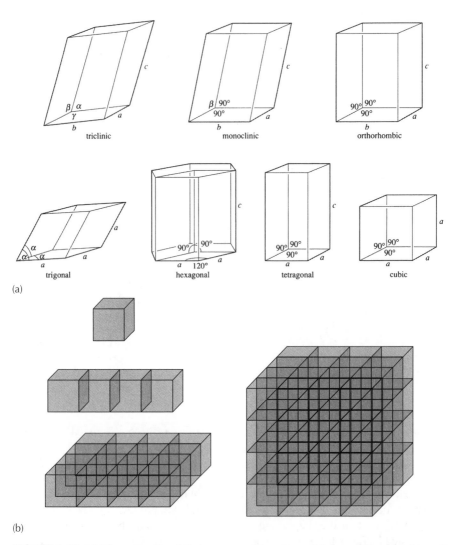

**FIGURE 1.23** (a) The unit cells of the seven crystal systems. (b) Assemblies of cubic unit cells in one, two, and three dimensions.

The symmetry of a crystal is a point group taken from a point at the centre of a perfect crystal. Only certain point groups are possible because of the constraint made by the fact that unit cells must be able to stack exactly with no spaces – so only one-, two-, three-, four-, and sixfold axes are possible. Combining this with planes of symmetry and centres of symmetry, we find 32 point groups that can describe the shapes of perfect crystals.

If we combine the 32 crystal point groups with the 14 Bravais lattices we find 230 three-dimensional **space groups** that crystal structures can adopt (i.e., 230

---

**TABLE 1.2**
**The seven crystal systems**

| System | Unit cell | Minimum symmetry requirements |
|---|---|---|
| Triclinic | $\alpha \neq \beta \neq \gamma \neq 90°$<br>$a \neq b \neq c$ | None |
| Monoclinic | $\alpha = \gamma = 90°$<br>$\beta \neq 90°$<br>$a \neq b \neq c$ | One twofold axis or one symmetry plane |
| Orthorhombic | $\alpha = \beta = \gamma = 90°$<br>$a \neq b \neq c$ | Any combination of three mutually perpendicular twofold axes or planes of symmetry |
| Trigonal/rhombohedral | $\alpha = \beta = \gamma \neq 90°$<br>$a = b = c$ | One threefold axis |
| Hexagonal | $\alpha = \beta = 90°$<br>$\gamma = 120°$<br>$a = b \neq c$ | One sixfold axis or one sixfold improper axis |
| Tetragonal | $\alpha = \beta = \gamma = 90°$<br>$a = b \neq c$ | One fourfold axis or one fourfold improper axis |
| Cubic | $\alpha = \beta = \gamma = 90°$<br>$a = b = c$ | Four threefold axes at 109° 28' to each other |

---

different space-filling patterns)! These are all documented in the International Tables for Crystallography (see Bibliography at end of the book).

It is important not to lose sight of the fact that the lattice points represent equivalent positions in a crystal structure and not atoms. In a real crystal, an atom, a complex ion, a molecule, or even a group of molecules could occupy a lattice point. The lattice points are used to simplify the repeating patterns within a structure, but they tell us nothing of the chemistry or bonding within the crystal — for that we have to include the atomic positions: this we will do later in the chapter when we look at some real structures.

It is instructive to note how much of a structure these various types of unit cell represent. We noted a difference between the centred and primitive two-dimensional unit cell where the centred cell contains two lattice points whereas the primitive cell contains only one. We can work out similar occupancies for the three-dimensional case. The number of unit cells sharing a particular molecule depends on its site. A corner site is shared by eight unit cells, an edge site by four, a face site by two and a molecule at the body-centre is not shared by any other unit cell (Figure 1.25). Using these figures, we can work out the number of molecules in each of the four types of cell in Figure 1.24, assuming that one molecule is occupying each lattice point. The results are listed in Table 1.4.

Solid State Chemistry: An Introduction

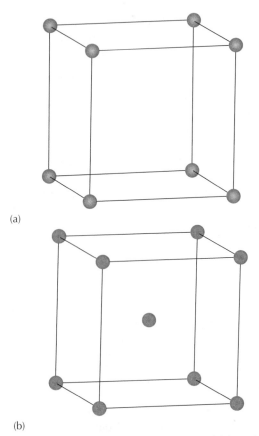

(a)

(b)

**FIGURE 1.24** Primitive (a), body-centred (b), face-centred (c), and face-centred (A, B, or C) (d), unit cells. (*–continued*)

### 1.5.5 MILLER INDICES

The faces of crystals, both when they grow and when they are formed by cleavage, tend to be parallel either to the sides of the unit cell or to planes in the crystal that contain a high density of atoms. It is useful to be able to refer to both crystal faces and to the planes in the crystal in some way — to give them a name — and this is usually done by using **Miller indices**.

First, we will describe how Miller indices are derived for lines in two-dimensional nets, and then move on to look at planes in three-dimensional lattices. Figure 1.26 is a rectangular net with several sets of lines, and a unit cell is marked on each set with the origin of each in the bottom left-hand corner corresponding to the directions of the $x$ and $y$ axes. A set of parallel lines is defined by two indices, $h$ and $k$, where $h$ and $k$ are the number of parts into which $a$ and $b$, the unit cell edges, are divided by the lines. Thus the indices of a line $hk$ are defined so that the line intercepts $a$ at $\frac{a}{h}$ and $b$ at $\frac{b}{k}$. Start by finding a line next to the one passing through the origin. In the set of lines marked A, the line next to the one passing through the origin

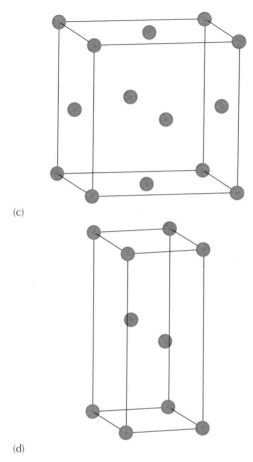

(c)

(d)

**FIGURE 1.24** (continued)

### TABLE 1.3
### Bravais lattices

| Crystal system | Lattice types |
|---|---|
| Cubic | P, I, F |
| Tetragonal | P, I |
| Orthorhombic | P, C, I, F |
| Hexagonal | P |
| Trigonal (Rhombohedral) | P/R[a] |
| Monoclinic | P, C |
| Triclinic | P |

[a]The primitive description of the rhombohedral lattice is normally given the symbol R.

**TABLE 1.4**
**Number of molecules in four types of cells**

| Name | Symbol | Number of molecules in unit cell |
|---|---|---|
| Primitive | P | 1 |
| Body-centred | I | 2 |
| Face-centred | A or B or C | 2 |
| All face-centred | F | 4 |

leaves $a$ undivided but divides $b$ into two; both intercepts lie on the positive side of the origin, therefore, in this case, the indices of the set of lines $hk$ are $12$ (referred to as the 'one-two' set). If the set of lines lies parallel to one of the axes then there is no intercept and the index becomes zero. If the intercepted cell edge lies on the negative side of the origin, then the index is written with a bar on the top (e.g., $\bar{2}$ ), known as 'bar-two'. Notice that if we had selected the line on the other side of the origin in A we would have indexed the lines as the $\bar{1}\ \bar{2}$; no difference exists between the two pairs of indices and always the $hk$ and the $\bar{h}\ \bar{k}$ lines are the same set of lines. Try Question 5 for more examples. Notice also, in Figure 1.26, that the lines with the lower indices are more widely spaced.

The Miller indices for planes in three-dimensional lattices are given by $hkl$, where $l$ is now the index for the $z$-axis. The principles are the same. Thus a plane is indexed $hkl$ when it makes intercepts $\dfrac{a}{h}$, $\dfrac{b}{k}$, and $\dfrac{c}{l}$ with the unit cell edges $a$, $b$, and $c$. Figure 1.27 depicts some cubic lattices with various planes shaded. The positive directions of the axes are marked, and these are orientated to conform to

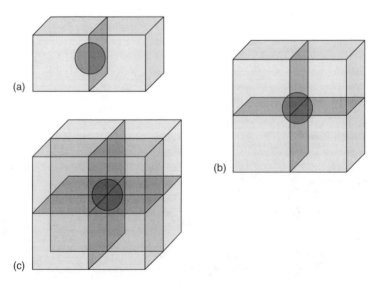

(a)

(b)

(c)

**FIGURE 1.25** Unit cells showing a molecule on (a) a face, (b) an edge, and (c) a corner.

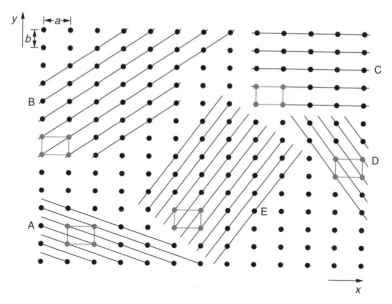

**FIGURE 1.26** A rectangular net showing five sets of lines, A–E, with unit cells marked.

the conventional right-hand rule as illustrated in Figure 1.28. In Figure 1.27(a), the shaded planes lie parallel to $y$ and $z$, but leave the unit cell edge $a$ undivided; the Miller indices of these planes are thus *100*. Again, take note that the *hkl* and $\overline{hkl}$ planes are the same.

### 1.5.6 Interplanar Spacings

It is sometimes useful to be able to calculate the perpendicular distance $d_{hkl}$ between parallel planes (Miller indices *hkl*). When the axes are at right angles to one another (orthogonal) the geometry is simple and for an orthorhombic system where $a \neq b \neq c$ and $\alpha = \beta = \gamma = 90°$, this gives:

$$\frac{1}{d_{hkl^2}} = \frac{h^2}{a^2} + \frac{k^2}{b^2} + \frac{l^2}{c^2}$$

Other relationships are summarized in Table 1.5.

### 1.5.7 Packing Diagrams

Drawing structures in three-dimensions is not easy and so crystal structures are often represented by two-dimensional plans or projections of the unit cell contents — in much the same way as an architect makes building plans. These projections are called **packing diagrams** because they are particularly useful in molecular structures

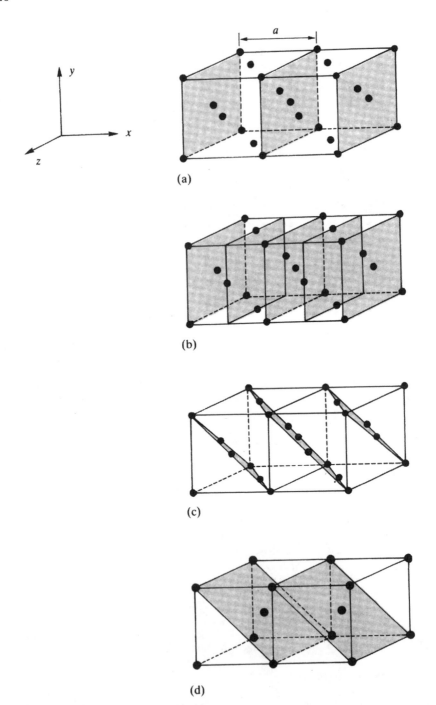

**FIGURE 1.27** (a)–(c) Planes in a face-centred cubic lattice. (d) Planes in a body-centred cubic lattice (two unit cells are shown).

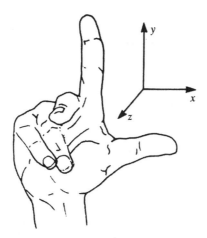

**FIGURE 1.28** The right-handed rule for labelling axes.

for showing how the molecules pack together in the crystal, and thus the intermolecular interactions.

The position of an atom or ion in a unit cell is described by its **fractional coordinates**; these are simply the coordinates based on the unit cell axes (known as the **crystallographic axes**), but expressed as *fractions of the unit cell lengths*. It has the simplicity of a universal system which enables unit cell positions to be compared from structure to structure regardless of variation in unit cell size.

---

**TABLE 1.5**
**d-spacings in different crystal systems**

| Crystal system | $d_{hkl}$ as a function of Miller indices and lattice parameters |
|---|---|
| Cubic | $\dfrac{1}{d^2} = \dfrac{h^2 + k^2 + l^2}{a^2}$ |
| Tetragonal | $\dfrac{1}{d^2} = \dfrac{h^2 + k^2}{a^2} + \dfrac{l^2}{c^2}$ |
| Orthorhombic | $\dfrac{1}{d^2} = \dfrac{h^2}{a^2} + \dfrac{k^2}{b^2} + \dfrac{l^2}{c^2}$ |
| Hexagonal | $\dfrac{1}{d^2} = \dfrac{4}{3}\left(\dfrac{h^2 + hk + k^2}{a^2}\right) + \dfrac{l^2}{c^2}$ |
| Monoclinic | $\dfrac{1}{d^2} = \dfrac{1}{\sin^2\beta}\left(\dfrac{h^2}{a^2} + \dfrac{k^2\sin^2\beta}{b^2} + \dfrac{l^2}{c^2} - \dfrac{2hl\cos\beta}{ac}\right)$ |

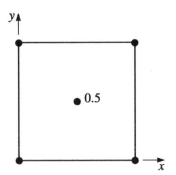

**FIGURE 1.29** Packing diagram for a body-centred unit cell.

To take a simple example, in a cubic unit cell with $a = 1000$ pm, an atom with an $x$ coordinate of 500 pm has a fractional coordinate in the $x$ direction of

$\dfrac{x}{a} = \dfrac{500}{1000} = 0.5$. Similarly, in the $y$ and $z$ directions, the fractional coordinates are

given by $\dfrac{y}{b}$ and $\dfrac{z}{c}$, respectively.

A packing diagram is shown in Figure 1.29 for the body-centred unit cell of Figure 1.8. The projection is shown on the $yx$ plane (i.e., we are looking at the unit cell straight down the $z$-axis). The $z$-fractional coordinate of any atoms/ions lying in the top or bottom face of the unit cell will be 0 or 1 (depending on where you take the origin) and it is conventional for this not to be marked on the diagram. Any $z$-coordinate that is not 0 or 1 is marked on the diagram in a convenient place. There is an opportunity to practice constructing these types of diagram in the questions at the end of the chapter.

## 1.6  CRYSTALLINE SOLIDS

We start this section by looking at the structures of some simple **ionic solids**. Ions tend to be formed by the elements in the Groups at the far left and far right of the Periodic Table. Thus, we expect the metals in Groups I and II to form cations and the nonmetals of Groups VI(16) and VII(17) and nitrogen to form anions, because by doing so they are able to achieve a stable noble gas configuration. Cations can also be formed by some of the Group III(13) elements, such as aluminium, $Al^{3+}$, by some of the low oxidation state transition metals and even occasionally by the high atomic number elements in Group IV(14), such as tin and lead, giving $Sn^{4+}$ and $Pb^{4+}$. Each successive ionization becomes more difficult because the remaining electrons are more strongly bound due to the greater effective nuclear charge, and so highly charged ions are rather rare.

An **ionic bond** is formed between two oppositely charged ions because of the electrostatic attraction between them. Ionic bonds are strong but are also non-directional; their strength decreases with increasing separation of the ions. Ionic crystals are therefore composed of infinite arrays of ions which have packed together in such a way as to maximize the coulombic attraction between oppositely charged ions and to minimize the repulsions between ions of the same charge. We expect to find ionic compounds in the halides and oxides of the Group I and II metals, and it is with these crystal structures that this section begins.

However, just because it is possible to form a particular ion, does not mean that this ion will always exist whatever the circumstances. In many structures, we find that the bonding is not purely ionic but possesses some degree of **covalency**: the electrons are shared between the two bonding atoms and not merely transferred from one to the other. This is particularly true for the elements in the centre of the Periodic Table. This point is taken up in Section 1.6.4 where we discuss the size of ions and the limitations of the concept of ions as hard spheres.

Two later sections (1.6.5 and 1.6.6) look at the crystalline structures of covalently bonded species. First, extended covalent arrays are investigated, such as the structure of **diamond** — one of the forms of elemental carbon — where each atom forms strong covalent bonds to the surrounding atoms, forming an infinite three-dimensional network of localized bonds throughout the crystal. Second, we look at molecular crystals, which are formed from small, individual, covalently-bonded molecules. These molecules are held together in the crystal by weak forces known collectively as **van der Waals forces**. These forces arise due to interactions between dipole moments in the molecules. Molecules that possess a permanent dipole can interact with one another (**dipole–dipole interaction**) and with ions (**charge–dipole interaction**). Molecules that do not possess a dipole also interact with each other because 'transient dipoles' arise due to the movement of electrons, and these in turn induce dipoles in adjacent molecules. The net result is a weak attractive force known as the **London dispersion force**, which falls off very quickly with distance.

Finally, in this section, we take a very brief look at the structures of some silicates — the compounds that largely form the earth's crust.

## 1.6.1 IONIC SOLIDS WITH FORMULA MX

### The Caesium Chloride Structure (CsCl)

A unit cell of the caesium chloride structure is shown in Figure 1.30. It shows a caesium ion, $Cs^+$, at the centre of the cubic unit cell, surrounded by eight chloride

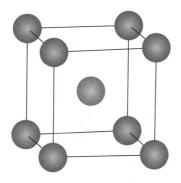

**FIGURE 1.30** The CsCl unit cell. Cs, blue sphere; Cl, grey spheres (or vice versa).

ions, Cl⁻, at the corners. It could equally well have been drawn the other way round with chloride at the centre and caesium at the corners because the structure consists of two interpenetrating primitive cubic arrays. Note the similarity of this unit cell to the body-centred cubic structure adopted by some of the elemental metals such as the Group I (alkali) metals. However, the caesium chloride structure is not body-centred cubic because the environment of the caesium at the centre of the cell is not the same as the environment of the chlorides at the corners: a body-centred cell would have chlorides at the corners (i.e., at [0, 0, 0], etc. *and* at the body-centre $[\frac{1}{2}, \frac{1}{2}, \frac{1}{2}]$). Each caesium is surrounded by eight chlorines at the corners of a cube and *vice versa*, so the coordination number of each type of atom is eight. The unit cell contains one formula unit of CsCl, with the eight corner chlorines each being shared by eight unit cells. With ionic structures like this, individual molecules are not distinguishable because individual ions are surrounded by ions of the opposite charge.

Caesium is a large ion (ionic radii are discussed in detail later in Section 1.6.4) and so is able to coordinate eight chloride ions around it. Other compounds with large cations that can also accommodate eight anions and crystallize with this structure include CsBr, CsI, TlCl, TlBr, TlI, and $NH_4Cl$.

## The Sodium Chloride (or Rock Salt) Structure (NaCl)

Common salt, or sodium chloride, is also known as rock salt. It is mined all over the world from underground deposits left by the dried-up remains of ancient seas, and has been so highly prized in the past that its possession has been the cause of much conflict, most notably causing the 'salt marches' organized by Gandhi, and helping to spark off the French Revolution. A unit cell of the sodium chloride structure is illustrated in Figure 1.31. The unit cell is cubic and the structure consists of two interpenetrating face-centred arrays, one of Na⁺ and the other of Cl⁻ ions. Each sodium ion is surrounded by six equidistant chloride ions situated at the corners

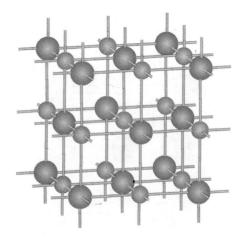

**FIGURE 1.31** The NaCl unit cell. Na, blue spheres; Cl, grey spheres (or vice versa).

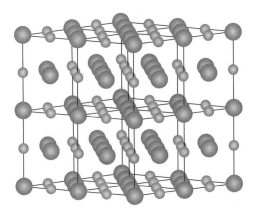

**FIGURE 1.32** The close-packed layers in NaCl. Na, blue spheres; Cl, grey spheres.

of an octahedron and in the same way each chloride ion is surrounded by six sodium ions: we say that the coordination is 6:6.

An alternative way of viewing this structure is to think of it as a cubic close-packed array of chloride ions with sodium ions filling all the octahedral holes. The conventional unit cell of a *ccp* array is an F face-centred cube (hence the cubic in *ccp*); the close-packed layers lie at right angles to a cube diagonal (Figure 1.32). Filling all the octahedral holes gives a Na:Cl ratio of 1:1 with the structure as illustrated in Figure 1.31. Interpreting simple ionic structures in terms of the close-packing of one of the ions with the other ion filling some or all of either the octahedral or tetrahedral holes, is extremely useful: it makes it particularly easy to see both the coordination geometry around a specific ion and also the available spaces within a structure.

As you might expect from their relative positions in Group I, a sodium ion is smaller than a caesium ion and so it is now only possible to pack six chlorides around it and not eight as in caesium chloride.

The sodium chloride unit cell contains four formula units of NaCl. If you find this difficult to see, work it out for yourself by counting the numbers of ions in the different sites and applying the information given in Table 1.4.

Table 1.6 lists some of the compounds that adopt the NaCl structure; more than 200 are known.

---

**TABLE 1.6**

**Compounds that have the NaCl (rock-salt) type of crystal structure**

Most alkali halides, MX, and AgF, AgCl, AgBr

All the alkali hydrides, MH

Monoxides, MO, of Mg, Ca, Sr, Ba

Monosulfides, MS, of Mg, Ca, Sr, Ba

---

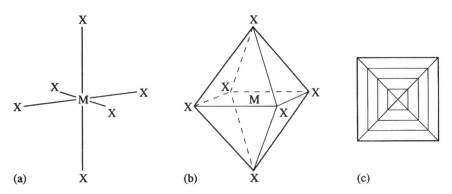

**FIGURE 1.33** (a) An [$MX_6$] octahedron, (b) a solid octahedron, and (c) plan of an octahedron with contours.

Many of the structures described in this book can be viewed as linked octahedra, where each octahedron consists of a metal atom surrounded by six other atoms situated at the corners of an octahedron (Figure 1.33(a) and Figure 1.33(b)). These can also be depicted as viewed from above with contours marked, as in Figure 1.33(c). Octahedra can link together via corners, edges, and faces, as seen in Figure 1.34. The

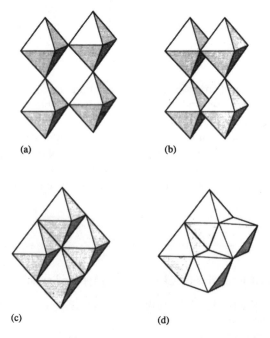

**FIGURE 1.34** The conversion of (a) corner-shared $MX_6$ octahedra to (b) edge-shared octahedra, and (c) edge-shared octahedra to (d) face-shared octahedra.

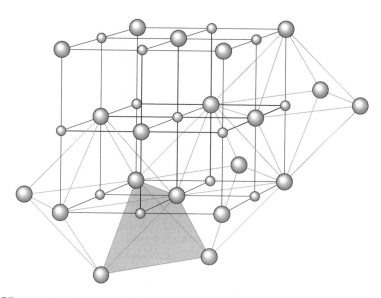

**FIGURE 1.35** NaCl structure showing edge-sharing of octahedra and the enclosed tetrahedral space (shaded).

linking of octahedra by different methods effectively eliminates atoms because some of the atoms are now shared between them: two $MO_6$ octahedra linked through a vertex has the formula, $M_2O_{11}$; two $MO_6$ octahedra linked through an edge has the formula, $M_2O_{10}$; two $MO_6$ octahedra linked through a face has the formula, $M_2O_9$.

The NaCl structure can be described in terms of $NaCl_6$ octahedra sharing edges. An octahedron has 12 edges, and each one is shared by two octahedra in the NaCl structure. This is illustrated in Figure 1.35, which shows a NaCl unit cell with three $NaCl_6$ octahedra shown in outline, and one of the resulting tetrahedral spaces is depicted by shading.

## The Nickel Arsenide Structure (NiAs)

The nickel arsenide structure is the equivalent of the sodium chloride structure in hexagonal close-packing. It can be described as an *hcp* array of arsenic atoms with nickel atoms occupying the octahedral holes. The geometry about the nickel atoms is thus octahedral. This is not the case for arsenic: each arsenic atom sits in the centre of a **trigonal prism** of six nickel atoms (Figure 1.36).

## The Zinc Blende (or Sphalerite) and Wurtzite Structures (ZnS)

Unit cells of these two structures are shown in Figure 1.37 and Figure 1.38, respectively. They are named after two different naturally occurring mineral forms of zinc sulfide. Zinc blende is often contaminated by iron, making it very dark in colour and thus lending it the name of 'Black Jack'. Structures of the same element or compound that differ only in their atomic arrangements are termed **polymorphs**.

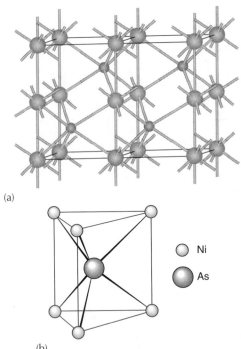

(a)

(b)

**FIGURE 1.36** (a) The unit cell of nickel arsenide, NiAs. (For undistorted *hcp* $c/a$ = 1.633, but this ratio is found to vary considerably.) Ni, blue spheres; As, grey spheres. (b) The trigonal prismatic coordination of arsenic in NiAs.

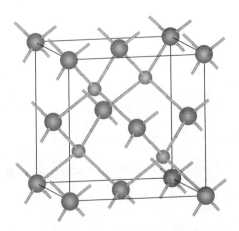

**FIGURE 1.37** The crystal structure of zinc blende or sphalerite, ZnS. Zn, blue spheres; S, grey spheres (or vice versa).

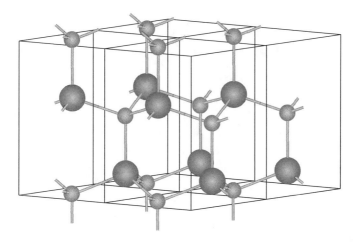

**FIGURE 1.38** The crystal structure of wurtzite, ZnS. Zn, blue spheres; S, grey spheres.

The zinc blende structure can be described as a *ccp* array of sulfide ions with zinc ions occupying every other tetrahedral hole in an ordered manner. Each zinc ion is thus tetrahedrally coordinated by four sulfides and vice versa. Compounds adopting this structure include the copper halides and Zn, Cd, and Hg sulfides. Notice that if all the atoms were identical, the structure would be the same as that of a diamond (see Section 1.6.5). Notice that the atomic positions are equivalent, and we could equally well generate the structure by swapping the zinc and sulfurs.

The wurtzite structure is composed of an *hcp* array of sulfide ions with alternate tetrahedral holes occupied by zinc ions. Each zinc ion is tetrahedrally coordinated by four sulfide ions and vice versa. Compounds adopting the structure include BeO, ZnO, and $NH_4F$.

Notice how the coordination numbers of the structures we have observed so far have changed. The coordination number for close-packing, where all the atoms are identical, is twelve. In the CsCl structure, it is eight; in NaCl, it is six; and in both of the ZnS structures, it is four. Generally, the larger a cation is, the more anions it can pack around itself (see Section 1.6.4).

## 1.6.2 Solids with General Formula MX$_2$

### The Fluorite and Antifluorite Structures

The fluorite structure is named after the mineral form of calcium fluoride, $CaF_2$, which is found in the U.K. in the famous Derbyshire 'Blue John' mines. The structure is illustrated in Figure 1.39. It can be described as related to a *ccp* array of calcium ions with fluorides occupying all of the tetrahedral holes. There is a problem with this as a description because calcium ions are rather smaller than fluoride ions, and so, physically, fluoride ions would not be able to fit into the tetrahedral holes of a calcium ion array. Nevertheless, it gives an exact description of the *relative* positions of the ions. The diagram in Figure 1.39(a) depicts the fourfold tetrahedral coordination

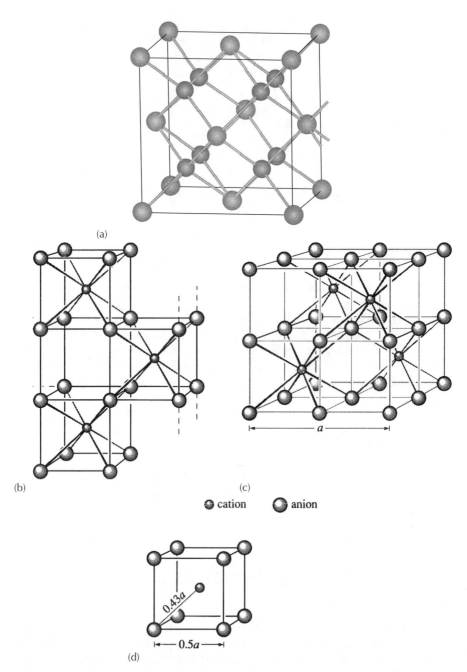

(a)

(b)                                                                (c)

● cation        ○ anion

(d)

**FIGURE 1.39** The crystal structure of fluorite, $CaF_2$. (a) Computer generated unit cell as a *ccp* array of cations: Ca, blue spheres; F, grey spheres. (b) and (c) The same structure redrawn as a primitive cubic array of anions. (d) Relationship of unit cell dimensions to the primitive anion cube (the octant).

of the fluoride ions very clearly. Notice also that the larger octahedral holes are vacant in this structure — one of them is located at the body-centre of the unit cell in Figure 1.39(a). This becomes a very important feature when we come to look at the movement of ions through defect structures in Chapter 5.

By drawing cubes with fluoride ions at each corner as has been done in Figure 1.39(b), you can see that there is an eightfold cubic coordination of each calcium cation. Indeed, it is possible to move the origin and redraw the unit cell so that this feature can be seen more clearly as has been done in Figure 1.39(c). The unit cell is now divided into eight smaller cubes called **octants**, with each alternate octant occupied by a calcium cation.

In the antifluorite structure, the positions of the cations and anions are merely reversed, and the description of the structure as cations occupying all the tetrahedral holes in a *ccp* array of anions becomes more realistic. In the example with the biggest anion and smallest cation, $Li_2Te$, the telluriums are approximately close-packed (even though there is a considerable amount of covalent bonding). For the other compounds adopting this structure, such as the oxides and sulfides of the alkali metals, $M_2O$ and $M_2S$, the description accurately shows the relative positions of the atoms. However, the anions could not be described as close-packed because they are not touching. The cations are too big to fit in the tetrahedral holes, and, therefore, the anion–anion distance is greater than for close-packing.

These are the only structures where 8:4 coordination is found. Many of the fast-ion conductors are based on these structures (see Chapter 5, Section 5.4).

## The Cadmium Chloride ($CdCl_2$) and Cadmium Iodide ($CdI_2$) Structures

Both of these structures are based on the close-packing of the appropriate anion with half of the octahedral holes occupied by cations. In both structures, the cations occupy all the octahedral holes in every other anion layer, giving an overall layer structure with 6:3 coordination. The cadmium chloride structure is based on a *ccp* array of chloride ions whereas the cadmium iodide structure is based on an *hcp* array of iodide ions. The cadmium iodide structure is shown in Figure 1.40, and in (a) we can see that an iodide anion is surrounded by three cadmium cations on one side but by three iodides on the other (i.e., it is not completely surrounded by ions of the opposite charge as we would expect for an ionic structure). This is evidence that the bonding in some of these structures is not entirely ionic, as we have tended to imply so far. This point is discussed again in more detail in Section 1.6.4.

## The Rutile Structure

The rutile structure is named after one mineral form of titanium oxide ($TiO_2$). Rutile has a very high refractive index, scattering most of the visible light incident on it, and so is the most widely used white pigment in paints and plastics. A unit cell is illustrated in Figure 1.41. The unit cell is tetragonal and the structure again demonstrates 6:3

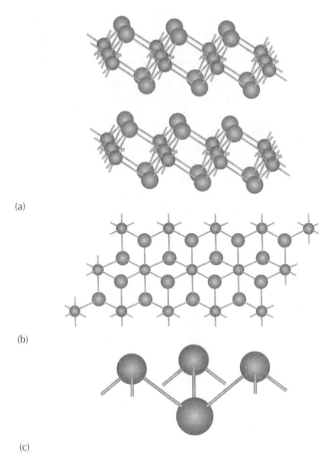

(a)

(b)

(c)

**FIGURE 1.40** (a) The crystal structure of cadmium iodide, $CdI_2$; (b) the structure of the layers in $CdI_2$ and $CdCl_2$: the halogen atoms lie in planes above and below that of the metal atoms; and (c) the coordination around one iodine atom in $CdI_2$. Cd, blue spheres; I, grey spheres.

coordination but is not based on close-packing: each titanium atom is coordinated by six oxygens at the corners of a (slightly distorted) octahedron and each oxygen atom is surrounded by three planar titaniums which lie at the corners of an (almost) equilateral triangle. It is not geometrically possible for the coordination around Ti to be a perfect octahedron *and* for the coordination around O to be a perfect equilateral triangle.

The structure can be viewed as chains of linked $TiO_6$ octahedra, where each octahedron shares a pair of opposite edges, and the chains are linked by sharing vertices: this is shown in Figure 1.41(b). Figure 1.41(c) shows a plan of the unit cell looking down the chains of octahedra so that they are seen in projection.

Occasionally the **antirutile structure** is encountered where the metal and non-metals have changed places, such as in $Ti_2N$.

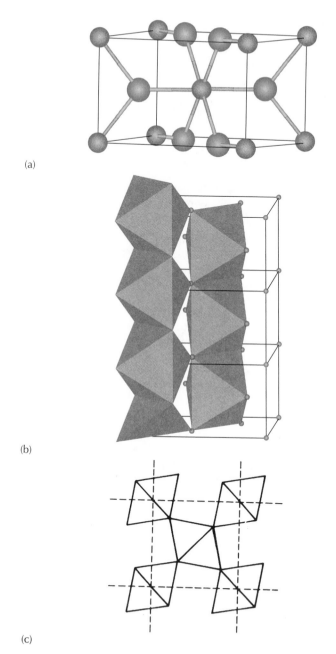

(a)

(b)

(c)

**FIGURE 1.41** The crystal structure of rutile, $TiO_2$. (a) Unit cell, (b) parts of two chains of linked $[TiO_6]$ octahedra, and (c) projection of structure on base of unit cell. Ti, blue spheres; O, grey spheres.

## The $\beta$-cristobalite Structure

The $\beta$-cristobalite structure is named after one mineral form of silicon dioxide, $SiO_2$. The silicon atoms are in the same positions as both the zinc and sulfurs in zinc blende (or the carbons in diamond, which we look at later in Section 1.6.5): each pair of silicon atoms is joined by an oxygen midway between. The only metal halide adopting this structure is beryllium fluoride, $BeF_2$, and it is characterized by 4:2 coordination.

## 1.6.3  OTHER IMPORTANT CRYSTAL STRUCTURES

As the valency of the metal increases, the bonding in these simple binary compounds becomes more covalent and the highly symmetrical structures characteristic of the simple ionic compounds occur far less frequently, with molecular and layer structures being common. Many thousands of inorganic crystal structures exist. Here we describe just a few of those that are commonly encountered and those that occur in later chapters.

## The Bismuth Triiodide Structure ($BiI_3$)

This structure is based on an *hcp* array of iodides with the bismuths occupying one-third of the octahedral holes. Alternate pairs of layers have two-thirds of the octahedral sites occupied.

## Corundum $\alpha$-$Al_2O_3$

This mineral is the basis for ruby and sapphire gemstones, their colour depending on the impurities. It is very hard – second only to diamond. This structure may be described as an *hcp* array of oxygen atoms with two-thirds of the octahedral holes occupied by aluminium atoms. As we have seen before, geometrical constraints dictate that octahedral coordination of the aluminiums precludes tetrahedral coordination of the oxygens. However, it is suggested that this structure is adopted in preference to other possible ones because the four aluminiums surrounding an oxygen approximate most closely to a regular tetrahedron. The structure is also adopted by $Ti_2O_3$, $V_2O_3$, $Cr_2O_3$, $\alpha$-$Fe_2O_3$, $\alpha$-$Ga_2O_3$, and $Rh_2O_3$.

## The Rhenium Trioxide Structure ($ReO_3$)

This structure (also called the aluminium fluoride structure) is adopted by the fluorides of Al, Sc, Fe, Co, Rh, and Pd; also by the oxides $WO_3$ (at high temperature) and $ReO_3$ (see Chapter 5, Section 5.8.1). The structure consists of $ReO_6$ octahedra linked together through each corner to give a highly symmetrical three-dimensional network with cubic symmetry. Part of the structure is given in Figure 1.42(a), the linking of the octahedra in (b), and the unit cell in (c).

## Mixed Oxide Structures

Three important mixed oxide structures exist: **spinel**, **perovskite**, and **ilmenite**.

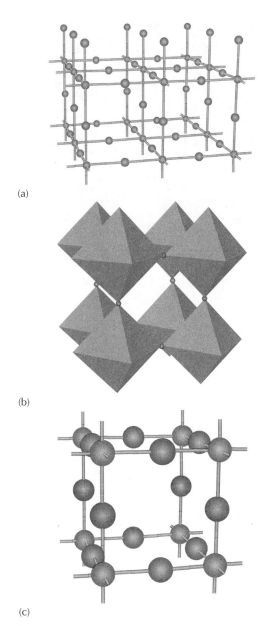

(a)

(b)

(c)

**FIGURE 1.42** (a) Part of the $ReO_3$ structure, (b) $ReO_3$ structure showing the linking of $[ReO_6]$ octahedra, and (c) unit cell. Re, blue spheres; O, grey spheres.

## The Spinel and Inverse-spinel Structures

The spinels have the general formula $AB_2O_4$, taking their name from the mineral spinel $MgAl_2O_4$: generally, A is a divalent ion, $A^{2+}$, and B is trivalent, $B^{3+}$. The structure can be described as being based on a cubic close-packed array of oxide ions, with $A^{2+}$ ions occupying tetrahedral holes and $B^{3+}$ ions occupying octahedral

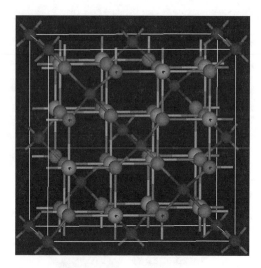

**FIGURE 1.43** The spinel structure, $CuAl_2O_4$ ($AB_2O_4$). See colour insert following page 196. Cu, blue spheres; Al, pink spheres; O, red spheres.

holes. A spinel crystal containing $n$ $AB_2O_4$ formula units has $8n$ tetrahedral holes and $4n$ octahedral holes; accordingly, one-eighth of the tetrahedral holes are occupied by $A^{2+}$ ions and one-half of the octahedral holes by the $B^{3+}$ ions. A unit cell is illustrated in Figure 1.43. The A ions occupy tetrahedral positions together with the corners and face-centres of the unit cell. The B ions occupy octahedral sites. Spinels with this structure include compounds of formula $MAl_2O_4$ where M is Mg, Fe, Co, Ni, Mn, or Zn.

When compounds of general formula $AB_2O_4$ adopt the inverse-spinel structure, the formula is better written as $B(AB)O_4$, because this indicates that half of the $B^{3+}$ ions now occupy tetrahedral sites, and the remaining half, together with the $A^{2+}$ ions, occupy the octahedral sites. Examples of inverse-spinels include magnetite, $Fe_3O_4$, (see Chapter 9, Section 9.7) $Fe(MgFe)O_4$, and $Fe(ZnFe)O_4$.

*The Perovskite Structure*

This structure is named after the mineral $CaTiO_3$. A unit cell is shown in Figure 1.44(a): This unit cell is known as the A-type because if we take the general formula $ABX_3$ for the perovskites, then theA atom is at the centre in this cell. The central Ca (A) atom is coordinated to 8 Ti atoms (B) at the corners and to 12 oxygens (X) at the midpoints of the cell edges. The structure can be usefully described in other ways. First, it can be described as a *ccp* array of A and X atoms with the B atoms occupying the octahedral holes (compare with the unit cell of NaCl in Figure 1.31 if you want to check this). Second, perovskite has the same octahedral framework as $ReO_3$ based on $BX_6$ octahedra with an A atom added in at the centre of the cell (Figure 1.42(b)). Compounds adopting this structure include $SrTiO_3$, $SrZrO_3$, $SrHfO_3$, $SrSnO_3$, and $BaSnO_3$. The structures of the high temperature superconductors are based on this structure (see Chapter 10, Section 10.3.1).

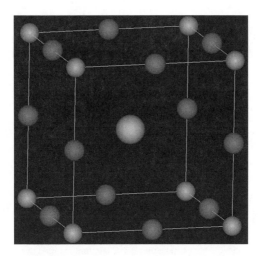

**FIGURE 1.44** The perovskite structure of compounds $ABX_3$, such as $CaTiO_3$. See colour insert following page 196. Ca, green sphere; Ti, silver spheres; O, red spheres.

### The Ilmenite Structure

The ilmenite structure is adopted by oxides of formula $ABO_3$ when A and B are similar in size and their total charge adds up to +6. The structure is named after the mineral of $Fe^{II}Ti^{IV}O_3$, and the structure is very similar to the corundum structure described previously, an *hcp* array of oxygens, but now two different cations are present occupying two-thirds of the octahedral holes.

The structures related to close-packing are summarized in Table 1.7.

### 1.6.4 Ionic Radii

We know from quantum mechanics that atoms and ions do not have precisely defined radii. However, from the foregoing discussion of ionic crystal structures we have seen that ions pack together in an extremely regular fashion in crystals, and their atomic positions, and thus their interatomic distances, can be measured very accurately. It is a very useful concept, therefore, particularly for those structures based on close-packing, to think of ions as hard spheres, each with a particular radius.

If we take a series of alkali metal halides, all with the rock salt structure, as we replace one metal ion with another, say sodium with potassium, we would expect the metal-halide internuclear distance to change by the same amount each time if the concept of an ion as a hard sphere with a particular radius holds true. Table 1.8 presents the results of this procedure for a range of alkali halides, and the change in internuclear distance on swapping one ion for another is highlighted.

From Table 1.8, we can see that the change in internuclear distance on changing the ion is not constant, but also that the variation is not great: this provides us with some experimental evidence that it is not unreasonable to think of the ions as having a fixed radius. We can see, however, that the picture is not precisely true, neither would we expect it to be, because atoms and ions are squashable entities and their

**TABLE 1.7**
**Structures related to close-packed arrangements of anions**

| | | | Examples | |
| --- | --- | --- | --- | --- |
| Formula | Cation:anion coordination | Type and number of holes occupied | Cubic close-packing | Hexagonal close-packing |
| MX | 6:6 | All octahedral | Sodium chloride: NaCl, FeO, MnS, TiC | Nickel arsenide: NiAs, FeS, NiS |
| | 4:4 | Half tetrahedral; every alternate site occupied | Zinc blende: ZnS, CuCl, $\gamma$-AgI | Wurtzite: ZnS, $\beta$-AgI |
| $MX_2$ | 8:4 | All tetrahedral; | Fluorite : $CaF_2$, $ThO_2$, $ZrO_2$, $CeO_2$ | None |
| | 6:3 | Half octahedral; alternate layers have fully occupied sites | Cadmium chloride : $CdCl_2$ | Cadmium iodide: $CdI_2$, $TiS_2$ |
| $MX_3$ | 6:2 | One-third octahedral; alternate pairs of layers have two-thirds of the octahedral sites occupied | | Bismuth iodide: $BiI_3$, $FeCl_3$, $TiCl_3$, $VCl_3$ |
| $M_2X_3$ | 6:4 | Two-thirds octahedral | | Corundum: $\alpha$-$Al_2O_3$, $\alpha$-$Fe_2O_3$, $V_2O_3$, $Ti_2O_3$, $\alpha$-$Cr_2O_3$ |
| $ABO_3$ | | Two-thirds octahedral | | Ilmenite: $FeTiO_3$ |
| $AB_2O_4$ | | One-eighth tetrahedral and one-half octahedral | Spinel: $MgAl_2O_4$ inverse spinel: $MgFe_2O_4$, $Fe_3O_4$ | Olivine: $Mg_2SiO_4$ |

size is going to be affected by their environment. Nevertheless, it is a useful concept to develop a bit further as it enables us to describe some of the ionic crystal structures in a simple pictorial way.

There have been many suggestions as to how individual ionic radii can be assigned, and the literature contains several different sets of values. Each set is named after the person(s) who originated the method of determining the radii. We will describe some of these methods briefly before listing the values most commonly used at present. It is most important to remember that you must not mix radii from more than one set of values. Even though the values vary considerably from set to set, each set is internally consistent (i.e., if you add together two radii from one set

**TABLE 1.8**
**Interatomic distances of some alkali halides, $r_{M-X}$/pm**

|       | F⁻  |    | Cl⁻ |    | Br⁻ |    | I⁻  |
| ----- | --- | -- | --- | -- | --- | -- | --- |
| Li⁺   | 201 | 56 | 257 | 18 | 275 | 27 | 302 |
|       | 30  |    | 24  |    | 23  |    | 21  |
| Na⁺   | 231 | 50 | 281 | 17 | 298 | 25 | 323 |
|       | 35  |    | 33  |    | 31  |    | 30  |
| K⁺    | 266 | 48 | 314 | 15 | 329 | 24 | 353 |
|       | 16  |    | 14  |    | 14  |    | 13  |
| Rb⁺   | 282 | 46 | 328 | 15 | 343 | 23 | 366 |

of values, you will obtain an approximately correct internuclear distance as determined from the crystal structure).

The internuclear distances can be determined by X-ray crystallography. In order to obtain values for individual ionic radii from these, the value of one radius needs to be fixed by some method. Originally in 1920, Landé suggested that in the alkali halide with the largest anion and smallest cation — LiI — the iodide ions must be in contact with each other with the tiny Li⁺ ion inside the octahedral hole: as the Li–I distance is known, it is then a matter of simple geometry to determine the iodide radius. Once the iodide radius is known then the radii of the metal cations can be found from the structures of the metal iodides — and so on. Bragg and Goldschmidt later extended the list of ionic radii using similar methods. It is very difficult to come up with a consistent set of values for the ionic radii, because of course the ions are not hard spheres, they are somewhat elastic, and the radii are affected by their environment such as the nature of the oppositely charged ligand and the coordination number. Pauling proposed a theoretical method of calculating the radii from the internuclear distances; he produced a set of values that is both internally consistent and shows the expected trends in the Periodic Table. Pauling's method was to take a series of alkali halides with isoelectronic cations and anions and assume that they are in contact: if you then assume that each radius is inversely proportional to the effective nuclear charge felt by the outer electrons of the ion, a radius for each ion can be calculated from the internuclear distance. Divalent ions undergo additional compression in a lattice and compensation has to be made for this effect in calculating their radii. With some refinements this method gave a consistent set of values that was widely used for many years; they are usually known as **effective ionic radii**.

It is also possible to determine accurate electron density maps for the ionic crystal structures using X-ray crystallography. Such a map is shown for NaCl and LiF in Figure 1.45. The electron density contours fall to a minimum — although not to zero — in between the nuclei and it is suggested that this minimum position should be taken as the radius position for each ion. These experimentally determined ionic radii are often called **crystal radii**; the values are somewhat different from the older sets and tend to make the anions smaller and the cations bigger than previously. The most comprehensive set of radii has been compiled by

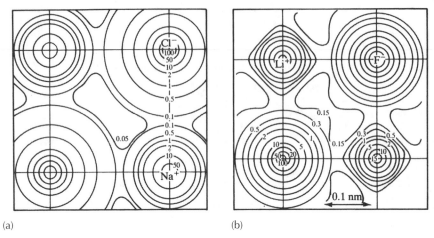

(a)                                                    (b)

**FIGURE 1.45** Electron density maps for (a) NaCl and (b) LiF.

Shannon and Prewitt using data from almost a thousand crystal structure determinations and based on conventional values of 126 pm and 119 pm for the radii of the $O^{2-}$ and $F^-$ ions, respectively. These values differ by a constant factor of 14 pm from traditional values but it is generally accepted that they correspond more closely to the actual physical sizes of ions in a crystal. A selection of this data is shown in Table 1.9.

Several important trends in the sizes of ions can be noted from the data in Table 1.9:

1. The radii of ions within a Group of the Periodic Table, such as the alkali metals, increase with atomic number, $Z$: as you go down a Group, more electrons are present, and the outer ones are further from the nucleus.
2. In a series of isoelectronic cations, such as $Na^+$, $Mg^{2+}$, and $Al^{3+}$, the radius decreases rapidly with increasing positive charge. The number of electrons is constant but the nuclear charge increases and so pulls the electrons in and the radii decrease.
3. For pairs of isoelectronic anions (e.g., $F^-$, $O^{2-}$), the radius increases with increasing charge because the more highly charged ion has a smaller nuclear charge.
4. For elements with more than one oxidation state (e.g., $Ti^{2+}$ and $Ti^{3+}$ ), the radii decrease as the oxidation state increases. In this case, the nuclear charge stays the same, but the number of electrons that it acts on decreases.
5. As you move across the Periodic Table for a series of similar ions, such as the first row transition metal divalent ions, $M^{2+}$, there is an overall decrease in radius. This is due to an increase in nuclear charge across the Table because electrons in the same shell do not screen the nucleus from each other very well. A similar effect is observed for the $M^{3+}$ ions of the lanthanides and this is known as the **lanthanide contraction**.

## TABLE 1.9
## Crystal radii (pm)[a]

| 1 | 2 | | | | | | | | | | | 13 | 14 | 15 | 16 | 17 | 18 |
|---|---|---|---|---|---|---|---|---|---|---|---|---|---|---|---|---|---|
| H 126(−1)[b] | | | | | | | | | | | | | | | | | He — |
| Li 90(+1) | Be 59(+2) | | | | | | | | | | | B 41(+3) | C 30(+4) | N 132(−3)[b] 30(+3) | O 126(−2) | F 119(−1) | Ne — |
| Na 116(+1) | Mg 86(+2) | | | | | | | | | | | Al 68(+3) | Si 54(+4) | P 58(+3) | S 170(−2) | Cl 167(−1) | Ar — |
| K 152(+1) | Ca 114(+2) | Sc 89(+3) | Ti 100(+2) 81(+3) 75(+4) | V 93(+2) 78(+3) | Cr 87/94(+2)[b] 76(+3) | Mn 81/97(+2)[b] 72/79(+3)[b] | Fe 75/92(+2)[b] 69/79(+3)[b] | Co 79/89(+2)[b] 69/75(+3)[b] | Ni 83(+2) 70/74(+3)[b] | Cu 91(+1) 87(+2) 68(+3) | Zn 88(+2) | Ga 76(+3) | Ge 87(+2) 67(+4) | As 72(+3) | Se 184(−2) | Br 182(−1) | Kr — |
| Rb 166(+1) | Sr 132(+2) | Y 104(+3) | Zr 86(+4) | | | | | | | Ag 129(+1) | Cd 109(+2) | In 94(+3) | Sn 136(+2) 83(+4) | Sb 90(+3) | Te 207(−2) | I 206(−1) | Xe 62(+8) |
| Cs 181(+1) | Ba 149(+2) | Lu 100(+3) | Hf 85(+4) | | | | | | | Au 151(+1) | Hg 116(+2) | Tl 103(+3) | Pb 133(+2) 92(+4) | Bi 117(+3) | Po — | At — | Rn — |
| Fr 194(+1) | Ra — | | | | | | | | | | | | | | | | |

[a] Values taken from R.D. Shannon, *Acta Cryst.*, A32 (1976), 751.
[b] Figures in parentheses indicate the charge on the ion.

6. For transition metals, the spin state affects the ionic radius.
7. The crystal radii increase with an increase in coordination number — see $Cu^+$ and $Zn^{2+}$ in Table 1.9. One can think of fewer ligands around the central ion as allowing the counterions to compress the central ion.

The picture of ions as hard spheres works best for fluorides and oxides, both of which are small and somewhat uncompressable ions. As the ions get larger, they are more easily compressed — the electron cloud is more easily distorted — and they are said to be more **polarizable**.

When we were discussing particular crystal structures in the previous section we noted that a larger cation such as $Cs^+$, was able to pack eight chloride ions around it, whereas the smaller $Na^+$ only accommodated six. If we continue to think of ions as hard spheres for the present, for a particular structure, as the ratio of the cation and anion radii changes there will come a point when the cation is so small that it will no longer be in touch with the anions. The lattice would not be stable in this state because the negative charges would be too close together for comfort and we would predict that the structure would change to one of lower coordination, allowing the anions to move further apart. If the ions are hard spheres, using simple geometry, it is possible to quantify the radius ratio (i.e., $\dfrac{r_{cation}}{r_{anion}} = \dfrac{r_+}{r_-}$ ) at which this happens, and this is illustrated for the octahedral case in Figure 1.46.

Taking a plane through the centre of an octahedrally coordinated metal cation, the stable situation is shown in Figure 1.46(a) and the limiting case for stability, when the anions are touching, in Figure 1.46(b). The anion radius, $r_-$ in Figure 1.46(b) is OC, and the cation radius, $r_+$, is (OA – OC). From the geometry of the right-angled triangle we can see that $\cos 45° = \dfrac{OC}{OA} = 0.707$. The radius ratio, $\dfrac{r_+}{r_-}$ is given by $\dfrac{(OA - OC)}{OC} = (\dfrac{OA}{OC} - 1) = (1.414 - 1) = 0.414$. Using similar calculations it is possible to calculate limiting ratios for the other geometries: Table 1.10 summarizes these.

On this basis, we would expect to be able to use the ratio of ionic radii to predict possible crystal structures for a given substance. But does it work? Unfortunately, only about 50% of the structures are correctly predicted. This is because the model is too simplistic — ions are not hard spheres, but are polarized under the influence of other ions instead. In larger ions, the valence electrons are further away from the nucleus, and shielded from its influence by the inner core electrons, and so the electron cloud is more easily distorted. The ability of an ion to distort an electron cloud — its polarizing power — is greater for small ions with high charge; the distortion of the electron cloud means that the bonding between two such ions becomes more directional in character. This means that the bonding involved is rarely truly ionic but frequently involves at least some degree of covalency. The higher the formal charge on a metal ion, the greater will be the proportion of covalent

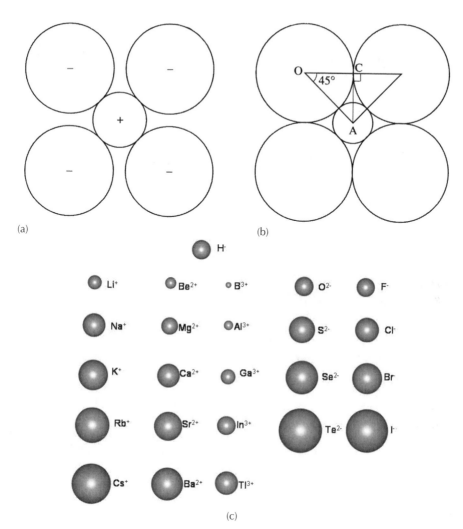

**FIGURE 1.46** (a) Anions packed around a cation on a horizontal plane, (b) anion–anion contact on a horizontal plane through an octahedron, and (c) relative sizes of the typical element ions.

bonding between the metal and its ligands. The higher the degree of covalency, the less likely is the concept of ionic radii and radius ratios to work. It also seems that there is little energy difference between the six-coordinate and eight-coordinate structures, and the six-coordinate structure is usually preferred — eight-coordinate structures are rarely found, for instance, no eight-coordinate oxides exist. The preference for the six-coordinate rock-salt structure is thought to be due to the small amount of covalent-bond contribution. In this structure, the three orthogonal p orbitals lie in the same direction as the vectors joining the cation to the surrounding six anions. Thus, they are well placed for good overlap of the orbitals necessary for

**TABLE 1.10**
**Limiting radius ratios for different coordination numbers**

| Coordination number | Geometry | Limiting radius ratio | Possible structures |
|---|---|---|---|
| | | 0.225 | |
| 4 | Tetrahedral | | Wurtzite, zinc blende |
| | | 0.414 | |
| 6 | Octahedral | | Rock-salt, rutile |
| | | 0.732 | |
| 8 | Cubic | | Caesium chloride, fluorite |
| | | 1.00 | |

σ bonding to take place. The potential overlap of the p orbitals in the caesium chloride structure is less favourable.

### 1.6.5 EXTENDED COVALENT ARRAYS

The last section noted that many 'ionic' compounds in fact possess some degree of covalency in their bonding. As the formal charge on an ion becomes greater we expect the degree of covalency to increase, so we would generally expect compounds of elements in the centre of the Periodic Table to be covalently bonded. Indeed, some of these elements themselves are covalently bonded solids at room temperature. Examples include elements such as Group III(3), boron; Group IV(14), carbon, silicon, germanium; Group V(15), phosphorus, arsenic; Group VI(16), selenium, tellurium; they form **extended covalent arrays** in their crystal structures.

Take, for instance, one of the forms of carbon: **diamond**. Diamond has a cubic crystal structure with an F-centred lattice (Figure 1.47); the positions of the atomic

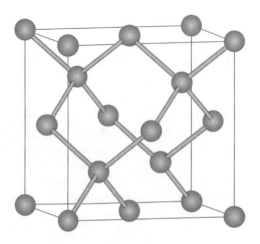

**FIGURE 1.47** A unit cell of the diamond structure.

centres are the same as in the zinc blende structure, with carbon now occupying both the zinc and the sulfur positions. Each carbon is equivalent and is tetrahedrally coordinated to four others, forming a covalently bonded **giant molecule** throughout the crystal: the carbon–carbon distances are all identical (154 pm). It is interesting to note how the different type of bonding has affected the coordination: here we have identical atoms all the same size but the coordination number is now restricted to four because this is the maximum number of covalent bonds that carbon can form. In the case of a metallic element such as magnesium forming a crystal, the structure is close-packed with each atom 12-coordinated (bonding in metals is discussed in Chapter 4). The covalent bonds in diamond are strong, and the rigid three-dimensional network of atoms makes diamond the hardest substance known; it also has a high melting temperature (m.t.) (3773 K). Silicon carbide (SiC) known as carborundum also has this structure with silicons and carbons alternating throughout: it, too, is very hard and is used for polishing and grinding.

Silica ($SiO_2$) gives us other examples of giant molecular structures. There are two crystalline forms of silica at atmospheric pressure: **quartz** and **cristobalite**. Each of these also exists in low and high temperature forms, α- and β-, respectively. We have already discussed the structure of β-cristobalite in terms of close-packing in Section 1.6.2. Quartz is commonly encountered in nature: the structure of β-quartz is illustrated in Figure 1.48, and consists of $SiO_4$ tetrahedra linked so that each oxygen atom is shared by two tetrahedra, thus giving the overall stoichiometry of $SiO_2$. Notice how once again the covalency of each atom dictates the coordination around itself, silicon having four bonds and oxygen two, rather than the larger coordination numbers that are found for metallic and some ionic structures. Quartz is unusual in that the linked tetrahedra form **helices** or spirals throughout the crystal, which are all either left- or right-handed, producing laevo- or dextrorotatory crystals, respectively; these are known as **enantiomorphs**.

For our final example in this section, we will look at the structure of another polymorph of carbon. Normal **graphite** is illustrated in Figure 1.49. (There are other

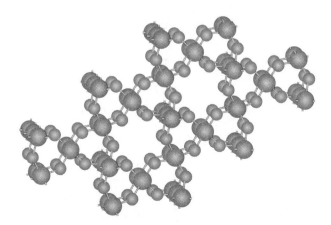

**FIGURE 1.48** The β-quartz structure.

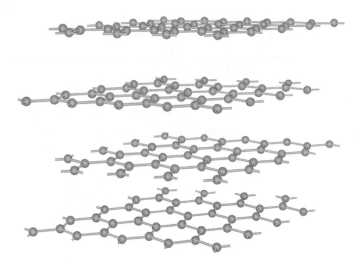

**FIGURE 1.49** SThe crystal structure of graphite.

graphite structures that are more complex.) The structure of normal graphite consists of two-dimensional layers of carbon atoms joined together in a hexagonal array. Within the layers, each carbon atom is strongly bonded to three others at a distance of 142 pm. This carbon–carbon distance is rather shorter than the one observed in diamond, due to the presence of some double bonding. (The planar hexagonal configuration of carbon atoms puts some of the 2p orbitals in a suitable position for $\pi$-overlap in a similar fashion to the $\pi$-bonding in benzene where the carbon–carbon distance is 139.7 pm.) The distance between the layers is much greater at 340 pm; this is indicative of weak bonding between the layers due to van der Waals forces. Graphite is a soft grey solid with a high m.t. and low density; its softness is attributed to the weak bonding between the layers which allows them to slide over one another. The weak bonding and large distance between the layers also explains the low density of graphite. In addition, graphite crystals shear easily parallel to the layers. It is a popular myth, however, that this ease of shearing makes graphite a useful lubricant. In fact, its lubricant properties are dependent on an adsorbed layer of nitrogen, and when this is lost under extreme conditions such as low pressure or high temperature, then the lubricant properties are also lost. When graphite is used as a lubricant under high vacuum conditions, such as pertain in outer space, then surface additives have to be incorporated to maintain the low-friction properties. The electrons in the delocalised $\pi$-orbital are mobile, and so graphite is electrically conducting in the layers like a two-dimensional metal, but is a poor conductor perpendicular to the layers.

### 1.6.6 BONDING IN CRYSTALS

Although bonding in solids will be discussed in detail in Chapter 4, it is convenient at this point to summarize the different types of bonding that we meet in crystal structures. In Section 1.2, we considered the structures of metallic crystals held together by metallic bonding. In Section 1.6, we looked at structures, such as NaCl and CsCl, that have ionic bonding, and later saw the influence of covalent bonding in the layer structures of $CdCl_2$ and $CdI_2$. In the graphite structure we see covalently bonded layers of carbon atoms held together by weak van der Waals forces, and will meet this again, together with hydrogen bonding in the next section on molecular crystal structures.

## Metallic Bonding

Metals consist of a regular array of metal cations surrounded by a 'sea' of electrons. These electrons occupy the space between the cations, binding them together, but are able to move under the influence of an external field, thus accounting for the electrical conductivity of metals.

## Ionic Bonding

An ionic bond forms between two oppositely charged ions due to the electrostatic attraction between them. The attractive force, $F$, is given by **Coulomb's Law**:

$F \propto \dfrac{q_1 q_2}{r^2}$ , where $q_1$ and $q_2$ are the charges on the two ions, and $r$ is the distance between them. A similar but opposite force is experienced by two ions of the same charge. Ionic bonds are strong and nondirectional; the energy of the interaction is given by force × distance, and is inversely proportional to the separation of the charges, $r$. Ionic forces are effective over large distances compared with other bonding interactions. Ions pack together in regular arrays in ionic crystals, in such a way as to maximize Coulombic attraction, and minimize repulsions.

## Covalent Bonding

In covalent bonds, the electrons are shared between two atoms resulting in a build-up of electron density between the atoms. Covalent bonds are strong and directional.

## Charge–Dipole and Dipole–Dipole Interactions

In a covalent bond, electronegative elements such as oxygen and nitrogen attract an unequal share of the bonding electrons, such that one end of the bond acquires a partial negative charge, δ–, and the other end a partial positive charge, δ+. The separation of negative and positive charge creates an **electric dipole**, and the molecule can align itself in an electric field. Such molecules are said to be **polar**. The partial electric charges on polar molecules can attract one another in a **dipole–dipole interaction**. The dipole–dipole interaction is about 100 times weaker than ionic interactions and falls off quickly with distance, as a function of $\dfrac{1}{r^3}$ .

Polar molecules can also interact with ions in a **charge–dipole interaction** which is about 10 to 20 times weaker than ion–ion interactions, and which decreases with distance as $\dfrac{1}{r^2}$.

## London Dispersion Forces

Even if molecules do not possess a permanent dipole moment weak forces can exist between them. The movement of the valence electrons creates 'transient dipoles', and these in turn induce dipole moments in adjacent molecules. The transient dipole in one molecule can be attracted to the transient dipole in a neighbouring molecule, and the result is a weak, short-range attractive force known as the **London dispersion force**. These dispersion forces drop off rapidly with distance, decreasing as a function of $\dfrac{1}{r^6}$.

The weak nonbonded interactions that occur between molecules are often referred to collectively as **van der Waals forces**.

## Hydrogen-Bonding

In one special case, polar interactions are strong enough for them to be exceptionally important in dictating the structure of the solid and liquid phases. Where hydrogen is bonded to a very electronegative element such as oxygen or fluorine, there is a partial negative charge, $\delta-$, on the electronegative element, and an equal and opposite $\delta+$ charge on the hydrogen. The positively charged $H^{\delta+}$ can also be attracted to the partial negative charge on a neighbouring molecule, forming a weak bond known as a **hydrogen-bond**, O–H---O, and pulling the three atoms almost into a straight line. A network of alternating weak and strong bonds is built up, and examples can be seen in water ($H_2O$) and in hydrogen fluoride (HF). The longer, weaker hydrogen bonds can be thought of as dipole–dipole interactions, and are particularly important in biological systems, and in any crystals that contain water.

## 1.6.7 ATOMIC RADII

An atom in a covalently bonded molecule can be assigned a **covalent radius**, $r_c$ and a non-bonded radius, known as the **van der Waals radius**.

Covalent radii are calculated from half the interatomic distance between two singly bonded like atoms. For diatomic molecules such as $F_2$, this is no problem, but for other elements, such as carbon, which do not have a diatomic molecule, an average value is calculated from a range of compounds that contain a C–C single bond.

The van der Waals radius is defined as a nonbonded distance of closest approach, and these are calculated from the smallest interatomic distances in crystal structures that are considered to be not bonded to one another. Again, these are average values compiled from many crystal structures. If the sum of the van der Waals radii of two adjacent atoms in a structure is greater than the measured distance between them,

**TABLE 1.11**
**Single-bond covalent radii and van der Waals radii (in parentheses) for the typical elements/pm**

| Group I | Group II | Group III | Group IV | Group V | Group VI | Group VII | Group VIII |
|---|---|---|---|---|---|---|---|
| | | | H<br>37 (120) | | | | He<br>— (140) |
| Li<br>135 | Be<br>90 | B<br>80 | C<br>77 (170) | N<br>74 (155) | O<br>73 (152) | F<br>71 (147) | Ne<br>— (154) |
| Na<br>154 | Mg<br>130 | Al<br>125 | Si<br>117 (210) | P<br>110 (180) | S<br>104 (180) | Cl<br>99 (175) | Ar<br>— (188) |
| K<br>200 | Ca<br>174 | Ga<br>126 | Ge<br>122 | As<br>121 (185) | Se<br>117 (190) | Br<br>114 (185) | Kr<br>— (202) |
| Rb<br>211 | Sr<br>192 | In<br>141 | Sn<br>137 | Sb<br>141 | Te<br>137 (206) | I<br>133 (198) | Xe<br>— (216) |
| Cs<br>225 | Ba<br>198 | Tl<br>171 | Pb<br>175 | Bi<br>170 | Po<br>140 | At<br>— | Rn<br>— |

then it is assumed that there is some bonding between them. Table 1.11 gives the covalent and van der Waals radii for the typical elements.

### 1.6.8 MOLECULAR STRUCTURES

Finally, we consider crystal structures that do not contain any extended arrays of atoms. The example of graphite in the previous section in a way forms a bridge between these structures and the structures with infinite three-dimensional arrays. Many crystals contain small, discrete, covalently bonded molecules that are held together only by weak forces.

Examples of molecular crystals are found throughout organic, organometallic, and inorganic chemistry. Low melting and boiling temperatures characterize the crystals. We will look at just two examples, carbon dioxide and water (ice), both familiar, small, covalently bonded molecules.

Gaseous carbon dioxide, $CO_2$, when cooled sufficiently forms a molecular crystalline solid, which is illustrated in Figure 1.50. Notice that the unit cell contains clearly discernible $CO_2$ molecules, which are covalently bonded, and these are held together in the crystal by weak van der Waals forces.

The structure of one form of ice (crystalline water) is depicted in Figure 1.51. Each $H_2O$ molecule is tetrahedrally surrounded by four others. The crystal structure is held together by the hydrogen bonds formed between a hydrogen atom on one water molecule and the oxygen of the next, forming a three-dimensional arrangement throughout the crystal. This open hydrogen-bonded network of water molecules makes ice less dense than water, so that it floats on the surface of water.

A summary of the various types of crystalline solids is given in Table 1.12, relating the type of structure to its physical properties. It is important to realise that

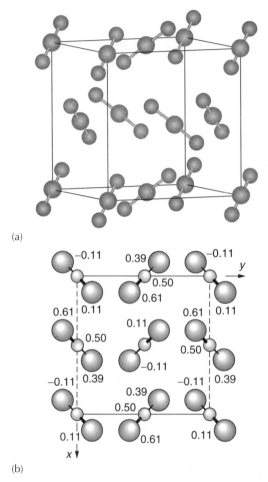

(a)

(b)

**FIGURE 1.50** (a) The crystal structure of $CO_2$, (b) packing diagram of the unit cell of $CO_2$ projected on to the $xy$ plane. The heights of the atoms are expressed as fractional coordinates of c. C, blue spheres; O, grey spheres.

this only gives a broad overview, and is intended as a guide only: not every crystal will fall exactly into one of these categories.

### 1.6.9 SILICATES

The **silicates** form a large group of crystalline compounds with rather complex but interesting structures. A great part of the earth's crust is formed from these complex oxides of silicon.

Silicon itself crystallizes with the same structure as diamond. Its normal oxide, silica, $SiO_2$, is polymorphic and in previous sections we have discussed the crystal structure of two of its polymorphs — β-cristobalite (Section 1.5.2); and β-quartz;

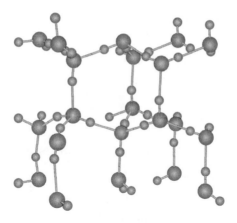

**FIGURE 1.51** The crystal structure of ice. H, blue spheres; O, grey spheres.

(Section 1.5.5). Quartz is one of the commonest minerals in the earth, occurring as sand on the seashore, as a constituent in granite and flint and, in less pure form, as agate and opal. The silicon atom in all these structures is tetrahedrally coordinated.

**TABLE 1.12**
**Classification of crystal structures**

| Type | Structural unit | Bonding | Characteristics | Examples |
|---|---|---|---|---|
| Ionic | Cations and anions | Electrostatic, non-directional | Hard, brittle, crystals of high m.t.; moderate insulators; melts are conducting | Alkali metal halides |
| Extended covalent array | Atoms | Mainly covalent | Strong hard crystals of high m.t.; insulators | Diamond, silica |
| Molecular | Molecules | Mainly covalent between atoms in molecule, van der Waals or hydrogen bonding between molecules | Soft crystals of low m.t. and large coefficient of expansion; insulators | Ice, organic compounds |
| Metallic | Metal atoms | Band model (see Chapter 4) | Single crystals are soft; strength depends on structural defects and grain; good conductors; m.t.s vary but tend to be high | Iron, aluminium, sodium |

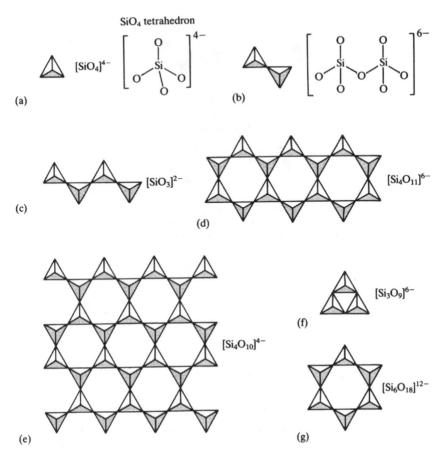

**FIGURE 1.52** A structural classification of mineral silicates.

The silicate structures are most conveniently discussed in terms of the $SiO_4^{4-}$ unit. The $SiO_4^{4-}$ unit has tetrahedral coordination of silicon by oxygen and is represented in these structures by a small tetrahedron as shown in Figure 1.52(a). The silicon-oxygen bonds possess considerable covalent character.

Some minerals, such as **olivines** (Figure 1.53) contain discrete $SiO_4^{4-}$ tetrahedra. These compounds do not contain Si—O—Si—O—Si— . . . chains, but there is considerable covalent character in the metal-silicate bonds. These are often described as **orthosilicates** — salts of orthosilicic acid, $Si(OH)_4$ or $H_4SiO_4$, which is a very weak acid. The structure of olivine itself, $(Mg,Fe)_2SiO_4$, which can be described as an assembly of $SiO_4^{4-}$ ions and $Mg^{2+}$ (or $Fe^{2+}$) ions, appears earlier in Table 1.7 because an alternative description is of an *hcp* array of oxygens with silicons occupying one-eighth of the tetrahedral holes and magnesium ions occupying one-half of the octahedral holes.

In most silicates, however, the $SiO_4^{4-}$ tetrahedra are linked by oxygen sharing through a vertex, such as is illustrated in Figure 1.52(b) for two linked tetrahedra to give $Si_2O_7^{6-}$. Notice that each terminal oxygen confers a negative charge on the

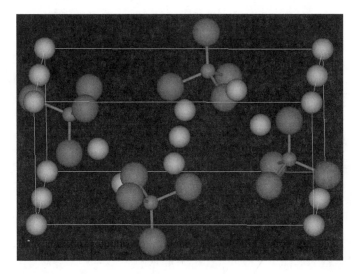

**FIGURE 1.53** The unit cell of olivine. See colour insert following page 196. Key: Mg,Fe, green; Si, grey; O, red.

anion and the shared oxygen is neutral. The diagrams of silicate structures showing the silicate frameworks, such as those depicted in Figure 1.52, omit these charges as they can be readily calculated. By the sharing of one or more oxygen atoms through the vertices, the tetrahedra are able to link up to form chains, rings, layers, etc. The negative charges on the silicate framework are balanced by metal cations in the lattice. Some examples are discussed next.

## Discrete $SiO_4^{4-}$ Units

Examples are found in: **olivine** (Table 1.7, Figure 1.53) an important constituent of basalt; **garnets**, $M_3^{II}M_2^{III}(SiO_4)_3$ (where $M^{II}$ can be $Ca^{2+}$, $Mg^{2+}$, or $Fe^{2+}$, and $M^{III}$ can be $Al^{3+}$, $Cr^{3+}$, or $Fe^{3+}$) and the framework of which is composed of $M^{III}O_6$ octahedra which are joined to six others via vertex-sharing $SiO_4$ tetrahedra — the $M^{II}$ ions are coordinated by eight oxygens in dodecahedral interstices; $Ca_2SiO_4$, found in mortars and Portland cement; and **zircon**, $ZrSiO_4$, which has eight-coordinate Zr.

## Disilicate Units ($Si_2O_7^{6-}$)

Structures containing this unit (Figure 1.52(b)) are not common but occur in **thortveitite**, $Sc_2Si_2O_7$, and **hemimorphite**, $Zn_4(OH)_2Si_2O_7$.

## Chains

$SiO_4^{4-}$ units share two corners to form infinite chains (Figure 1.52(c)). The repeat unit is $SiO_3^{2-}$. Minerals with this structure are called **pyroxenes** (e.g., **diopside** ($CaMg(SiO_3)_2$) and **enstatite** ($MgSiO_3$)). The silicate chains lie parallel to one another and are linked together by the cations that lie between them.

**FIGURE 1.54** The structure of an amphibole. See colour insert following page 196. Key: Mg, green; Si, grey; O, red; Na, purple.

## Double Chains

Here alternate tetrahedra share two and three oxygen atoms, respectively, as in Figure 1.52(d). This class of minerals is known as the **amphiboles** (Figure 1.54), an example of which is **tremolite**, $Ca_2Mg_5(OH)_2(Si_4O_{11})_2$. Most of the asbestos minerals fall in this class. The repeat unit is $Si_4O_{11}{}^{6-}$.

## Infinite Layers

The tetrahedra all share three oxygen atoms (Figure 1.52(e)). The repeat unit is $Si_4O_{10}{}^{4-}$. Examples are the **mica** group (e.g., $KMg_3(OH)_2Si_3AlO_{10}$) of which **biotite**, $(K(Mg,Fe)_3(OH)_2Si_3AlO_{10})$ (Figure 1.55) and **talc** $(Mg_3(OH)_2Si_4O_{10})$ are members, and contains a sandwich of two layers with octahedrally coordinated cations between the layers and clay minerals such as **kaolin**, $Al_4(OH)_8Si_4O_{10}$.

## Rings

Each $SiO_4{}^{4-}$ unit shares two corners as in the chains. Figure 1.52(f) and Figure 1.52(g) show three and six tetrahedra linked together; these have the general formula $[SiO_3]_n{}^{-2n}$; rings also may be made from four tetrahedra. An example of a six-tetrahedra ring is **beryl** (emerald), $Be_3Al_2Si_6O_{18}$; here the rings lie parallel with metal ions between them. Other examples include **dioptase** $(Cu_6Si_6O_{18}.6H_2O)$ and **benitoite** $(BaTiSi_3O_9)$.

## Three-dimensional Structures

If $SiO_4{}^{4-}$ tetrahedra share all four oxygens, then the structure of silica, $SiO_2$ is produced. However, if some of the silicon atoms are replaced by the similarly sized atoms of the Group III element aluminium (i.e., if $SiO_4{}^{4-}$ is replaced by $AlO_4{}^{5-}$), then other cations must be introduced to balance the charges. Such minerals include the **feldspars** (general formula, $M(Al,Si)_4O_8$) the most abundant of the rock-forming minerals; the **zeolites**, which are used as ion exchangers, molecular sieves, and catalysts (these are discussed in detail in Chapter 7); the **ultramarines**, which are

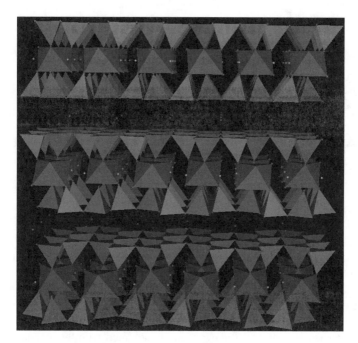

**FIGURE 1.55** The structure of biotite. See colour insert following page 196. Key: Mg, green; Si, grey; O, red; Na, purple; Al, pink; Fe, blue.

coloured silicates manufactured for use as pigments, **lapis lazuli** being a naturally occurring mineral of this type.

As one might expect there is an approximate correlation between the solid state structure and the physical properties of a particular silicate. For instance, cement contains discrete $SiO_4^{4-}$ units and is soft and crumbly; asbestos minerals contain double chains of $SiO_4^{4-}$ units and are characteristically fibrous; mica contains infinite layers of $SiO_4^{4-}$ units, the weak bonding between the layers is easily broken, and micas show cleavage parallel to the layers; and granite contains feldspars that are based on three-dimensional $SiO_4^{4-}$ frameworks and are very hard.

## 1.7 LATTICE ENERGY

The **lattice energy**, $L$, of a crystal is the standard enthalpy change when one mole of the solid is formed from the *gaseous ions* (e.g., for NaCl, $L = \Delta H_m^{\ominus}$ for the reaction in Equation (1.1) = $-787$ kJ mol$^{-1}$).

$$Na^+(g) + Cl^-(g) = NaCl(s) \tag{1.1}$$

Because it is not possible to measure lattice energies directly, they can be determined experimentally from a **Born–Haber cycle**, or they can be calculated as we see in Section 1.7.2.

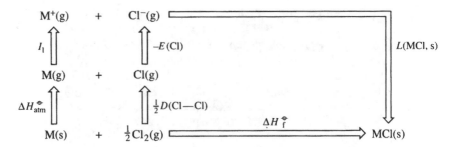

**FIGURE 1.56** The Born–Haber cycle for a metal chloride (MCl).

### 1.7.1 THE BORN–HABER CYCLE

A Born–Haber cycle is the application of **Hess's Law** to the enthalpy of formation of an ionic solid at 298 K. Hess's law states that the enthalpy of a reaction is the same whether the reaction takes place in one step or in several. A Born–Haber cycle for a metal chloride (MCl) is depicted in Figure 1.56; the metal chloride is formed from the constituent elements in their standard state in the equation at the bottom, and by the clockwise series of steps above. From Hess's law, the sum of the enthalpy changes for each step around the cycle can be equated with the standard enthalpy of formation, and we get that:

$$\Delta H_f^{\ominus}\ (\text{MCl,s}) = \Delta H_{\text{atm}}^{\ominus}\ (\text{M,s}) + I_1(\text{M}) + \frac{1}{2}\,D_{\text{m}}(\text{Cl–Cl}) - E(\text{Cl}) + L(\text{MCl,s})\ (1.2)$$

By rearranging this equation, we can write an expression for the lattice energy in terms of the other quantities, which can then be calculated if the values for these are known. The terms in the Born–Haber cycle are defined in Table 1.13 together with some sample data. Notice that the way in which we have defined lattice energy gives **negative** values; you may find Equation (1.1) written the other way round in some texts, in which case they will quote positive lattice energies. Notice also that electron affinity is defined as the heat *evolved* when an electron is added to an atom; as an enthalpy change refers to the heat *absorbed*, the electron affinity and the enthalpy change for that process, will have opposite signs.

Cycles such as this can be constructed for other compounds such as oxides (MO), sulfides (MS), higher valent metal halides ($MX_n$), etc. The difficulty in these cycles sometimes comes in the determination of values for the electron affinity, $E$. In the case of an oxide, it is necessary to know the double electron affinity for oxygen (the negative of the enthalpy change of the following reaction):

$$2e^-(g) + O(g) = O^{2-}(g) \qquad (1.3)$$

This can be broken down into two stages:

$$e^-(g) + O(g) = O^-(g) \qquad (1.4)$$

## TABLE 1.13
## Terms in the Born–Haber cycle

| Term | Definition of the reaction to which the term applies | NaCl/ kJ mol$^{-1}$ | AgCl/ kJ mol$^{-1}$ |
|---|---|---|---|
| $\Delta H_{atm}^{\ominus}$ (M) | M(s) = M(g) <br> Standard enthalpy of atomization of metal M | 107.8 | 284.6 |
| $I_1$(M) | M(g) = M$^+$(g) + e$^-$(g) <br> First ionization energy of metal M | 494 | 732 |
| $\frac{1}{2}$ D(Cl–Cl) | $\frac{1}{2}$ Cl$_2$(g) = Cl(g) <br><br> Half the dissociation energy of Cl$_2$ | 122 | 122 |
| –E(Cl) | Cl(g) + e$^-$(g) = Cl$^-$(g) <br> The enthalpy change of this reaction is defined as <br> minus the electron affinity of chlorine | –349 | –349 |
| L(MCl,s) | M$^+$(g) + Cl$^-$(g) = MCl(s) <br> Lattice energy of MCl(s) | | |
| $\Delta H_f^{\ominus}$ (MCl,s) | M(s) + $\frac{1}{2}$ Cl$_2$(g) = MCl(s) <br><br> Standard enthalpy of formation of MCl(s) | –411.1 | –127.1 |

and

$$e^-(g) + O^-(g) = O^{2-}(g) \tag{1.5}$$

It is impossible to determine the enthalpy of reaction for Equation (1.5) exper-imentally, and so this value can only be found if the lattice energy is known — a Catch-22 situation! To overcome problems such as this, methods of calculating (instead of measuring) the lattice energy have been devised and they are described in the next section.

### 1.7.2 CALCULATING LATTICE ENERGIES

For an ionic crystal of known structure, it should be a simple matter to calculate the energy released on bringing together the ions to form the crystal, using the equations of simple electrostatics. The energy of an ion pair, M$^+$, X$^-$ (assuming they are point charges), separated by distance, $r$, is given by **Coulomb's Law**:

$$E = - \frac{e^2}{4\pi\varepsilon_0 r} \tag{1.6}$$

and where the *magnitudes* of the charges on the ions are $Z_+$ and $Z_-$, for the cation and anion, respectively, by

$$E = - \frac{Z_+ Z_- e^2}{4\pi\varepsilon_0 r} \tag{1.7}$$

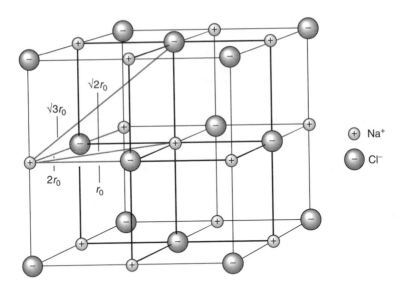

**FIGURE 1.57** Sodium chloride structure showing internuclear distances.

($e$ is the electronic charge, $1.6 \times 10^{-19}$ C, and $\varepsilon_0$ is the permittivity of a vacuum, $8.854 \times 10^{-12}$ F m$^{-1}$).

The energy due to coulombic interactions in a crystal is calculated for a particular structure by summing all the ion–pair interactions, thus producing an infinite series. The series will include terms due to the attraction of the opposite charges on cations and anions and repulsion terms due to cation/cation and anion/anion interactions. Figure 1.57 depicts some of these interactions for the NaCl structure. The Na$^+$ ion in the centre is immediately surrounded by 6 Cl$^-$ ions at a distance of $r$, then by 12 cations at a distance of $\sqrt{2}\,r$, then by eight anions at $\sqrt{3}\,r$, followed by a further 6 cations at $2r$, and so on. The coulombic energy of interaction is given by the summation of all these interactions:

$$E_C = -\frac{e^2}{4\pi\varepsilon_0 r}\left(6 - \frac{12}{\sqrt{2}} + \frac{8}{\sqrt{3}} - \frac{6}{2} + \frac{24}{\sqrt{5}} \dots\dots\right)$$

or

$$E_C = -\frac{e^2}{4\pi\varepsilon_0 r}\left(\frac{6}{\sqrt{1}} - \frac{12}{\sqrt{2}} + \frac{8}{\sqrt{3}} - \frac{6}{\sqrt{4}} + \frac{24}{\sqrt{5}} \dots\dots\right) \tag{1.8}$$

The term inside the brackets is known as the **Madelung constant, $A$,** in this case for the NaCl structure. The series is slow to converge, but values of the Madelung constant have been computed, not only for NaCl, but also for most of the simple ionic structures. For one mole of NaCl, we can write:

$$E_C = -\frac{N_A A e^2}{4\pi\varepsilon_0 r} \tag{1.9}$$

**TABLE 1.14**
**Madelung constants for some common ionic lattices**

| Structure | Madelung constant, $A$ | Number of ions in formula unit, $v$ | $\dfrac{A}{v}$ | Coordination |
|---|---|---|---|---|
| Caesium chloride, CsCl | 1.763 | 2 | 0.882 | 8:8 |
| Sodium chloride, NaCl | 1.748 | 2 | 0.874 | 6:6 |
| Fluorite, CaF$_2$ | 2.519 | 3 | 0.840 | 8:4 |
| Zinc blende, ZnS | 1.638 | 2 | 0.819 | 4:4 |
| Wurtzite, ZnS | 1.641 | 2 | 0.821 | 4:4 |
| Corundum, Al$_2$O$_3$ | 4.172 | 5 | 0.835 | 6:4 |
| Rutile, TiO$_2$ | 2.408 | 3 | 0.803 | 6:3 |

where $N_A$ is the **Avogadro number**, $6.022 \times 10^{23}$ mol$^{-1}$. (Note that the expression is multiplied by $N_A$ and not by $2N_A$, even though $N_A$ cations and $N_A$ anions are present; this avoids counting every interaction twice!) The value of the Madelung constant is dependent only on the geometry of the lattice, and not on its dimensions; values for various structures are given in Table 1.14.

Ions, of course, are not point charges, but consist of positively charged nuclei surrounded by electron clouds. At small distances, these electron clouds repel each other, and this too needs to be taken into account when calculating the lattice energy of the crystal. At large distances, the repulsion energy is negligible, but as the ions approach one another closely, it increases very rapidly. Max Born suggested that the form of this repulsive interaction could be expressed by:

$$E_R = \frac{B}{r^n} \tag{1.10}$$

where B is a constant and $n$ (known as the **Born exponent**) is large and also a constant.

We can now write an expression for the lattice energy in terms of the energies of the interactions that we have considered:

$$L = E_C + E_R = -\frac{N_A A Z_+ Z_- e^2}{4\pi\varepsilon_0 r} + \frac{B}{r^n} \tag{1.11}$$

The lattice energy will be a minimum when the crystal is at equilibrium (i.e., when the internuclear distance is at the equilibrium value of $r_0$). If we minimize the lattice energy (see Box), we get:

$$L = -\frac{N_A A Z_+ Z_- e^2}{4\pi\varepsilon_0 r_0}\left(1 - \frac{1}{n}\right) \tag{1.12}$$

This is known as the **Born–Landé equation**: the values of $r_0$ and $n$ can be obtained from X-ray crystallography and from compressibility measurements, respectively. The other terms in the equation are well-known constants, and when values for these are substituted, we get:

$$L/\text{kJ mol}^{-1} = - \frac{1.389 \times 10^5 \, AZ_+Z_-}{r_0 \, / \, \text{pm}} \left(1 - \frac{1}{n}\right) \tag{1.13}$$

If the units of $r_0$ are pm, then the units of $L$ will be kJ mol$^{-1}$.

---

### Derivation of the Born–Landé Equation

We can minimize the lattice energy function by using the standard mathematical technique of differentiating with respect to $r$ and then equating to zero:

$$\frac{dL}{dr} = \frac{N_A AZ_+Z_- e^2}{4\pi\varepsilon_0 r^2} - \frac{nB}{r^{n+1}}$$

but $\dfrac{dL}{dr} = 0$ when $r = r_0$

so $\dfrac{nB}{r_0^{n+1}} = \dfrac{N_A AZ_+Z_- e^2}{4\pi\varepsilon_0 r_0^{2}}$

$$B = \frac{N_A AZ_+Z_- e^2 r_0^{n+1}}{n4\pi\varepsilon_0 r_0^{2}} = \frac{N_A AZ_+Z_- e^2 r_0^{n-1}}{n4\pi\varepsilon_0}$$

$$L = -\frac{N_A AZ_+Z_- e^2}{4\pi\varepsilon_0 r_0} + \frac{N_A AZ_+Z_- e^2 r_0^{n-1}}{n4\pi\varepsilon_0 r_0^{n}}$$

$$L = -\frac{N_A AZ_+Z_- e^2}{4\pi\varepsilon_0 r_0} \left(1 - \frac{1}{n}\right) \tag{1.12}$$

---

Pauling demonstrated that the values of $n$ could be approximated with reasonable accuracy for compounds of ions with noble gas configurations, by averaging empirical constants for each ion. The values of these constants are given in Table 1.15. For example, $n$ for rubidium chloride, RbCl, is 9.5 (average of 9 and 10) and for strontium chloride, SrCl$_2$, is 9.33 (the average of 9, 9, and 10).

Notice what a dramatic effect the charge on the ions has on the value of the lattice energy. A structure containing one doubly charged ion has a factor of two in the equation ($Z_+Z_- = 2$), whereas one containing *two* doubly charged ions is multiplied by a factor of *four* ($Z_+Z_- = 4$). Structures containing multiply charged ions tend to have much larger (numerically) lattice energies.

---

**TABLE 1.15**
**Constants used to calculate *n***

| Ion type | Constant |
|----------|----------|
| [He] | 5 |
| [Ne] | 7 |
| [Ar] | 9 |
| [Kr] | 10 |
| [Xe] | 12 |

---

A Russian chemist, A.F. Kapustinskii, noted that if the Madelung constants, $A$, for a number of structures are divided by the number of ions in one formula unit of the structure, $v$, the resulting values are almost constant (see Table 1.14) varying only between approximately 0.88 and 0.80. This led to the idea that it would be possible to set up a general lattice energy equation that could be applied to any crystal regardless of its structure. We can now set up a general equation and use the resulting equation to calculate the lattice energy of an unknown structure. First replace the Madelung constant, $A$, in the Born–Landé Equation (1.12) with value from the NaCl structure, $0.874v$, and $r_0$ by $(r_+ + r_-)$, where $r_+$ and $r_-$ are the cation and anion radii for six-coordination, giving:

$$L/\text{kJ mol}^{-1} = -\frac{1.214 \times 10^5 v Z_+ Z_-}{r_+ + r_- / \text{pm}}(1 - \frac{1}{n}) \qquad (1.14)$$

If $n$ is assigned an average value of 9, we arrive at:

$$L/\text{kJ mol}^{-1} = -\frac{1.079 \times 10^5 v Z_+ Z_-}{r_+ + r_- / \text{pm}} \qquad (1.15)$$

These equations are known as the **Kapustinskii equations**.

Some lattice energy values that have been calculated by various methods are shown in Table 1.16 for comparison with experimental values that have been computed using a Born–Haber cycle. Remarkably good agreement is achieved, considering all the approximations involved. The largest discrepancies are for the large polarizable ions, where, of course, the ionic model is not expected to be perfect. The equations can be improved to help with these discrepancies by including the effect of van der Waals forces, zero point energy (the energy due to the vibration of the ions at 0 K), and heat capacity. The net effect of these corrections is only of the order of 10 kJ mol$^{-1}$ and the values thus obtained for the lattice energy are known as **extended-calculation values**.

It is important to note that the good agreement achieved between the Born–Haber and calculated values for lattice energy, do not in any way prove that the ionic model is valid. This is because the equations possess a self-compensating feature in that they use *formal charges* on the ions, but take *experimental internuclear distances*.

## TABLE 1.16
## Lattice energies of some alkali and alkaline earth metal halides at 0 K

| Compound | Structure | Born–Haber cycle[a] | Born–Landé Equation 1.12[b] | Extended calculation[c] | Kapustinskii Equation 1.15 |
|---|---|---|---|---|---|
| | | | L/kJ mol$^{-1}$ | | |
| LiF | NaCl | −1025 | — | −1033 | −1033 |
| LiI | NaCl | −756 | — | −738 | −729 |
| NaF | NaCl | −910 | −904 | −906 | −918 |
| NaCl | NaCl | −772 | −757 | −770 | −763 |
| NaBr | NaCl | −736 | −720 | −735 | −724 |
| NaI | NaCl | −701 | −674 | −687 | −670 |
| KCl | NaCl | −704 | −690 | −702 | −677 |
| KI | NaCl | −646 | −623 | −636 | −603 |
| CsF | NaCl | −741 | −724 | −734 | −719 |
| CsCl | CsCl | −652 | −623 | −636 | −620 |
| CsI | CsCl | −611 | −569 | −592 | −558 |
| MgF$_2$ | Rutile | −2922 | −2883 | −2914 | −3158 |
| CaF$_2$ | Fluorite | −2597 | −2594 | −2610 | −2779 |
| CaCl$_2$ | Deformed rutile | −2226 | — | −2223 | −2304 |

[a] D.A. Johnson (1982) *Some Thermodynamic Aspects of Inorganic Chemistry*, 2nd edn, Cambridge University Press, Cambridge.

[b] D.F.C. Morris (1957) *J. Inorg. Nucl. Chem.*, **4**, 8.

[c] D. Cubiociotti (1961) *J. Chem. Phys.*, **34**, 2189; T.E. Brackett and E.B. Brackett (1965) *J. Phys. Chem.*, **69**, 3611; H.D.B. Jenkins and K.F. Pratt (1977) *Proc. Roy. Soc.*, (A) **356**, 115.

The values of $r_0$ are the result of *all* the various types of bonding in the crystal, not just of the ionic bonding, and so are rather shorter than one would expect for purely ionic bonding.

As we end this section, let us reconsider ionic radii briefly. Many ionic compounds contain complex or polyatomic ions. Clearly, it is going to be extremely difficult to measure the radii of ions such as ammonium, $NH_4^+$, or carbonate, $CO_3^{2-}$, for instance. However, Yatsimirskii has devised a method which determines a value of the radius of a polyatomic ion by applying the Kapustinskii equation to lattice energies determined from thermochemical cycles. Such values are called **thermochemical radii**, and Table 1.17 lists some values.

### 1.7.3 CALCULATIONS USING THERMOCHEMICAL CYCLES AND LATTICE ENERGIES

It is not yet possible to measure lattice energy directly, which is why the best experimental values for the alkali halides, as listed in Table 1.16, are derived from a thermochemical cycle. This in itself is not always easy for compounds other than the alkali halides because, as we noted before, not all of the data is necessarily available. Electron affinity values are known from experimental measurements for

**TABLE 1.17**

**Thermochemical radii of polyatomic ions[a]**

| Ion | pm | Ion | pm | Ion | pm |
|---|---|---|---|---|---|
| $NH_4^+$ | 151 | $ClO_4^-$ | 226 | $MnO_4^{2-}$ | 215 |
| $Me_4N^+$ | 215 | $CN^-$ | 177 | $O_2^{2-}$ | 144 |
| $PH_4^+$ | 171 | $CNS^-$ | 199 | $OH^-$ | 119 |
| $AlCl_4^-$ | 281 | $CO_3^{2-}$ | 164 | $PtF_6^{2-}$ | 282 |
| $BF_4^-$ | 218 | $IO_3^-$ | 108 | $PtCl_6^{2-}$ | 299 |
| $BH_4^-$ | 179 | $N_3^-$ | 181 | $PtBr_6^{2-}$ | 328 |
| $BrO_3^-$ | 140 | $NCO^-$ | 189 | $PtI_6^{2-}$ | 328 |
| $CH_3COO^-$ | 148 | $NO_2^-$ | 178 | $SO_4^{2-}$ | 244 |
| $ClO_3^-$ | 157 | $NO_3^-$ | 165 | $SeO_4^{2-}$ | 235 |

[a] J.E. Huheey (1983) *Inorganic Chemistry*, 3rd edn, Harper & Row, New York; based on data from H.D.B. Jenkins and K.P. Thakur (1979), *J. Chem. Ed.*, **56**, 576.

most of the elements, but when calculating lattice energy for a sulfide, for example, then we need to know the enthalpy change for the reaction in Equation (1.16):

$$2e^-(g) + S(g) = S^{2-}(g) \tag{1.16}$$

which is minus a **double electron affinity** and is not so readily available. This is where lattice energy calculations come into their own because we can use one of the methods discussed previously to calculate a value of $L$ for the appropriate sulfide and then plug it into the thermochemical cycle to calculate the enthalpy change of Equation (1.16).

**Proton affinities** can be found in a similar way: a proton affinity is defined as the enthalpy change of the reaction shown in Equation (1.17), where a proton is lost by the species, A.

$$AH^+(g) = A(g) + H^+(g) \tag{1.17}$$

This value can be obtained from a suitable thermochemical cycle providing the lattice energy is known. Take as an example the formation of the ammonium ion, $NH_4^+(g)$:

$$NH_3(g) + H^+(g) = NH_4^+(g) \tag{1.18}$$

The enthalpy change of the reaction in Equation (1.18) is minus the proton affinity of ammonia, $-P(NH_3,g)$. This could be calculated from the thermochemical cycle shown in Figure 1.58, provided the lattice energy of ammonium chloride is known.

Thermochemical cycles can also be used to provide us with information on the thermodynamic properties of compounds with metals in unusual oxidation states that have not yet been prepared. For instance, we can use arguments to determine whether it is possible to prepare a compound of sodium in a higher oxidation state

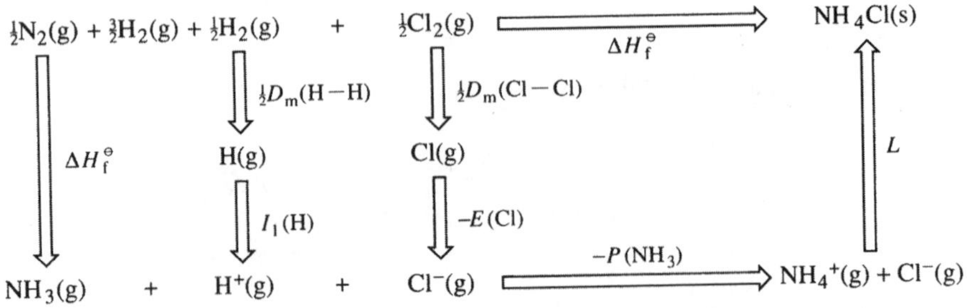

**FIGURE 1.58** Thermochemical cycle for the calculation of the proton affinity of ammonia.

than normal, $NaCl_2$. To calculate a value for the enthalpy of formation and thence ($NaCl_2$,s), we need to set up a Born–Haber cycle for $NaCl_2(s)$ of the type shown in Figure 1.59. The only term in this cycle that is not known is the lattice energy of $NaCl_2$, and for this we make the approximation that it will be the same as that of the isoelectronic compound $MgCl_2$, which *is* known. The summation becomes:

$$\Delta H_f^{\ominus} (NaCl_2,s) = \Delta H_{atm}^{\ominus} (Na,s) + I_1(Na) + I_2(Na) + D_m(Cl–Cl) – 2E(Cl)$$
$$+ L(NaCl_2,s) \tag{1.19}$$

Table 1.18 lists the relevant values.

We noted earlier that solids containing doubly charged ions tend to have large negative lattice energies, yet the calculation for $NaCl_2$ has shown a large positive enthalpy of formation. Why? A glance at the figures in Table 1.18 reveals the answer: the second ionization energy for sodium is huge (because this relates to the removal of an inner shell electron) and this far outweighs the larger lattice energy. Therefore, $NaCl_2$ does not exist because the extra stabilization of the lattice due to a doubly charged ion is not enough to compensate for the large second ionization energy.

The calculation we have just performed shows that for the reaction:

$$NaCl_2(s) = Na(s) + Cl_2(g): \quad \Delta H_m^{\ominus} = -2\ 190 \text{ kJ mol}^{-1}$$

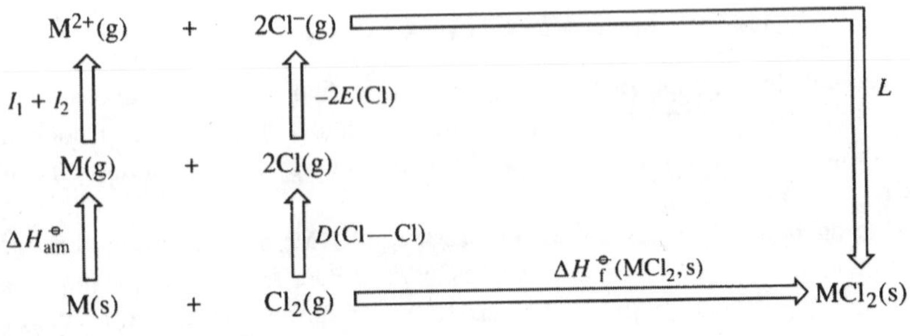

**FIGURE 1.59** Born–Haber cycle for a metal dichloride, $MCl_2$.

**TABLE 1.18**

**Values of the Born–Haber cycle terms for NaCl$_2$ and MgCl$_2$/kJ mol$^{-1}$**

|  | Na | Mg |
|---|---|---|
| $\Delta H_{\text{atm}}^{\ominus}$ | 108 | 148 |
| $I_1$ | 494 | 736 |
| $I_2{}^a$ | 4565 | 1452 |
| $D$(Cl-Cl) | 244 | 244 |
| $-2E$(Cl) | −698 | −698 |
| $L$(MCl$_2$,s) | −2523 | −2523 |
| $\Delta H_f^{\ominus}$ (MCl$_2$,s) | 2 190 | −641 |

$^a$The second ionization energy, $I_2$, refers to the energy change of the reaction: $M^+(g) - e^-(g) = M^{2+}(g)$.

To be sure of the stability of the compound, we need a value for $\Delta G_m^{\ominus}$. For the analogous reaction of MgCl$_2$(s),

$$\Delta S_m^{\ominus} = 166.1 \text{ J K}^{-1} \text{ mol}^{-1}$$

and so

$$\Delta G_m^{\ominus} = \Delta H_m^{\ominus} - T\Delta S_m^{\ominus} = -2\ 190 - (298.15 \times 0.166) = -2\ 239 \text{ kJ mol}^{-1}$$

suggesting that NaCl$_2$ would indeed be unstable with respect to sodium and chlorine.

Interestingly, it was arguments and calculations of this sort that led Neil Bartlett to the discovery of the first noble gas compound, XePtF$_6$. Bartlett had prepared a new complex O$_2$PtF$_6$, which by analogy with the diffraction pattern of KPtF$_6$, he formulated as containing the dioxygenyl cation, [O$_2$$^+$][PtF$_6$$^-$]. He realised that the ionization energies of oxygen and xenon are very similar and that although the radius of the Xe$^+$ ion is slightly different, because the PtF$_6$$^-$ anion is very large the lattice energy of [Xe$^+$][PtF$_6$$^-$] should be very similar to that of the dioxygenyl complex and therefore should exist! Accordingly, he mixed xenon and PtF$_6$ and obtained the orange-yellow solid of xenon hexafluoroplatinate — the first noble gas compound. (Although in fact, the compound turned out not to have the structure that Bartlett predicted because at room temperature the XePtF$_6$ reacts with another molecule of PtF$_6$ to give a product containing [XeF]$^+$[PtF$_6$]$^-$ and [PtF$_5$]$^-$.)

Several examples of problems involving these types of calculation are included in the questions at the end of the chapter.

## 1.8  CONCLUSION

This opening chapter has introduced many of the principles and ideas that lie behind a discussion of the crystalline solid state. We have discussed in detail the structure of a number of important ionic crystal structures and shown how they can be linked to a simple view of ions as hard spheres that pack together as closely as possible, but can also be viewed as the linking of octahedra or tetrahedra in various ways. Taking these ideas further, we have investigated the size of these ions in terms of their radii, and

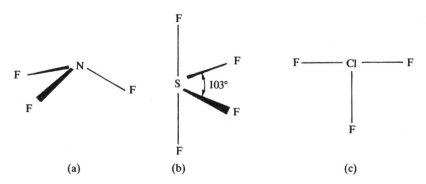

**FIGURE 1.60** (a) $NF_3$, (b) $SF_4$, and (c) $ClF_3$.

thence the energy involved in forming a lattice with ionic bonding. We also noted that covalent bonding is present in many structures and that when only covalent bonding is present we tend to see a rather different type of crystal structure.

## QUESTIONS*

1. Figure 1.60 depicts several molecules. Find and draw all the symmetry elements possessed by each molecule.
2. Does the $CF_4$ molecule in Figure 1.14 possess a centre of inversion? What other rotation axis is coincident with the $\bar{4}$ ($S_4$)?
3. How many centred cells are drawn in Figure 1.17?
4. Name the symmetry properties found in these five commonly occurring space groups: P $\bar{1}$, $P2_1/c$, C2/m, Pbca, and F $\bar{4}$ 3m.
5. Index the sets of lines in Figure 1.26 marked B, C, D, and E.
6. Index the sets of planes in Figure 1.27(b), (c), and (d).
7. Figure 1.24(c) shows a unit cell of a face-centred cubic structure. If a single atom is placed at each lattice point then this becomes the unit cell of the *ccp* (cubic close-packed) structure. Find the *100*, *110*, and the *111* planes and calculate the density of atoms per unit area for each type of plane. (Hint: Calculate the area of each plane assuming a cell length $a$. Decide the fractional contribution made by each atom to the plane.)
8. Using Figures 1.30, 1.31, and 1.37 together with models if necessary, draw unit cell projections for (a) CsCl, (b) NaCl, and (c) ZnS (zinc blende).
9. How many formula units, ZnS, are there in the zinc blende unit cell?
10. Draw a projection of a unit cell for both the *hcp* and *ccp* structures, seen perpendicular to the close-packed layers (i.e., assume that the close-packed layer is the *ab* plane, draw in the $x$ and $y$ coordinates of the atoms in their correct positions and mark the third coordinate $z$ as a fraction of the corresponding repeat distance $c$).
11. The unit cell dimension $a$, of NaCl is 564 pm, calculate the density of NaCl in kg m$^{-3}$.

---

**TABLE 1.19**
**Values of the Born–Haber cycle terms for CaCl$_2$/kJ mol$^{-1}$**

| Term | Value |
| --- | --- |
| $\Delta H_{atm}^{\ominus}$ | 178 |
| $I_1$ | 590 |
| $I_2$ | 1146 |
| $D$(Cl–Cl) | 244 |
| $-2E$(Cl) | −698 |
| $\Delta H_f^{\ominus}$ (CaCl$_2$,s) | −795.8 |

12. A compound AgX has the same crystal structure as NaCl, a density of 6477 kg m$^{-3}$, and a unit cell dimension of 577.5 pm. Identify X.

13. Estimate a value for the radius of the iodide ion. The distance between the lithium and iodine nuclei in lithium iodide is 300 pm.

14. Calculate a radius for F$^-$ from the data in Table 1.8 for NaI and NaF. Repeat the calculation using RbI and RbF.

15. The diamond structure is given in Figure 1.47. Find the *100, 110,* and *111* planes and determine the relative atomic densities per unit area. (See Question 7.)

16. Use the Born–Haber cycle in Figure 1.59 and the data in Table 1.19 to calculate the lattice energy of solid calcium chloride, CaCl$_2$:

17. Calculate the value of the Madelung constant for the structure in Figure 1.61. All bond lengths are equal and all bond angles are 90°. Assume that no ions exist other than those shown in the figure, and that the charges on the cations and anion are +1 and −1, respectively.

18. Calculate a value for the lattice energy of potassium chloride using Equation (1.15). Compare this with the value you calculate from the thermodynamic data in Table 1.20.

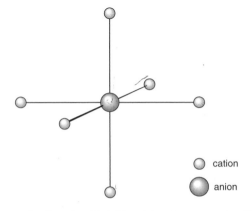

cation

anion

**FIGURE 1.61** Structure for Question 17. All bond lengths are equal and bond angles are 90°.

**TABLE 1.20**
**Values of the Born–Haber cycle terms for KCl/kJ mol$^{-1}$**

| Term | Value |
|---|---|
| $\Delta H_{atm}^{\ominus}$ | 89.1 |
| $I_1$ | 418 |
| $\frac{1}{2} D_m(\text{Cl–Cl})$ | 122 |
| $-E(\text{Cl})$ | $-349$ |
| $\Delta H_f^{\ominus}$ (KCl,s) | $-436.7$ |

**TABLE 1.21**
**Values of the Born–Haber cycle terms for FeS/kJ mol$^{-1}$**

| Term | Value |
|---|---|
| $\Delta H_{atm}^{\ominus}$ (Fe,s) | 416.3 |
| $I_1(\text{Fe})$ | 761 |
| $I_2(\text{Fe})$ | 1 561 |
| $\Delta H_{atm}^{\ominus}$ (S,s) | 278.8 |
| $\Delta H_f^{\ominus}$ (FeS,s) | $-100.0$ |
| $E(\text{S})$ | 200 |

**TABLE 1.22**
**Values of the Born–Haber cycle terms for MgO/kJ mol$^{-1}$**

| Term | Value |
|---|---|
| $\Delta H_{atm}^{\ominus}$ (Mg,s) | 147.7 |
| $I_1(\text{Mg})$ | 736 |
| $I_2(\text{Mg})$ | 1 452 |
| $\frac{1}{2} D_m(\text{O–O})$ | 249 |
| $\Delta H_f^{\ominus}$ (MgO,s) | $-601.7$ |
| $E(\text{O})$ | 141 |

**TABLE 1.23**
**Values of the Born–Haber cycle terms for NH$_4$Cl/kJ mol$^{-1}$**

| Term | Value |
|---|---|
| $\Delta H_f^{\ominus}$ (NH$_3$) | −46.0 |
| $\Delta H_f^{\ominus}$ (NH$_4$Cl,s) | −314.4 |
| $\frac{1}{2}D_m$(H–H) | 218 |
| $\frac{1}{2}D_m$(Cl–Cl) | 122 |
| $I$(H) | 1314 |
| $E$(Cl) | 349 |
| $r_+$(NH$_4^+$) | 151 pm |

19. Calculate a value for the electron affinity of sulfur for two electrons. Take iron(II) sulfide, FeS, as a model and devise a suitable cycle. Use the data given in Table 1.21.

20. Calculate a value for the electron affinity of oxygen for two electrons. Take magnesium oxide, MgO, as a model and devise a suitable cycle. Use the data given in Table 1.22.

21. Calculate a value for the proton affinity of ammonia using the cycle in Figure 1.58 and data in Table 1.23.

22. Compounds of aluminium and magnesium in the lower oxidation states, Al(I) and Mg(I), do not exist under normal conditions. If we make an assumption that the radius of Al$^+$ or Mg$^+$ is the same as that of Na$^+$ (same row of the Periodic Table), then we can also equate the lattice energies, MCl. Use this information in a Born–Haber cycle to calculate a value of the enthalpy of formation, $\Delta H_f^{\ominus}$, for AlCl(s) and MgCl(s), using the data in Table 1.24.

**TABLE 1.24**
**Values of the Born–Haber cycle terms for NaCl, MgCl, and AlCl/kJ mol$^{-1}$**

| Term | Na | Mg | Al |
|---|---|---|---|
| $\Delta H_{atm}^{\ominus}$ | 108 | 148 | 326 |
| $I_1$ | 494 | 736 | 577 |
| $\frac{1}{2}D_m$(Cl–Cl) | 122 | 122 | 122 |
| $-E$(Cl) | −349 | −349 | −349 |
| $\Delta H_f^{\ominus}$ (NaCl,s) | −411 | | |
| $L$(MCl,s) | | | |

# 2 Physical Methods for Characterizing Solids

## 2.1 INTRODUCTION

A vast array of physical methods are used to investigate the structures of solids, each technique with its own strengths and weaknesses — some techniques are able to investigate the local coordination around a particular atom or its electronic properties, whereas others are suited to elucidating the long-range order of the structure. No book could do justice to all the techniques on offer, so here we describe just some of the more commonly available techniques, and try to show what information can be gleaned from each one, and the limitations. We start with X-ray diffraction by powders and single crystals. Single crystal X-ray diffraction is used to determine atomic positions precisely and therefore the bond lengths (to a few tens of picometres*) and bond angles of molecules within the unit cell. It gives an overall, average picture of a long-range ordered structure, but is less suited to giving information on the structural positions of defects, dopants, and non-stoichiometric regions. It is often very difficult to grow single crystals, but most solids can be made as a crystalline powder. Powder X-ray diffraction is probably the most commonly employed technique in solid state inorganic chemistry and has many uses from analysis and assessing phase purity to determining structure. Single crystal X-ray diffraction techniques have provided us with the structures upon which the interpretation of most powder data is based.

## 2.2 X-RAY DIFFRACTION

### 2.2.1 GENERATION OF X-RAYS

The discovery of X-rays was made by a German physicist, Wilhelm Röntgen, in 1895 for which he was awarded the first Nobel Prize in Physics in 1901. The benefits of his discovery in terms of medical diagnosis and treatment, and in investigating molecular and atomic structure are immeasurable, and yet Röntgen was a man of such integrity that he refused to make any financial gain out of his discovery, believing that scientific research should be made freely available.

An electrically heated filament, usually tungsten, emits electrons, which are accelerated by a high potential difference (20–50 kV) and allowed to strike a metal target or anode which is water cooled (Figure 2.1(a)). The anode emits a continuous spectrum of 'white' X-radiation but superimposed on this are sharp, intense X-ray

---

* Many crystallographers still work in Ångstroms, Å; 1 Å = 100 pm.

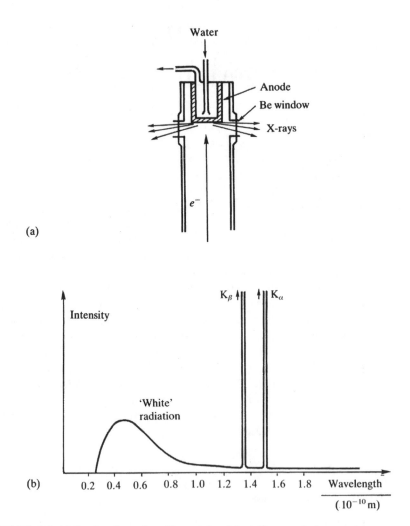

**FIGURE 2.1** (a) Section through an X-ray tube; (b) an X-ray emission spectrum.

peaks ($K_\alpha$, $K_\beta$) as depicted in Figure 2.1(b). The frequencies of the $K_\alpha$ and $K_\beta$ lines are characteristic of the anode metal; the target metals most commonly used in X-ray crystallographic studies are copper and molybdenum, which have $K_\alpha$ lines at 154.18 pm and 71.07 pm, respectively. These lines occur because the bombarding electrons knock out electrons from the innermost K shell ($n = 1$) and this in turn creates vacancies which are filled by electrons descending from the shells above. The decrease in energy appears as radiation; electrons descending from the L shell ($n = 2$) give the $K_\alpha$ lines and electrons from the M shell ($n = 3$) give the $K_\beta$ lines. (These lines are actually very closely spaced doublets — $K_{\alpha 1}$, $K_{\alpha 2}$ and $K_{\beta 1}$, $K_{\beta 2}$ — which are usually not resolved.) As the atomic number, Z, of the target increases, the lines shift to shorter wavelength.

Normally in X-ray diffraction, monochromatic radiation (single wavelength or a very narrow range of wavelengths) is required. Usually the $K_\alpha$ line is selected and the $K_\beta$ line is filtered out by using a filter made of a thin metal foil of the element adjacent $(Z - 1)$ in the Periodic Table; thus, nickel effectively filters out the $K_\beta$ line of copper, and niobium is used for molybdenum. A monochromatic beam of X-rays can also be selected by reflecting the beam from a plane of a single crystal, normally graphite (the reasons why this works will become obvious after you have read the next section).

## 2.2.2 DIFFRACTION OF X-RAYS

By 1912, the nature of X-rays — whether they were particles or waves — was still unresolved; a demonstration of X-ray diffraction effects was needed to demonstrate their wave nature. This was eventually achieved by Max von Laue using a crystal of copper sulfate as the diffraction grating, work which earned him the Nobel Prize for Physics in 1914. Crystalline solids consist of regular arrays of atoms, ions or molecules with interatomic spacings of the order of 100 pm. For diffraction to take place, the wavelength of the incident light has to be of the same order of magnitude as the spacings of the grating. Because of the periodic nature of the internal structure, it is possible for crystals to act as a three-dimensional diffraction grating to light of a suitable wavelength: a Laue photograph is pictured in Figure 2.2.

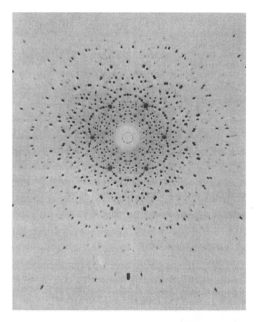

**FIGURE 2.2** X-ray diffraction by a crystal of beryl using the Laue method. (From W.J. Moore (1972) *Physical Chemistry*, 5th edn, Longman, London.)

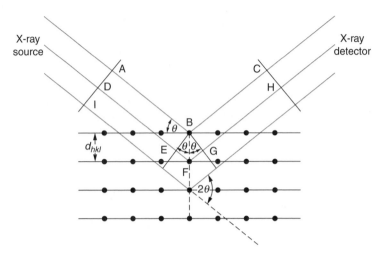

**FIGURE 2.3** Bragg reflection from a set of crystal planes with a spacing $d_{hkl}$.

This discovery was immediately noted by W.H. and W.L. Bragg (father and son), and they started experiments on using X-ray crystal diffraction as a means of structure determination. In 1913 they first determined the crystal structure of NaCl, and they went on to determine many structures including those of KCl, ZnS, $CaF_2$, $CaCO_3$, and diamond. W.L. (Lawrence) Bragg noted that X-ray diffraction behaves like 'reflection' from the planes of atoms within the crystal and that only at specific orientations of the crystal with respect to the source and detector are X-rays 'reflected' from the planes. It is not like the reflection of light from a mirror, as this requires that the angle of incidence equals the angle of reflection, and this is possible for all angles. With X-ray diffraction, the reflection only occurs when the conditions for constructive interference are fulfilled.

Figure 2.3 illustrates the Bragg condition for the reflection of X-rays by a crystal. The array of black points in the diagram represents a section through a crystal and the lines joining the dots mark a set of parallel planes with Miller indices $hkl$ and interplanar spacing $d_{hkl}$. A parallel beam of monochromatic X-rays ADI is incident to the planes at an angle $\theta_{hkl}$. The ray A is scattered by the atom at B and the ray D is scattered by the atom at F. For the reflected beams to emerge as a single beam of reasonable intensity, they must reinforce, or arrive in phase with one another. This is known as constructive interference, and for constructive interference to take place, the path lengths of the interfering beams must differ by an integral number of wavelengths. If BE and BG are drawn at right angles to the beam, the difference in path length between the two beams is given by:

$$\text{difference in path length} = EF + FG$$

but

$$EF = FG = d_{hkl}\sin\theta_{hkl}$$

so

$$\text{difference in path length} = 2d_{hkl}\sin\theta_{hkl} \tag{2.1}$$

This must be equal to an integral number, $n$, of wavelengths. If the wavelength of the X-rays is $\lambda$, then

$$n\lambda = 2d_{hkl}\sin\theta_{hkl} \tag{2.2}$$

This is known as the **Bragg equation**, and it relates the spacing between the crystal planes, $d_{hkl}$, to the particular Bragg angle, $\theta_{hkl}$ at which reflections from these planes are observed (mostly the subscript $hkl$ is dropped from the Bragg angle $\theta$ without any ambiguity as the angle is unique for each set of planes).

When $n = 1$, the reflections are called first order, and when $n = 2$ the reflections are second order and so on. However, the Bragg equation for a second order reflection from a set of planes $hkl$ is

$$2\lambda = 2d_{hkl}\sin\theta$$

which can be rewritten as

$$\lambda = 2\,\frac{d_{hkl}}{2}\,\sin\theta \tag{2.3}$$

Equation 2.3 represents a first order reflection from a set of planes with interplanar spacing $\frac{d_{hkl}}{2}$. The set of planes with interplanar spacing $\frac{d_{hkl}}{2}$ has Miller indices $2h\ 2k\ 2l$. Therefore, the second order reflection from $hkl$ is indistinguishable from the first order reflection from $2h\ 2k\ 2l$, and the Bragg equation may be written more simply as

$$\lambda = 2d_{hkl}\sin\theta \tag{2.4}$$

## 2.3 POWDER DIFFRACTION

### 2.3.1 POWDER DIFFRACTION PATTERNS

A finely ground crystalline powder contains a very large number of small crystals, known as crystallites, which are oriented randomly to one another. If such a sample is placed in the path of a monochromatic X-ray beam, diffraction will occur from planes in those crystallites which happen to be oriented at the correct angle to fulfill the Bragg condition. The diffracted beams make an angle of $2\theta$ with the incident beam. Because the crystallites can lie in all directions while still maintaining the Bragg condition, the reflections lie on the surface of cones whose *semi*-apex angles

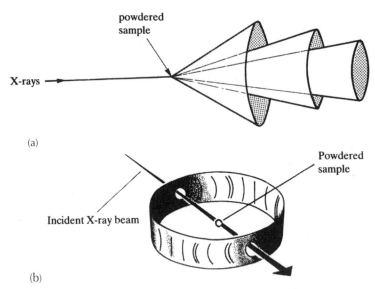

(a)

(b)

**FIGURE 2.4** (a) Cones produced by a powder diffraction experiment; (b) experimental arrangement for a Debye–Scherrer photograph.

are equal to the deflection angle $2\theta$ (Figure 2.4(a)). In the **Debye–Scherrer** photographic method, a strip of film was wrapped around the inside of a X-ray camera (Figure 2.4(b)) with a hole to allow in the collimated incident beam and a beam-stop to absorb the undiffracted beam. The sample was rotated to bring as many planes as possible into the diffracting condition, and the cones were recorded as arcs on the film. Using the radius of the camera and the distance along the film from the centre, the Bragg angle $2\theta$, and thus the $d_{hkl}$ spacing for each reflection can be calculated. Collection of powder diffraction patterns is now almost always performed by automatic diffractometers (Figure 2.5(a)), using a scintillation or CCD detector to record the angle and the intensity of the diffracted beams, which are plotted as intensity against $2\theta$ (Figure 2.5(b)). The resolution obtained using a diffractometer is better than photography as the sample acts like a mirror helping to refocus the X-ray beam. The data, both position and intensity, are readily measured and stored on a computer for analysis.

The difficulty in the powder method lies in deciding which planes are responsible for each reflection; this is known as 'indexing the reflections' (i.e., assigning the correct $hkl$ index to each reflection). Although this is often possible for simple compounds in high symmetry systems, as we shall explain in Section 2.4.3, it is extremely difficult to do for many larger and/or less symmetrical systems.

### 2.3.2 ABSENCES DUE TO LATTICE CENTRING

First, consider a primitive cubic system. From equation 2.4, we see that the planes giving rise to the reflection with the smallest Bragg angle will have the largest $d_{hkl}$ spacing. In the primitive cubic system the *100* planes have the largest separation

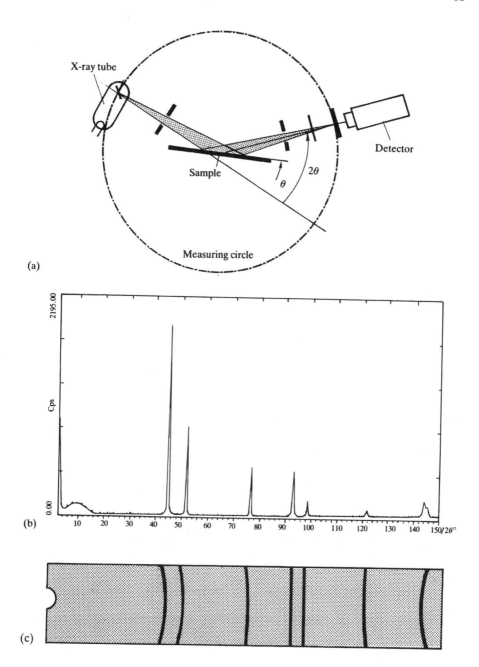

**FIGURE 2.5** (a) Diagram of a powder diffractometer; (b) a powder diffraction pattern for Ni powder compared with (c) the Debye–Scherrer photograph of Ni powder.

and thus give rise to this reflection, and as $a = b = c$ in a cubic system the $010$ and the $001$ also reflect at this position. For the cubic system, with a unit cell dimension $a$, the spacing of the reflecting planes is given by Equation 2.4

$$d_{hkl} = \frac{a}{\sqrt{(h^2 + k^2 + l^2)}} \tag{2.4}$$

Combining this with the Bragg equation gives

$$\lambda = \frac{2a\sin\theta_{hkl}}{\sqrt{(h^2 + k^2 + l^2)}}$$

and rearranging gives

$$\sin^2\theta_{hkl} = \frac{\lambda^2}{4a^2}(h^2 + k^2 + l^2) \tag{2.5}$$

For the primitive cubic class all integral values of the indices $h$, $k$, and $l$ are possible. Table 2.1 lists the values of $hkl$ in order of increasing value of ($h^2 + k^2 + l^2$) and therefore of increasing $\sin\theta$ values.

One value in the sequence, 7, is missing because no possible integral values exists for ($h^2 + k^2 + l^2$) = 7. Other higher missing values exist where ($h^2 + k^2 + l^2$) cannot be an integer: 15, 23, 28, etc., but note that this is only an arithmetical phenomenon and is nothing to do with the structure.

Taking Equation 2.5, if we plot the intensity of diffraction of the powder pattern of a primitive cubic system against $\sin^2\theta_{hkl}$ we would get six equi-spaced lines with the 7th, 15th, 23rd, etc., missing. Consequently, it is easy to identify a primitive cubic system and by inspection to assign indices to each of the reflections.

The cubic unit cell dimension $a$ can be determined from any of the indexed reflections using Equation 2.5. The experimental error in measuring the Bragg angle is constant for all angles, so to minimize error, either, the reflection with the largest Bragg angle is chosen, or more usually, a least squares refinement to all the data is used.

The pattern of observed lines for the two other cubic crystal systems, body-centred and face-centred is rather different from that of the primitive system. The differences arise because the centring leads to destructive interference for some reflections and these extra missing reflections are known as **systematic absences**.

---

**TABLE 2.1**
**Values of ($h^2 + k^2 + l^2$)**

| $hkl$ | 100 | 110 | 111 | 200 | 210 | 211 | 220 | 300 = 221 |
|---|---|---|---|---|---|---|---|---|
| ($h^2 + k^2 + l^2$) | 1 | 2 | 3 | 4 | 5 | 6 | 8 | 9 |

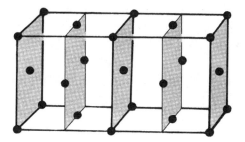

**FIGURE 2.6** Two F-centred unit cells with the *200* planes shaded.

Consider the *200* planes that are shaded in the F face-centred cubic unit cells depicted in Figure 2.6; if *a* is the cell dimension, they have a spacing $\frac{a}{2}$. Figure 2.7 illustrates the reflections from four consecutive planes in this structure. The reflection from the *200* planes is exactly out of phase with the *100* reflection. Throughout the crystal, equal numbers of the two types of planes exist, with the result that complete destructive interference occurs and no *100* reflection is observed. Examining reflections from all the planes for the F face-centred system in this way, we find that in order for a reflection to be observed, the indices must be either *all odd* or *all even*.

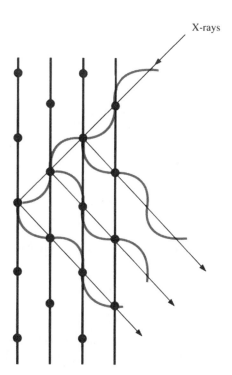

**FIGURE 2.7** The *100* reflection from an F-centred cubic lattice.

**TABLE 2.2**
**Allowed values of ($h^2 + k^2 + l^2$) for cubic crystals**

| Forbidden numbers | Primitive, P | Face-centred, F | Body-centred, I | Corresponding *hkl* values |
|---|---|---|---|---|
| | 1 | | | *100* |
| | 2 | | 2 | *110* |
| | 3 | 3 | | *111* |
| | 4 | 4 | 4 | *200* |
| | 5 | | | *210* |
| | 6 | | 6 | *211* |
| 7 | | | | *—* |
| | 8 | 8 | 8 | *220* |
| | 9 | | | *221, 300* |
| | 10 | | 10 | *310* |
| | 11 | 11 | | *311* |
| | 12 | 12 | 12 | *222* |
| | 13 | | | *320* |
| | 14 | | 14 | *321* |
| 15 | | | | *—* |
| | 16 | 16 | 16 | *400* |

A similar procedure for the body-centred cubic system finds that for reflections to be observed the *sum of the indices must be even*.

It is possible to characterize the type of Bravais lattice present by the pattern of systematic absences. Although our discussion has centred on cubic crystals, these absences apply to all crystal systems, not just to cubic, and are summarized in Table 2.3 at the end of the next section. The allowed values of $h^2 + k^2 + l^2$ are listed in Table 2.2 for each of the cubic lattices.

Using these pieces of information and Equation (2.5), we can see that if the observed $\sin^2\theta$ values for a pattern are in the ratio $1 : 2 : 3 : 4 : 5 : 6 : 8 \ldots$, then the unit cell is likely to be primitive cubic, and the common factor is $\dfrac{\lambda^2}{4a^2}$.

A face-centred cubic unit cell can also be recognized: if the first two lines have a common factor, A, then dividing all the observed $\sin^2\theta$ values by A gives a series of numbers, $3, 4, 8, 11, 12, 16 \ldots$, and A is equal to $\dfrac{\lambda^2}{4a^2}$.

A body-centred cubic system gives the values of $\sin^2\theta$ in the ratio $1 : 2 : 3 : 4 : 5 : 6 : 7 : 8 \ldots$ with the values 7 and 15 apparently *not* missing, but now the common factor is $\dfrac{2\lambda^2}{4a^2}$.

**TABLE 2.3**
**Systematic absences due to translational symmetry elements**

| Symmetry element | | Affected reflection | Condition for reflection to be present |
|---|---|---|---|
| Primitive lattice | P | $hkl$ | none |
| Body-centred lattice | I | $hkl$ | $h + k + l$ = even |
| Face-centred lattice | A | $hkl$ | $k + l$ = even |
| | B | | $h + l$ = even |
| | C | | $h + k$ = even |
| Face-centred lattice | F | | $h\ k\ l$ all odd or all even |
| | | | |
| twofold screw, $2_1$ along | | | |
| fourfold screw, $4_2$ along | $a$ | $h00$ | $h$ = even |
| sixfold screw, $6_3$ along | | | |
| threefold screw, $3_1$, $3_2$ along | | $00l$ | $l$ divisible by 3 |
| sixfold screw, $6_2$, $6_4$ along | $c$ | | |
| fourfold screw $4_1$, $4_3$ along | $a$ | $h00$ | $h$ divisible by 4 |
| sixfold screw, $6_1$, $6_5$ along | $c$ | $00l$ | $l$ divisible by 6 |
| | | | |
| Glide plane perpendicular to | $b$ | | |
| Translation $\frac{a}{2}$ ($a$ glide) | | | $h$ = even |
| Translation $\frac{c}{2}$ ($c$ glide) | | $h0l$ | $l$ = even |
| $\frac{b}{2} + \frac{c}{2}$ ($n$ glide) | | | $h + l$ = even |
| $\frac{b}{4} + \frac{c}{4}$ ($d$ glide) | | | $h + l$ divisible by 4 |

## 2.3.3 SYSTEMATIC ABSENCES DUE TO SCREW AXES AND GLIDE PLANES

The presence of translational symmetry elements in a crystal structure can be detected because they each lead to a set of systematic absences in the $hkl$ reflections. Figure 1.20 and Figure 1.21 illustrate how a twofold screw ($2_1$) along $z$ introduces a plane of atoms exactly halfway between the $001$ planes: reflections from these planes will destructively interfere with reflections from the $001$ planes and the $001$ reflection will be absent, as will any reflection for which $l$ is odd. The effect of a glide plane (Figure 1.19) is to introduce a plane of atoms halfway along the unit cell in the direction of the glide. For an $a$ glide perpendicular to $b$, the $10l$ reflection will be absent, and in general the $h0l$ reflections will only be present when $h$ is even. Systematic absences are summarized in Table 2.3.

Fairly powerful computer programmes for indexing are now in existence, and powder diffraction patterns can be indexed readily for the high symmetry crystal classes such as cubic, tetragonal, and hexagonal. For the other systems, the pattern often consists of a large number of overlapping lines, and indexing can be much more difficult or even impossible.

From the cubic unit cell dimension $a$, we can calculate the volume of the unit cell, $V$. If the density, $\rho$, of the crystals are known, then the mass of the contents of the unit cell, $M$, can also be calculated

$$\rho = \frac{M}{V} \qquad (2.6)$$

From a knowledge of the molecular mass, the number of molecules, $Z$, in the unit cell can be calculated. Examples of these calculations are in the questions at the end of the chapter.

The density of crystals can be determined by preparing a mixture of liquids (in which the crystals are insoluble!) such that the crystals neither float nor sink: the crystals then have the same density as the liquid. The density of the liquid can be determined in the traditional way using a density bottle.

### 2.3.4  USES OF POWDER X-RAY DIFFRACTION

#### Identification of Unknowns and Phase Purity

Powder diffraction is difficult to use as a method of determining crystal structures for anything other than simple high symmetry crystals because as the structures become more complex the number of lines increases so that overlap becomes a serious problem and it is difficult to index and measure the intensities of the reflections. It is usefully used as a fingerprint method for detecting the presence of a known compound or phase in a product. This is made possible by the existence of a huge library of powder diffraction patterns that is regularly updated, known as the Joint Committee for Powder Diffraction Standards (JCPDS) files, which are available on CD-ROM. When the powder diffraction pattern of your sample has been measured and both the $d_{hkl}$ spacings and intensity of the lines recorded, these can be matched against the patterns of known compounds in the files. With modern diffractometers, the computer matches the recorded pattern of the sample to the patterns stored in the JCPDS files (Figure 2.8).

The identification of compounds using powder diffraction is useful for qualitative analysis, such as mixtures of small crystals in geological samples. It also gives a rough check of the purity of a sample – but note that powder diffraction does not detect amorphous products or impurities of less than about 5%.

Powder diffraction can confirm whether two similar compounds, where one metal substitutes for another for instance, have an isomorphous structure.

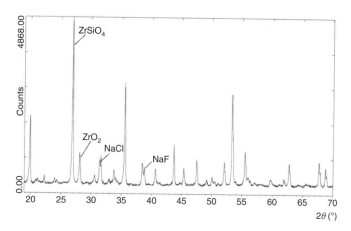

**FIGURE 2.8** X-ray diffraction pattern for the preparation of zircon, $ZrSiO_4$, from zirconia, $ZrO_2$, silica, $SiO_2$, and sodium halide mineralisers. The peaks demonstrate zircon to be the main product containing traces of all the starting materials.

## Crystallite Size

As the crystallite size decreases, the width of the diffraction peak increases. To either side of the Bragg angle, the diffracted beam will destructively interfere and we expect to see a sharp peak. However, the destructive interference is the resultant of the summation of all the diffracted beams, and close to the Bragg angle it takes diffraction from very many planes to produce complete destructive interference. In small crystallites not enough planes exist to produce complete destructive interference, and so we see a broadened peak.

The Debye–Scherrer formula enables the thickness of a crystallite to be calculated from the peak widths:

$$T = \frac{C\lambda}{B\cos\theta} = \frac{C\lambda}{(B_M^{\;2} - B_S^{\;2})^{\frac{1}{2}}\cos\theta} \tag{2.7}$$

where $T$ is the crystallite thickness, $\lambda$ the wavelength of the X-rays ($T$ and $\lambda$ have the same units), $\theta$ the Bragg angle, and $B$ is the full-width at half-maximum (FWHM) of the peak (radians) corrected for instrumental broadening. ($B_M$ and $B_S$ are the FWHMs of the sample and of a standard, respectively. A highly crystalline sample with a diffraction peak in a similar position to the sample is chosen and this gives the measure of the broadening due to instrumental effects.)

This method is particularly useful for plate-like crystals with distinctive shear planes (e.g., the *111*) as measuring the peak width of this reflection gives the thickness of the crystallites perpendicular to these planes.

It is a common feature of solid state reactions that reaction mixtures become more crystalline on heating as is evidenced by the X-ray diffraction pattern becoming sharper.

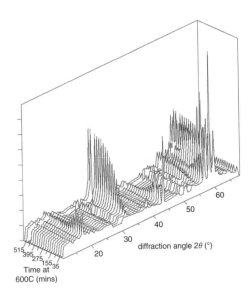

**FIGURE 2.9** Powder XRD patterns illustrate the phase changes in ferrosilicon with time, when heated at 600°C. (Courtesy of Professor F.J. Berry, Open University.)

## Following Reactions and Phase Diagrams

Powder X-ray diffraction is also a useful method for following the progress of a solid state reaction and determining mechanisms, and for determining phase diagrams. By collecting an X-ray pattern at regular intervals as the sample is heated on a special stage in the diffractometer, evolving phases can be seen as new lines which appear in the pattern, with a corresponding decrease in the lines due to the starting material(s). Figure 2.9 follows the phase transition of ferrosilicon from the non-stoichiometric alpha phase ($Fe_xSi_2$, $x = 0.77$–$0.87$) to the stoichiometric beta phase $FeSi_2$, when it is held at 600°C for a specific period. In Figure 2.10, we see three powder diffraction patterns taken at different temperatures from a solid state preparation of an Fe-doped zircon from powdered zirconia and silica according to the equation: $ZrO_2 + SiO_2 = ZrSiO_4$. Sodium halides are used to bring the reaction temperature down, and ferrous sulfate was the source of iron; as the temperature of the reaction mixture is increased, the peaks due to zirconia and silica decrease while those of zircon increase, until at 1060°C, this is the major component. During the reaction, peaks due to intermediates such as $Na_2SO_4$ and $Fe_2O_3$ also evolve.

A careful comparison of the intensities of particular lines using standards, not only enable the different phases be identified but also the proportions of different phases to be determined so that a phase diagram can be constructed.

### The Rietveld Method

In a high symmetry crystal system, very few peaks occur in the powder pattern, and they are often well resolved and well separated. It is then possible to measure their position and intensity with accuracy, and by the methods we described earlier, index

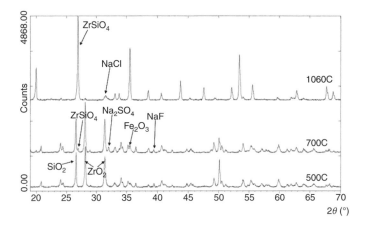

**FIGURE 2.10** The phase evolution of iron-doped zircon, $ZrSiO_4$, from zirconia, $ZrO_2$, silica, $SiO_2$, ferrous sulphate, $FeSO_4$, and sodium halide mineralisers.

the reflections and solve the structure. For larger and less symmetrical structures, far more reflections overlap considerably, and it becomes impossible to measure the intensities of individual peaks with any accuracy.

A method known as **Rietveld analysis** has been developed for solving crystal structures from powder diffraction data. The Rietveld method involves an interpretation of not only the line position but also of the line intensities, and because there is so much overlap of the reflections in the powder patterns, the method developed by Rietveld involves analysing the overall line profiles. Rietveld formulated a method of assigning each peak a gaussian shape and then allowing the gaussians to overlap so that an overall line profile could be calculated. The method was originally developed for neutron diffraction. In favourable cases, the Rietveld method can be used to solve a structure from the powder diffraction data. It starts by taking a trial structure, calculating a powder diffraction profile from it and then comparing it with the measured profile. The trial structure can then be gradually modified by changing the atomic positions and refined until a best-fit match with the measured pattern is obtained. The validity of the structure obtained is assessed by an $R$ factor, and by a difference plot of the two patterns (which should be a flat line). The method tends to work best if a good trial structure is already known, for instance if the unknown structure is a slight modification of a known structure, with perhaps one metal changed for another (Figure 2.11).

## 2.4  SINGLE CRYSTAL X-RAY DIFFRACTION

From a single crystal, it is possible to measure the position and intensity of the *hkl* reflections accurately and from this data determine not only the unit cell dimensions and space group, but also the precise atomic positions. In most cases, this can be done with speed and accuracy, and it is one of the most powerful structural techniques available to a chemist.

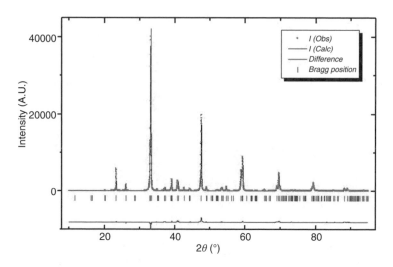

**FIGURE 2.11** Rietveld analysis of perovskite with partial substitution of Ti with Ca. (Courtesy of the Royal Society of Chemistry.)

### 2.4.1 THE IMPORTANCE OF INTENSITIES

So far, we have only discussed the effects of using crystals as three-dimensional diffraction gratings for X-rays. But you may have wondered why one goes to all this trouble. If we want to magnify an object to see its structure in more detail, why not use a lens system as in a microscope or a camera? Here a lens system focuses the light that is scattered from an object (which, if left alone, would form a diffraction pattern) and forms an image. Why not use a lens to focus X-rays and avoid all the complications? The problem is that there is no suitable way in which X-rays can be focussed, and so the effect of the lens has to be simulated by a mathematical calculation on the information contained in the diffracted beams. Much of this information is contained in the intensity of each beam, but as always, there is a snag! The recording methods do not record *all* of the information in the beam because they only record the intensity and are insensitive to the *phase*. Intensity is proportional to the *square* of the amplitude of the wave, and the phase information is lost. Unfortunately, it is this information which derives from the atomic positions in a structure. When a lens focuses light, this information is retained.

So far, we have seen that if we measure the Bragg angle of the reflections and successfully index them, then we get information on the size of the unit cell and, if it possesses any translational symmetry elements, also on the symmetry. In addition, we have seen that the intensity of each reflection is different and this too can be measured. In early photographic work, the relative intensities of the spots on the film were assessed by eye with reference to a standard, and later a scanning microdensitometer was used. In modern diffractometers, the beam is intercepted by a detector, either a charge coupled device (CCD) plate or a scintillation counter, and the intensity of each reflection is recorded electronically.

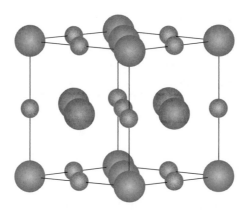

**FIGURE 2.12** NaCl unit cell depicting the close-packed *111* and *222* planes.

The interaction which takes place between X-rays and a crystal involves the electrons in the crystal: the more electrons an atom possesses, the more strongly will it scatter the X-rays. The effectiveness of an atom in scattering X-rays is called the **scattering factor** (or **form factor**), and given the symbol $f_0$. The scattering factor depends not only on the atomic number, but also on the Bragg angle $\theta$ and the wavelength of the X-radiation: as the Bragg angle increases, the scattering power drops off. The decrease in scattering power with angle is due to the finite size of an atom; the electrons are distributed around the nucleus and as $\theta$ increases, the X-rays scattered by an electron in one part of the atom are increasingly out of phase with those scattered in a different part of the electron cloud (see Figure 2.13a).

Why are intensities important? A simple example demonstrates this clearly. We know that the heavier an atom is, the better it is at scattering X-rays. On the face of it we might think that the planes containing the heavier atoms will give the more intense reflections. While this is true, the overall picture is more complicated than that because there are interactions with the reflected beams from other planes to take into account, which may produce *destructive interference*. Consider the diffraction patterns produced by NaCl and KCl crystals which both have the same structure (Figure 2.12). The structure can be thought of as two interlocking *ccp* arrays of $Na^+$ and $Cl^-$ ions. The unit cell depicted in the figure has close-packed layers of $Cl^-$ ions that lie parallel to a body diagonal, with indices *111*. Lying exactly halfway in between the *111* layers, and parallel to them, are close-packed layers of $Na^+$ ions; this means that a reflection from the $Cl^-$ close-packed layers is exactly out of phase with that from the equivalent $Na^+$ layers. Because a chloride ion has 18 electrons it scatters the X-rays more strongly than a sodium ion with 10 electrons, the reflections partially cancel and the intensity of the *111* reflection will be weak. The *222* layers contain the close-packed layers of both $Na^+$ and $Cl^-$, and this will be a strong reflection because the reflected waves will now reinforce one another. When we come to look at the equivalent situation in KCl, the reflection from the *111* layers containing $K^+$ ions is exactly out of phase with the reflection from the $Cl^-$ close-packed layers. But, $K^+$ and $Cl^-$ are isoelectronic, and so their scattering factors for X-rays are virtually identical and the net effect is that the two reflections cancel and the *111* reflection appears to be absent. Similarly, this means that the first observed reflection in the diffraction pattern from KCl is the *200* and it would be very easy to make the mistake that this was the *100* reflection from a primitive cubic cell with a unit cell length half that of the real face-centred cell.

The resultant of the waves scattered by all the atoms in the unit cell, in the direction of the *hkl* reflection, is called the **structure factor**, $F_{hkl}$, and is dependent on both the position of each atom and its scattering factor. It is given by the general expression for *j* atoms in a unit cell

$$F_{hkl} = \sum_j f_j e^{2\pi i(hx_j + ky_j + lz_j)} \tag{2.8}$$

where $f_j$ is the scattering factor of the *j*th atom and $x_j$ $y_j$ $z_j$ are its fractional coordinates. Series such as this can also be expressed in terms of sines and cosines, more obviously reflecting the periodic nature of the wave; they are known as **Fourier series**. In a crystal which has a centre of symmetry and *n* unique atoms in the unit cell (the unique set of atoms is known as the **asymmetric unit**), Equation (2.8) simplifies to

$$F_{hkl} = 2 \sum_n f_n \cos 2\pi (hx_n + ky_n + lz_n) \tag{2.9}$$

The electron density distribution within a crystal can be expressed in a similar way as a three-dimensional Fourier series:

$$\rho(x,y,z) = \frac{1}{V} \sum_h \sum_k \sum_l F_{hkl} e^{-2\pi i(hx + ky + lz)} \tag{2.10}$$

where $\rho(x,y,z)$ is the electron density at a position *x y z* in the unit cell and *V* is the volume of the unit cell. Notice the similarity between the expressions in Equation (2.8) and Equation (2.10). In mathematical terms, the electron density is said to be the **Fourier transform** of the structure factors and vice versa. This relationship means that if the structure factors are known, then it is possible to calculate the electron density distribution in the unit cell, and thus the atomic positions.

The intensity of the *hkl* reflections, $I_{hkl}$, are measured as described previously and form the data set for a particular crystal. The intensity of a reflection is proportional to the square of the structure factor:

$$I_{hkl} \propto F_{hkl}^2 \tag{2.11}$$

Taking the square root of the intensity gives a value for the *magnitude* of the structure factor (mathematically this is known as the *modulus* of the structure factor denoted by the vertical bars either side).

$$|F_{hkl}| \propto \sqrt{I_{hkl}} \tag{2.12}$$

Before this information can be used, the data set has to undergo some routine corrections, this process is known as **data reduction**. The **Lorentz correction, $L$**, relates to the geometry of the collection mode; the **polarization correction**, $p$, allows for the fact that the nonpolarized X-ray beam, may become partly polarized on reflection, and an **absorption correction** is often applied to data, particularly for inorganic structures, because the heavier atoms absorb some of the X-ray beam, rather than just scatter it. Corrections can also be made for **anomalous dispersion**, which affects the scattering power of an atom when the wavelength of the incident X-ray is close to its absorption edge. These corrections are applied to the scattering factor, $f_0$, of the atom.

The structure factor (and thus the intensity of a reflection) is dependent on both the position of each atom and its scattering factor. The structure factor can be *calculated*, therefore, from a knowledge of the types of atoms and their positions using Equation (2.8) or Equation (2.9). It is the great problem of X-ray crystallography that we need to be able to do the reverse of this calculation — we have the measured magnitudes of the structure factors, and from them we want to calculate the atomic positions. But there is the snag, which we mentioned earlier, known as **the phase problem**. When we take the square root of the intensity, we only obtain the modulus of the structure factor, and so we only know its magnitude and not its sign. The phase information is unfortunately lost, and we need it to calculate the electron density distribution and thus the atomic positions.

### 2.4.2  SOLVING SINGLE CRYSTAL STRUCTURES

It would seem to be an unresolvable problem — to calculate the structure factors we need the atomic positions and to find the atomic positions we need both the amplitude and the phase of the resultant waves, and we only have the amplitude. Fortunately, many scientists over the years have worked at finding ways around this problem, and have been extremely successful, to the extent that for many systems the solving of the structure has become a routine and fast procedure.

Single crystal X-ray diffraction data is nowadays collected using a computer controlled diffractometer, which measures the Bragg angle $\theta$ and the intensity $I$ for each *hkl* reflection. Many modern diffractometers employ a flat-plate detector (CCD), so that all the reflections can be collected and measured at the same time. A full data set, which can be thousands of reflections, can be accumulated in hours rather than the days or weeks of earlier times.

To summarize what we know about a structure:

- The size and shape of the unit cell is determined, usually from rotation photographs and scanning routines directly on the diffractometer.
- The reflections are indexed, and from the systematic absences the Bravais lattice and the translational symmetry elements of the structure determined: this information often determines the space group unequivocally, or narrows the possibilities down to a choice of two or three.
- The intensities of the indexed reflections are measured and stored as a data file.

- Correction factors are applied to the raw intensity data.
- Finally, the square roots of the corrected data are taken to give a set of observed structure factors. These are known as $F_{obs}$ or $F_o$.
- To calculate the electron density distribution in the unit cell, we need to know not only the magnitudes of the structure factors, but also their phase.

Crystal structures are solved by creating a set of trial phases for the structure factors. Two main methods are used to do this. The first is known as the **Patterson** method, and it relies on the presence of at least one (but not many) heavy atoms in the unit cell and so is useful for solving many inorganic molecular structures. The second is called **direct methods**, and it is best used for structures where the atoms have similar scattering properties. Direct methods calculate mathematical probabilities for the phase values and hence an electron density map of the unit cell; theoreticians have produced packages of accessible computer programs for solving and refining structures.

Once the atoms in a structure have been located, a calculated set of structure factors, $F_{calc}$ or $F_c$ is determined for comparison with the $F_{obs}$ magnitudes, and the positions of the atoms are refined using **least-squares methods**, for which standard computer programs are available. In practice, atoms vibrate about their equilibrium positions; this is often called **thermal motion**, although it depends not only on the temperature, but also on the mass of the atom and the strengths of the bonds holding it. The higher the temperature, the bigger the amplitude of vibration and the electron density becomes spread out over a larger volume, thus causing the scattering power of the atom to fall off more quickly. Part of the refinement procedure is to allow the electron density of each atom to refine in a sphere around the nucleus. Structure determinations usually quote an adjustable parameter known as the **isotropic displacement parameter**, $B$ (also called the **isotropic temperature factor**). The electron density of each atom can also be refined within an ellipsoid around its nucleus, when an **anisotropic displacement parameter** correction is applied which has six adjustable parameters.

The residual index, or $R$ **factor**, gives a measure of the difference between the observed and calculated structure factors and therefore of how well the structure has refined. It is defined as

$$R = \frac{\sum |(|F_o| - |F_c|)|}{\sum |F_o|} \tag{2.13}$$

and is used to give a guide to the correctness and precision of a structure. In general, the lower the $R$ value, the better the structure determination. $R$ values have to be used with caution because it is not unknown for structures to have a low $R$ value and still be wrong, although fortunately this does not happen often. No hard and fast rules exist for the expected value of $R$, and interpreting them is very much a matter of experience. It is usually taken as a rule of thumb for small molecule structures, that a correct structure for a reasonable quality data set would refine to below an $R$ of 0.1, anything above should be viewed with some degree of suspicion. That said, most structures nowadays, if collected from good quality crystals on modern diffractometers, would usually refine to below $R$ 0.05 and often to below $R$ 0.03.

A good structure determination, as well as having a low $R$ value, will also have low standard deviations on both the atomic positions and the bond lengths calculated from these positions. This is probably a more reliable guide to the quality of the refinement.

When a single crystal of a solid can be produced, X-ray diffraction provides an accurate, definitive structure, with bond lengths determined to tenths of a picometre. In recent years, the technique has been transformed from a very slow method reserved only for the most special structures, to a method of almost routine analysis: with modern machines, suites of computer programs and fast computers are used to solve several crystal structures per week.

## 2.5  NEUTRON DIFFRACTION

The vast majority of crystal structures published in the literature have been solved using X-ray diffraction. However, it is also possible to use neutron diffraction for crystallographic studies. It is a much less commonly used technique because very few sources of neutrons are available, whereas X-ray diffractometers can be housed in any laboratory. It does have advantages for certain structures, however.

The de Broglie relationship states that any beam of moving particles will display wave properties according to the formula

$$\lambda = \frac{h}{\rho} \tag{2.14}$$

where $\lambda$ is the wavelength, $\rho$ is the momentum of the particles ($\rho = mv$, mass $\times$ velocity), and h is Planck's constant. Neutrons are released in atomic fission processes from a uranium target, when they have very high velocities and a very small wavelength. The neutrons generated in a nuclear reactor can be slowed using heavy water so that they have a wavelength of about 100 pm and are thus suitable for structural diffraction experiments. The neutrons generated have a spread of wavelengths, and a monochromatic beam is formed using reflection from a plane of a single-crystal monochromator at a fixed angle (according to Bragg's law). Structural studies need a high flux of neutrons and this usually means that the only appropriate source is a high-flux nuclear reactor such as at Brookhaven and Oak Ridge in the United States, and Grenoble in France.

Alternative spallation sources are also available, such as the Rutherford laboratory in the United Kingdom, where the neutrons are produced by bombarding metal targets with high-energy protons. The diffraction experiments we have seen so far are set up with X-rays of a single wavelength $\lambda$, so that in order to collect all the diffracted beams, the Bragg angle $\theta$ is varied (Bragg equation $\lambda = 2d \sin\theta$). With the spallation source, the whole moderated beam with all its different wavelengths is used at a fixed angle, and the diffraction pattern is recorded as the function of the time of flight of the neutrons. (If we substitute $v = D/t$ [velocity = distance $\div$ time] in the de Broglie relationship, we see that the wavelength of the neutrons is propor-

tional to $t$: $\lambda = \dfrac{ht}{Dm}$ .) Because this method uses all of the beam, it has the advantage

of greater intensity.

The difference between the X-ray and neutron diffraction techniques lies in the scattering process: X-rays are scattered by the electrons around the nucleus, whereas neutrons are scattered by the nucleus. The scattering factor for X-rays increases linearly with the number of electrons in the atom, so that heavy atoms are much more effective at scattering than light atoms. However, because of the size of the atoms relative to the wavelength of the X-rays, the scattering from different parts of the cloud is not always in phase, so the scattering factor decreases with $\sin\theta/\lambda$ due to the destructive interference (Figure 2.13(a)). Because the nucleus is very small, neutron scattering factors do not decrease with $\sin\theta/\lambda$; and because nuclei are similar in size they are all similar in value (hydrogen is anomalously large due to the nuclear spin). Neutron scattering factors are also affected randomly by **resonance scattering**, when the neutron is absorbed by the nucleus and released later. This means that neutron scattering factors cannot be predicted but have to be determined experimentally and they vary for different atoms and indeed for different isotopes (Figure 2.13(b)).

Note that because of the different scattering mechanisms, the bond lengths determined by X-ray and neutron studies will be different. The neutron determination will give the true distance between the nuclei, whereas the X-ray values are distorted by the size of the electron cloud and so are shorter.

### 2.5.1 Uses of Neutron Diffraction

#### Locating Light Atoms

The fact that neutron scattering factors are similar for all elements means that light atoms scatter neutrons as effectively as heavy atoms and can therefore be located in the crystal structure; for example the X-ray scattering factors for deuterium and tungsten are 1 and 74, respectively, whereas the equivalent neutron values are 0.667 and 0.486. This property is particularly useful for locating hydrogen atoms in a structure, which can sometimes be difficult to do in an X-ray determination, especially if the hydrogen atoms are in the presence of a heavy metal atom. Accordingly, many neutron studies in the literature have been done with the express purpose of locating hydrogen atoms, or of exploring hydrogen bonding.

#### Heavy Atoms

The crystals do not absorb neutrons, so they are also useful for studying systems containing heavy atoms that absorb X-rays very strongly.

#### Similar Atomic Numbers and Isotopes

Atoms near each other in the Periodic Table have very similar X-ray scattering factors and cannot always be distinguished in an X-ray structure determination,

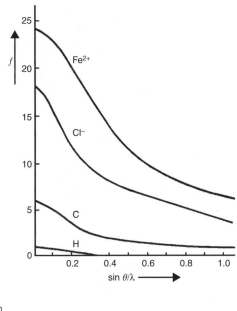

(a)

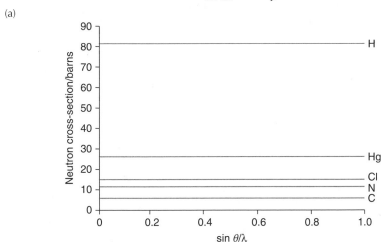

(b)

**FIGURE 2.13** (a) X-ray scattering factors for hydrogen, carbon, chloride and ferrous ions; (b) the neutron scattering cross sections for several elements, as a function of $\sin\theta/\lambda$.

oxygen and fluorine for instance, or similar metals in an alloy. A neutron structure determination may be able to identify atoms with similar atomic numbers.

## Magnetic Properties

As well as the scattering of the neutrons by the nuclei, there is additional magnetic scattering of the neutrons from paramagnetic atoms. This arises because a neutron has spin and so possesses a magnetic moment which can interact with the magnetic moment of an atom. The atomic magnetic moment is due to the alignment of the

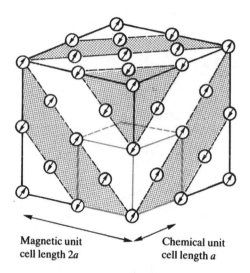

Magnetic unit                    Chemical unit
cell length 2*a*                 cell length *a*

**FIGURE 2.14** The magnetic ordering in NiO. The Ni planes only are pictured, and alternate close-packed layers have opposing magnetic moments. Note that the magnetic unit cell length is double that of the normal unit cell.

electron spins, and so this interaction, like the scattering of X-rays, falls off with increasing Bragg angle due to the size of the electron cloud. As will be discussed in Chapter 9, the magnetic moments of a paramagnetic crystal are arranged randomly, but in ferromagnetic, ferrimagnetic, and antiferromagnetic substances the atomic magnetic moments are arranged in an ordered fashion. In ferromagnetic substances, the magnetic moments are arranged so that they all point in the same direction and so reinforce one another; in antiferromagnetic substances, the magnetic moments are ordered so that they completely cancel one another out, and in ferrimagnetic substances the ordering leads to a partial cancellation of magnetic moments. Magnetic scattering of a polarized beam of neutrons from these ordered magnetic moments gives rise to magnetic Bragg peaks. For instance, the structure of NiO, as determined by X-ray diffraction is the same as NaCl. In the neutron study, however, below 120 K, extra peaks appear due to the magnetic interactions; these give a magnetic unit cell which has a cell length *twice* that of the standard cell. This arises (Figure 2.14) because the alternate close-packed layers of Ni atoms have their magnetic moments aligned in opposing directions, giving rise to antiferromagnetic behaviour.

### Rietveld Analysis

The technique of Rietveld profile analysis has already been mentioned in the context of X-ray powder diffraction, but it was with neutron powder diffraction that this technique originated. The fact that the neutron scattering factors are almost invariant with $\sin\theta/\lambda$ means that the intensity of the data does not drop off at high angles of $\theta$ as is the case with X-ray patterns, and so a neutron powder pattern tends to yield up considerably more data.

## Single Crystal Studies

The flux of a monochromatic source of neutrons is small, and this necessitates the use of large single crystals and long counting times for the experiment, in order to get sufficient intensity. Crystals typically have needed to be at least 1 mm in each direction, and it can be extremely difficult if not impossible to grow such large, perfect crystals. However, new high energy neutron sources are becoming available, such as the one at Grenoble, and the need for these large single crystals in neutron studies is receding.

## 2.6 ELECTRON MICROSCOPY

Optical microscopy has the advantages of cheapness and ease of sample preparation. A conventional optical microscope uses visible radiation (wavelength 400–700 nm) and so unfortunately cannot resolve images of objects which are smaller than half the wavelength of the light. However, a new technique known as **near-field scanning optical microscopy (NSOM)** uses a sub-wavelength-sized aperture rather than a lens to direct the light on to the sample. Moving the aperture and the sample relative to one another with sub-nanometre precision forms an image, and a spatial resolution of 10 to 100 nm can be achieved. This technique, still in its infancy, has been used successfully to study optical and optoelectronic properties of biological and nanometre scale materials.

Electron microscopy is widely used in the characterization of solids to study structure, morphology, and crystallite size, to examine defects and to determine the distribution of elements. An electron microscope is similar in principle to an optical microscope. The electron beam is produced by heating a tungsten filament, and focused by magnetic fields in a high vacuum (the vacuum prevents interaction of the beam with any extraneous particles in the atmosphere). The very short wavelength of the electrons allows resolution down to 0.1 nm.

### 2.6.1 SCANNING ELECTRON MICROSCOPY (SEM)

In this technique, the electrons from a finely focused beam are rastered across the surface of the sample. Electrons reflected by the surface of the sample and emitted secondary electrons are detected to give a map of the surface topography of samples such as catalysts, minerals, and polymers. It is useful for looking at particle size, crystal morphology, magnetic domains, and surface defects (Figure 2.15). A wide range of magnification can be used, the best achievable being about 2 nm. The samples may need to be coated with gold or graphite to stop charge building up on the surface.

### 2.6.2 TRANSMISSION ELECTRON MICROSCOPY (TEM)

In TEM, a thin sample (200 nm) is used and subjected to a high energy, high intensity beam of electrons; those which pass through the sample are detected forming a two-dimensional projection of the sample (Figure 2.16). The electrons may be elastically or inelastically scattered. The instrument can be operated to select either the direct

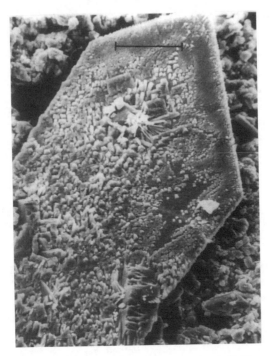

**FIGURE 2.15** SEM illustrating crystals of $VSbO_4$ growing out of $\beta$-$Sb_2O_4$ following reaction with $V_2O_5$. (Bar = 40 µm) (Courtesy of Professor Frank Berry, Open University.)

beam (bright field image) or the diffracted beam (dark field image). In high resolution instruments (sometimes called **high resolution electron microscopy [HREM]**), a very high potential field (up to $10^6$ V) accelerates the electrons, increasing their momentum to give very short wavelengths. Because the electrons pass through the sample, TEM/HREM images the bulk structure, and so can detect crystal defects such as phase boundaries, shear planes, and so on (Figure 2.17). Depending on the instrument, resolution of 0.5 nm can be achieved.

### 2.6.3 SCANNING TRANSMISSION ELECTRON MICROSCOPY (STEM)

These instruments combine the scanning ability of SEM with the high resolution achieved in TEM, a much smaller probe is used (10–15 nm) which scans across the sample.

### 2.6.4 ENERGY DISPERSIVE X-RAY ANALYSIS (EDAX)

As discussed in Section 2.2.2, an electron beam incident on a metal gives rise to the emission of characteristic X-rays from the metal. In electron microscopy, the elements present in the sample also emit characteristic X-rays. These are separated by a silicon-lithium detector, and each signal collected, amplified and corrected for

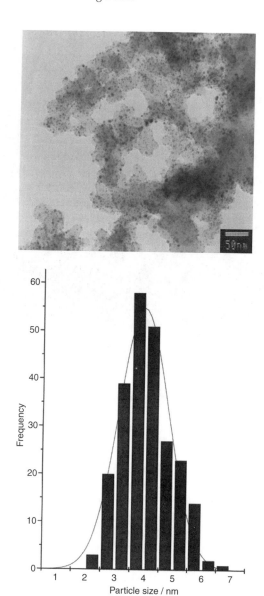

(a)

(b)

**FIGURE 2.16** (a) TEM image of a supported Pt/Cr bimetallic catalyst on C; (b) analysis of the metal particle sizes of this catalyst.

absorption and other effects, to give both qualitative and quantitative analysis of the elements present (for elements of atomic number greater than 11) in the irradiated particle, a technique known as **energy dispersive analysis of X-rays (EDAX or EDX)** (Figure 2.18).

**FIGURE 2.17** HREM image showing the atomic sites on the *111* plane of a Si crystal. (Courtesy of Dipl.–Ing. Michael Stöger-Pollach, Vienna University of Technology.)

## 2.7 X-RAY ABSORPTION SPECTROSCOPY

### 2.7.1 EXTENDED X-RAY ABSORPTION FINE STRUCTURE (EXAFS)

In high energy accelerators, electrons are injected into an electron storage ring (approximately 30 m in diameter) captured, and accelerated around this circular path by a series of magnets. When the electrons are accelerated to kinetic energies above the MeV range, they are travelling close to the speed of light and they emit so-called **synchrotron radiation** (Figure 2.19). For an accelerator in the GeV range (the synchrotron at Daresbury, United Kingdom, operates at 2 GeV, and its successor, DIAMOND at the Rutherford Laboratory at 3 GeV) the peak power is radiated at about $10^{18}$ Hz (approximately 10 keV and 1 Å) in the X-ray region of the electromagnetic spectrum. Unlike X-radiation from a conventional generator, the synchrotron radiation is of uniform intensity across a broad band of wavelengths and several orders of magnitude ($10^4$–$10^6$) higher in intensity (Figure 2.20). The shortest X-ray wavelengths emerge as almost fully collimated, polarized beams.

In an EXAFS experiment, the X-radiation is absorbed by a bound electron in a core shell (usually the K shell) and ejected as a photoelectron. If you measure the absorption coefficient of the sample as a function of the X-ray frequency, a sharp rise, or **absorption edge** is observed at the K shell threshold energy (Figure 2.21). Each element has its own characteristic K shell energy, and this makes it possible to study one type of atom in the presence of many others, by **tuning** the X-ray energy to its absorption edge. The appropriate frequency X-radiation from the continuous synchrotron radiation is selected by using the Bragg reflection from a

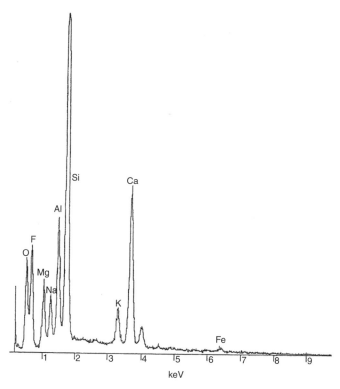

**FIGURE 2.18** EDAX analysis of a glaze.

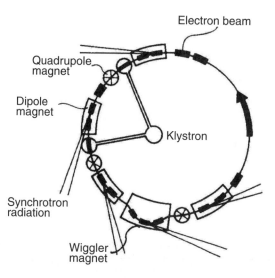

**FIGURE 2.19** Diagram of an electron storage ring for producing synchrotron radiation.

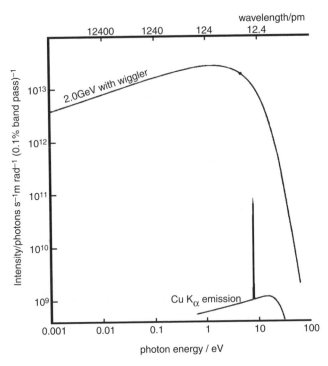

**FIGURE 2.20** Synchrotron radiation profile at 2 GeV compared with Cu K$_\alpha$ emission.

single plane of a carefully cut crystal such as Si (*220*); often, two crystals are used, as illustrated in the schematic diagram of a double crystal monochromator in Figure 2.22. By changing the Bragg angle of reflection, the frequency of the X-rays selected may be changed, and thus the absorption edges of a wide range of elements can be studied.

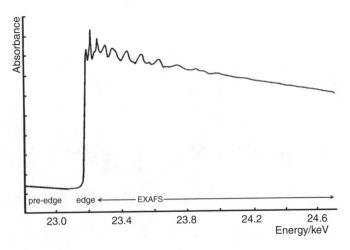

**FIGURE 2.21** The Rh absorption edge and EXAFS.

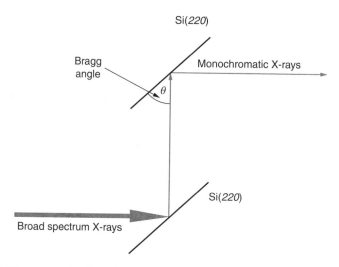

**FIGURE 2.22** Bragg reflections from a double-crystal monochromator. From the Bragg equation, $n\lambda = 2d\sin\theta$, $d$ for the planes of the crystal stays constant, so changing the angle changes the wavelength of the X-rays reflected. Two crystals are used to make the exit beam parallel to the entrance beam. Curved crystals focus the X-rays.

The waves of the ejected photoelectron from the K shell can be thought of as a spherical wave emanating from the nucleus of the absorbing atom; this encounters neighbouring atoms and is partially scattered by them producing a phase shift (Figure 2.23). Depending on the phase shift experienced by the electron, the reflected waves can then interfere constructively or destructively with the outgoing wave, producing a net interference pattern at the nucleus of the original atom. Absorption by the original atom is now modified, and the effect is seen as sinusoidal oscillations or **fine structure** superimposed on the absorption edge (Figure 2.21) extending out to several hundred eV after the edge. The extent to which the outgoing wave is reflected by a neighbouring atom, and so the intensity of the reflected wave, is partly dependent

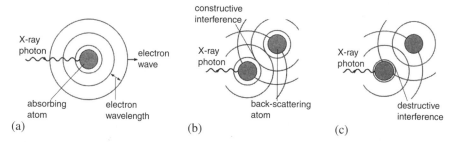

**FIGURE 2.23** The EXAFS process: (a) the photoelectron is ejected by X-ray absorption, (b) the outgoing photoelectron wave (solid line) is backscattered constructively by the surrounding atoms (dashed line), and (c) destructive interference between the outgoing and the backscattered wave.

on the scattering factor of that atom. The interference pattern making up the EXAFS thus depends on the *number*, and the *type* of neighbouring atoms, and their *distance* from the absorbing atom.

The EXAFS function is obtained from the X-ray absorption spectrum by subtracting the absorption due to the free atom. A Fourier transform of the EXAFS data gives a **radial distribution function** which shows the distribution of the neighbouring atoms as a function of internuclear distance from the absorbing atom. Shells of neighbours, known as **coordination shells**, surround the absorbing atom. Finally, the radial distribution function is fitted to a series of trial structural models until a structure which best fits the data is obtained, and the data is refined as a series of coordination shells surrounding the absorbing atom. The final structure will refine the number and types of atoms, and their distance from the absorbing atom. It is difficult to differentiate atoms of similar atomic number, and important to note that EXAFS only gives data on distance — no angular information is available. Depending on the quality of the data obtained, distances can be refined to about 1 pm, and in favourable cases several coordination shells out to about 600 pm can be refined.

In the example in Figure 2.24, a clay (a layered double hydroxide [LDH]) was intercalated with a transition metal complex $(NH_4)_2MnBr_4$. The EXAFS data in Figure 2.24(a) shows the Mn K-edge EXAFS of the pure complex, and we see one coordination sphere of four Br atoms at a distance of 2.49 Å, corresponding well to the tetrahedral coordination found in the X-ray crystal structure. However, after intercalation, the complex reacts with the layers in the clay, and the coordination changes to distorted octahedral where Mn is now surrounded by four O atoms at a distance of 1.92 Å and two Br atoms at a distance of 2.25 Å.

We saw earlier (Figure 2.9) that powder X-ray diffraction can be used to follow phase changes over time with heating; Figure 2.25 presents the corresponding iron K-edge EXAFS analysis for the same ferrosilicon sample. As the alpha ferrosilicon changes into the beta phase, a shell of eight silicon atoms at about 2.34 Å is found to surround Fe in both forms, but the beta form is found to have only two Fe atoms coordinated to Fe (2.97 Å), compared with 3.4 Fe atoms at 2.68 Å in the alpha phase.

X-rays are very penetrating, so EXAFS, like X-ray crystallography, examines the structure of the bulk of a solid. It has the disadvantage that it only provides information on interatomic distances, but has the considerable advantage that it is not confined to crystalline samples, and can be used on amorphous solids, glasses, and liquids. Not only that, but by using different absorption edges, it can investigate the coordination around more than one type of atom in the sample.

### 2.7.2 X-RAY ABSORPTION NEAR-EDGE STRUCTURE (XANES)

The precise position of the absorption edge varies with the chemical state of the absorbing atom, and this together with structure in the pre-edge region (Figure 2.21) can give information on the oxidation state of the atom and on its chemical environment. The example in Figure 2.26 depicts XANES spectra for manganese in different oxidation states.

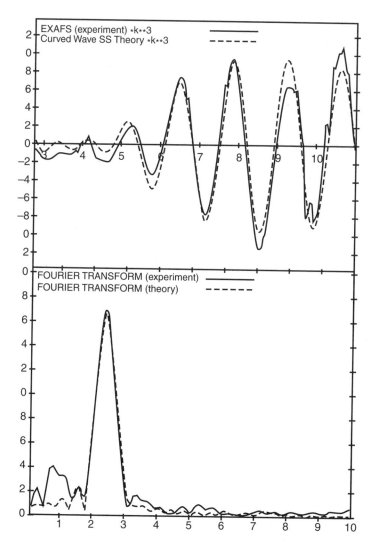

(a)

**FIGURE 2.24** EXAFS data for (a) $(NH_4^+)_2MnBr_4^{2-}$ : (upper) extracted EXAFS data; (lower) the radial distribution function, solid line experimental, dotted line calculated. (*–continued*)

## 2.8 SOLID STATE NUCLEAR MAGNETIC RESONANCE SPECTROSCOPY (MAS NMR)

In solution NMR spectroscopy, dipolar interactions and anisotropic effects are averaged out by the molecular motion, but this is not so in the solid state, and the NMR spectra of solids tend to be broadened by three main effects:

1. **Magnetic dipolar interactions** can be removed by the application of a high power decoupling field at the resonance frequency.

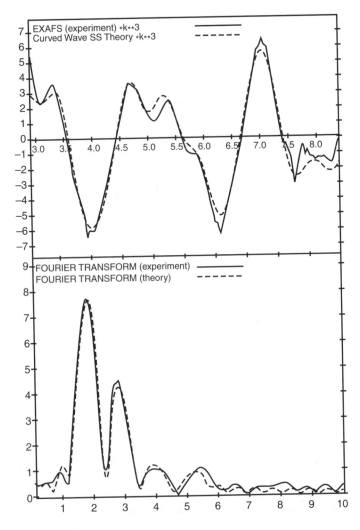

(b)

**FIGURE 2.24 (continued)** EXAFS data for (b) the intercalation of $MnBr_4^{2-}$ in a layered double hydroxide clay: (upper) extracted EXAFS data; (lower) the radial distribution function, solid line experimental, dotted line calculated.

2. Isotopes in low abundance have long spin-lattice relaxation times which give rise to poor signal-to-noise ratios. Sensitivity can be improved by using a technique known as **cross polarization** where a complex pulse sequence transfers polarization from an abundant nucleus to the dilute spin thereby enhancing the intensity of its signal.

3. The chemical shift of a particular atom varies with the orientation of the molecule to the field. In a solid this gives a range of values, an effect known as the **chemical shielding anisotropy**, which broadens the band.

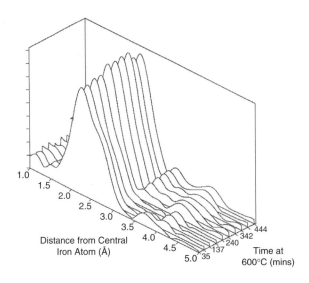

**FIGURE 2.25** EXAFS patterns as a function of time illustrating the phase evolution of beta ferrosilicon from the alpha form when heated at 600°C. (Courtesy of Professor F.J. Berry, Open University.)

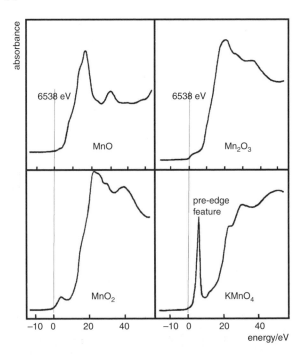

**FIGURE 2.26** Mn K-edge XANES data for various manganese oxides. The vertical dashed line is the position of the metal edge. The position of the edge changes, and the pre-edge feature increases, with oxidation state. (Courtesy of Dr. Neville Greaves, University of Aberystwyth.)

The line broadening is due to anisotropic interactions, all of which contain a $(3\cos^2\theta - 1)$ term. This term becomes zero when $3\cos^2\theta = 1$, or $\cos\theta = (1/3)^{1/2}$, (i.e., $\theta = 54°\ 44'$). **Magic angle spinning spectroscopy (MAS NMR)** spins the sample about an axis inclined at this so-called **magic angle** to the direction of the magnetic field and eliminates these sources of broadening, improving the resolution in chemical shift of the spectra. The spinning speed has to be greater than the frequency spread of the signal, if it is less, as may be the case for very broad bands, then a set of so-called "spinning side-bands" are observed, and care is needed in assigning the central resonance.

MAS NMR is often used nowadays as an umbrella term to imply the application of any or all of these techniques in obtaining a solid state NMR spectrum. High resolution spectra can be measured for most spin $I = {}^1/_2$ isotopes. MAS NMR has proved very successful in elucidating zeolite structures. Zeolites are three-dimensional framework silicate structures where many of the silicon sites are occupied by aluminium. Because Al and Si are next to each other in the Periodic Table they have similar X-ray atomic scattering factors, and consequently are virtually indistinguishable based on X-ray crystallographic data. It is possible to build up a picture of the overall shape of the framework with accurate atomic positions but not to decide which atom is Si and which is Al.

$^{29}$Si has a nuclear spin $I = {}^1/_2$ and so gives sharp spectral lines with no quadrupole broadening or asymmetry; the sensitivity is quite high and $^{29}$Si has a natural abundance of 4.7%. Pioneering work using MAS NMR on zeolites was carried by E. Lippmaa and G. Engelhardt in the late 1970s. They demonstrated that up to five peaks could be observed for the $^{29}$Si spectra of various zeolites and that these corresponded to the five different Si environments that can exist. Each Si is coordinated by four oxygen atoms, but each oxygen can then be attached either to a Si or to an Al atom giving the five possibilities: $Si(OAl)_4$, $Si(OAl)_3(OSi)$, $Si(OAl)_2(OSi)_2$, $Si(OAl)(OSi)_3$, and $Si(OSi)_4$. Most importantly, they also demonstrated that characteristic ranges of these shifts could be assigned to each coordination type. These ranges could then be used in further structural investigations of other zeolites (Figure 2.27). A MAS NMR spectrum of the zeolite known as analcite is depicted in Figure 2.28. Analcite has all five possible environments. Even with this information, it is still an extremely complicated procedure to decide where each linkage occurs in the structure.

$^{27}$Al has a 100% natural abundance and a nuclear spin $I = \frac{5}{2}$, resulting in a strong resonance which is broadened and rendered asymmetric by second-order quadrupolar effects. However, determining the $^{27}$Al MAS NMR spectrum of a zeolite can still have great diagnostic value because it distinguishes different types of aluminium coordination: octahedrally coordinated $[Al(H_2O)_6]^{3+}$ is frequently trapped as a cation in the pores of zeolites and gives a peak at about 0 ppm ($[Al(H_2O)_6]^{3+}$(aq) is used as the reference); tetrahedral $Al(OSi)_4$ gives rise to a single resonance with characteristic Al chemical shift values for individual zeolites in the range of 50 to 65 ppm; and $AlCl_4^-$, which may be present as a residue from the preparative process, has a resonance at about 100 ppm.

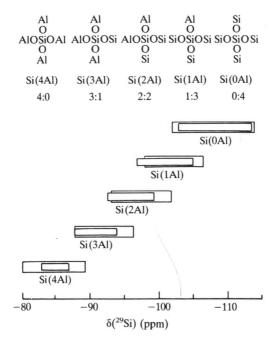

**FIGURE 2.27** The five possible local environments of a silicon atom together with their characteristic chemical shift ranges. The inner boxes represent the $^{29}$Si shift ranges suggested in the earlier literature. The outer boxes represent the extended $^{29}$Si shift ranges which are more unusual.

## 2.9  THERMAL ANALYSIS

Thermal analysis methods investigate the properties of solids as a function of a change in temperature. They are useful for investigating phase changes, decomposition, loss of water or oxygen, and for constructing phase diagrams.

### 2.9.1  DIFFERENTIAL THERMAL ANALYSIS (DTA)

A phase change produces either an absorption or an evolution of heat. The sample is placed in one chamber, and a solid that will not change phase over the temperature range of the experiment in the other. Both chambers are heated at a controlled uniform rate in a furnace, and the difference in temperature between the two is monitored and recorded against time. Any reaction in the sample will be represented as a peak in the plot of differential temperature; exothermic reactions give an increase in temperature, and endothermic a decrease, so the peaks appear in opposite directions. Figure 2.29 depicts three exotherms in the DTA of $KNO_3$, due to (i) a phase change from tetragonal to trigonal at 129°C, (ii) melting at 334°C, and (iii) decomposition above 550°C.

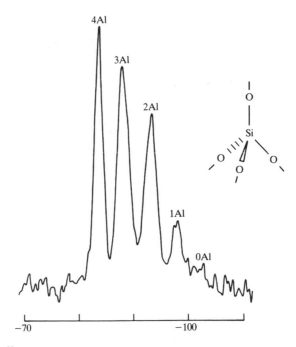

**FIGURE 2.28** $^{29}$Si MAS NMR spectrum at 79.6 MHz of analcite, illustrating five absorptions characteristic of the five possible permutations of Si and Al atoms attached at the corners of the $SiO_4$ tetrahedron as indicated.

### 2.9.2 THERMOGRAVIMETRIC ANALYSIS (TGA)

In this experiment, the weight of a sample is monitored as a function of time as the temperature is increased at a controlled uniform rate. The loss of water of crystallization or volatiles such as oxygen shows up as a weight loss, as does decomposition. Oxidation or adsorption of gas shows up as a weight gain. The TGA plot for $KNO_3$ in Figure 2.29 depicts a small weight loss up to about 550°C, probably due to the loss of adsorbed water, followed by a dramatic weight loss when the sample decomposes.

### 2.9.3 DIFFERENTIAL SCANNING CALORIMETRY (DSC)

DSC measures the amount of heat released by a sample as the temperature is increased or decreased at a controlled uniform rate, and so can investigate chemical reactions and measure heats of reaction for phase changes (Figure 2.30).

## 2.10 SCANNING TUNNELLING MICROSCOPY (STM) AND ATOMIC FORCE MICROSCOPY (AFM)

In **scanning tunnelling microscopy (STM)**, a sharp metal tip is brought sufficiently close to the surface of the solid sample (0.5–1 nm) that their electron-wave functions

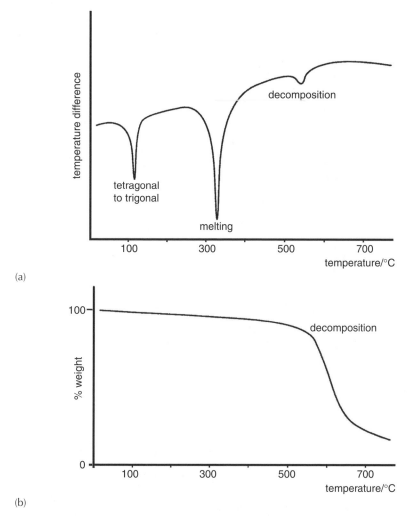

(a)

(b)

**FIGURE 2.29** (a) The DTA trace for $KNO_3$; (b) the TGA trace for $KNO_3$.

can overlap and electrons tunnel between the two. When a potential is applied to the solid surface, the electrons flow between the tip and the solid to give a tunnelling current in the range of pico to nano amperes. The magnitude of the current is very sensitive to the size of the gap, changing by a factor of 10 when the distance changes by 100 pm. The metal tip is scanned backward and forward across the solid, and the steep variation of the tunnelling current with distance gives an image of the atoms on the surface. The image is usually formed by keeping a constant tunnelling current and measuring the distance, thus creating contours of constant density of states on the surface. By changing the sign of the potential, the tunnelling direction reverses, and thus STM can map either occupied or unoccupied density of states. The map thus illustrates features due to both the topography and to the electronic structure, and can illustrate the positions of individual atoms (Figure 2.31(a)).

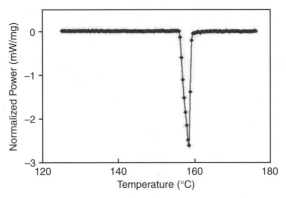

(a)

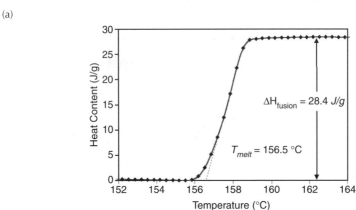

(b)

**FIGURE 2.30** (a) DSC trace for the melting of indium metal; (b) integration of the power data to give the heat of fusion for In. (Courtesy of Dr. Albert Sacco Jr., Northeastern University, Boston, Massachusetts.)

**AFM** is based on the detection of very small (of the order of nano newtons) forces between a sharp tip and atoms on a surface (Figure 2.31(b–d)). The tip is scanned across the surface at subnanometer distances, and the deflections due to attraction or repulsion by the underlying atoms detected. The technique produces atomic scale maps of the surface.

## 2.11 TEMPERATURE PROGRAMMED REDUCTION (TPR)

Temperature programmed reduction measures the reaction of hydrogen with a sample at various temperatures. The results are interpreted in terms of the different species present in the sample and their amenability to reduction. Therefore, these results can give information on the presence of different oxidation states or the effect of a dopant in a lattice. It is useful for measuring the temperature necessary for the complete reduction of a catalyst and is commonly used to investigate the interaction of a metal catalyst with its support, or of the effect of a promoter on a metal catalyst.

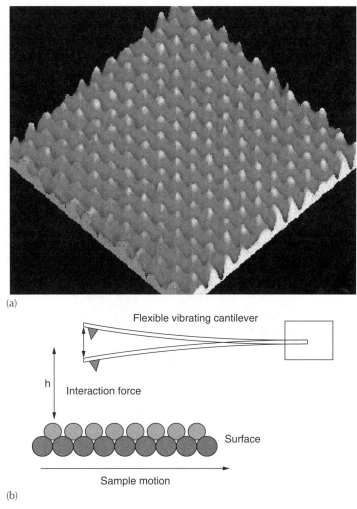

(a)

(b)

**FIGURE 2.31** (a) STM image of a periodic array of C atoms on HOPG. (Courtesy of Prof. R. Reifenberger, Purdue University.) (b) Diagram of AFM. *(–continued)*

The sample is heated over time in a furnace under a flowing gas mixture, typically 10% $H_2$ in $N_2$. Hydrogen has a high electrical conductivity, so a decrease in hydrogen concentration is marked by a decrease in conductivity of the gas mixture. This change is measured by a thermal conductivity cell (katharometer) and is recorded against either time or temperature.

The peaks in a TPR profile (Figure 2.32) measure the temperature at which various reduction steps take place. The amount of hydrogen used can be determined from the area under the peak. In the example, we see two peaks, indicating that the reduction of $Fe_2O_3$ takes place in two stages, first to $Fe_3O_4$ and then to metallic Fe.

Other temperature-programmed techniques include Temperature Programmed Oxidation and Sulfidation (TPO and TPS) for investigating oxidation and sulfidation

(c)

(d)

**FIGURE 2.31 (continued)** (c) AFM diamond tip on a silicon cantilever, and (d) AFM image of gold cluster on an oxidized surface. (Courtesy of Prof. R. Reifenberger, Purdue University.)

behaviour, and Temperature Programmed Desorption (TPD) (also called Thermal Desorption Spectroscopy [TDS]), which analyses gases desorbed from the surface of a solid or a catalyst on heating.

## 2.12  OTHER TECHNIQUES

In an introductory general text such as this, it is impossible to do justice to the plethora of physical techniques which are currently available to the solid state chemist. We have therefore concentrated on those which are most widely available, and on those which examine bulk material rather than surfaces. For these reasons, we have not covered Mössbauer spectroscopy, electron spectroscopies (XPS, UVPS, AES, and EELS) or low energy electron diffraction (LEED); for these, you will need to refer to more specialized texts. We have also not covered vibrational spectroscopy (IR and

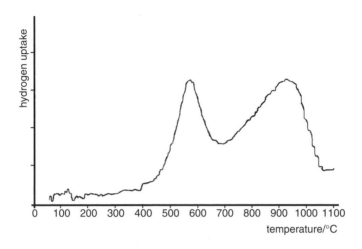

**FIGURE 2.32** TPR trace for $\alpha$-$Fe_2O_3$.

Raman) as good treatments of this subject are to be found in most undergraduate physical chemistry texts.

## QUESTIONS

1. What are the spacings of the *100*, *110*, and *111* planes in a cubic crystal system of unit cell dimension *a*? In what sequence would you expect to find these reflections in a powder diffraction photograph?
2. What is the sequence of the following reflections in a primitive cubic crystal: *220*, *300*, and *211*?
3. Nickel crystallizes in a cubic crystal system. The first reflection in the powder pattern of nickel is the *111*. What is the Bravais lattice?
4. The $\sin^2\theta$ values for $Cs_2TeBr_6$ are listed in Table 2.4 for the observed reflections. To which cubic class does it belong? Calculate its unit-cell length assuming Cu-$K_\alpha$ radiation of wavelength 154.2 pm.

**TABLE 2.4**
**$\sin^2\theta$ values for $Cs_2TeBr_6$**

| $\sin^2\theta$ |
| --- |
| .0149 |
| .0199 |
| .0399 |
| .0547 |
| .0597 |
| .0799 |
| .0947 |

**TABLE 2.5**
**$\theta$ values for NaCl**

$\theta_{hkl}$

13°41'
15°51'
22°44'
26°56'
28°14'
33°7'
36°32'
37°39'
42°0'
45°13'
50°36'
53°54'
55°2'
59°45'

5. X-ray powder data for NaCl is listed in Table 2.5. Determine the Bravais lattice, assuming that it is cubic.

6. Use the data given in Question 5 to calculate a unit-cell length for the NaCl unit cell.

7. If the unit cell length of NaCl is $a = 563.1$ pm and the density of NaCl is measured to be $2.17 \times 10^3$ kg m$^{-3}$; calculate Z, the number of formula units in the unit cell. (The atomic masses of sodium and chlorine are 22.99 and 35.45, respectively.)

8. A powder diffraction pattern establishes that silver crystallizes in a face-centred cubic unit cell. The *111* reflection is observed at $\theta = 19.1°$, using Cu-K$_\alpha$ radiation. Determine the unit cell length, $a$. If the density of silver is $10.5 \times 10^3$ kg m$^{-3}$ and $Z = 4$, calculate the value of the Avogadro constant. (The atomic mass of silver is 107.9.)

9. Calcium oxide crystallizes with a face-centred cubic lattice, $a = 481$ pm and a density $\rho = 3.35 \times 10^3$ kg m$^{-3}$: calculate a value for Z. (Atomic masses of Ca and O are 40.08 and 15.999, respectively.)

10. Thorium diselenide, ThSe$_2$, crystallizes in the orthorhombic system with $a = 442.0$ pm, $b = 761.0$ pm, $c = 906.4$ pm and a density $\rho = 8.5 \times 10^3$ kg m$^{-3}$: calculate Z. (Atomic masses of Th and Se are 232.03 and 78.96, respectively.)

11. Cu crystallizes with a cubic close-packed structure. The Bragg angles of the first two reflections in the powder pattern collected using Cu-K$_\alpha$ radiation are 21.6° and 25.15°. Calculate the unit cell length $a$, and estimate a radius for the Cu atom.

12. Arrange the following atoms in order of their ability to scatter X-rays: Na, Co, Cd, H, Tl, Pt, Cl, F, O.

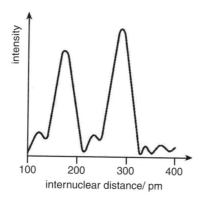

**FIGURE 2.33** Co edge EXAFS radial distribution function for Co(CO)$_4$.

13. In a cubic crystal we observe the *111* and *222* reflections, but not *001*. What is the Bravais lattice?

14. Interpret the EXAFS radial distribution function for Co(CO)$_4$ shown in Figure 2.33.

15. Figure 2.34 illustrates the $^{79}$Si MAS NMR spectrum of the zeolite faujasite. Use this figure to determine which Si environments are most likely to be present.

16. Figure 2.35 illustrates the $^{29}$Si spectrum of the same sample of faujasite as Figure 2.34, but after treatment with SiCl$_4$ and washing with water. What has happened to it?

17. A sample of faujasite was treated with SiCl$_4$ and four $^{27}$Al MAS NMR spectra were taken at various stages afterwards (Figure 2.36). Describe carefully what has happened during the process.

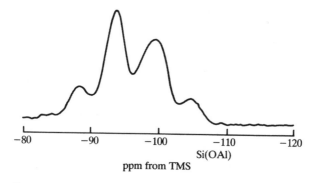

**FIGURE 2.34** $^{29}$Si MAS NMR spectrum at 79.6 MHz of faujasite of Si/Al = 2.61 (zeolite-Y).

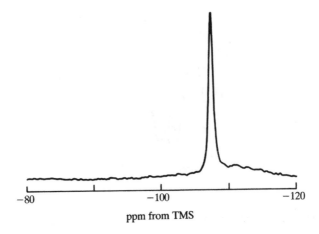

**FIGURE 2.35** $^{29}$Si MAS NMR spectrum at 79.6 MHz of faujasite of Si/Al = 2.61 after successive dealumination with $SiCl_4$ and washings.

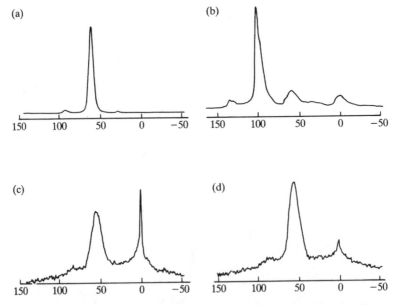

**FIGURE 2.36** $^{27}$Al MAS NMR spectra at 104.2 MHz obtained on faujasite samples at various stages of the $SiCl_4$ dealumination procedure: (a) starting faujasite sample, (b) intact sample after reaction with $SiCl_4$ before washing, (c) sample (b) after washing with distilled water, and (d) after several washings.

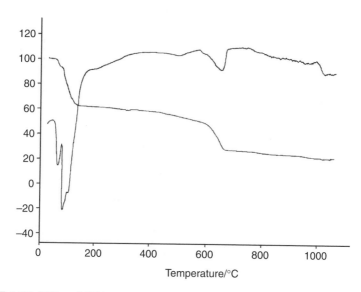

**FIGURE 2.37** DTA and TGA traces for ferrous sulfate heptahydrate.

18. A TGA trace for 25 mg of a hydrate of manganese oxalate, $MnC_2O_4.xH_2O$, revealed a weight loss of 20 mg at 100°C. What was the composition of the hydrate? A further weight loss occurs at 250°C, but then a weight *gain* at 900°C. What processes might be taking place?

19. Figure 2.37 illustrates the DTA and TGA traces for ferrous sulfate heptahydrate. Describe what processes are taking place.

# 3 Preparative Methods

## 3.1 INTRODUCTION

The interest in the properties of solids and the development of new materials has given rise to the development of a huge variety of methods for preparing them. The method chosen for any solid will depend not only on the composition of the solid but also on the form it is required in for its proposed use. For example, silica glass for fibre optics needs to be much freer of impurities than silica glass used to make laboratory equipment. Some methods may be particularly useful for producing solids in forms that are not the stable form under normal conditions; for example, the synthesis of diamond employs high pressures. Other methods may be chosen because they favour the formation of unusual oxidation states, for example, the preparation of chromium dioxide by the hydrothermal method, or because they promote the production of fine powders or, by contrast, large single crystals. For industrial use, a method that does not employ high temperatures could be favoured because of the ensuing energy savings.

In the preparation of solids, care usually has to be taken to use stoichiometric quantities, pure starting materials, and to ensure that the reaction has gone to completion because it is usually not possible to purify a solid once it has formed.

We do not have space here to discuss all the ingenious syntheses that have been employed over the past few years, so we shall concentrate on those that are commonly used with a few examples of techniques used for solids with particularly interesting properties. The preparation of organic solid state compounds and polymers is not covered because, generally, it involves organic synthesis techniques which is a whole field in itself, and is covered in many organic textbooks.

It is difficult to impose a logical order on such a diverse subject. The chapter begins by considering the most basic, and most commonly used, method of preparing solids, **the ceramic method**: this grand title disguises the fact that it simply means grinding up the reactant solids and heating them hard until they react! We then go on to look at refinements of this method, and ways of improving the uniformity of the reaction and reducing the reaction temperature. The following sections on microwave heating and combustion synthesis discuss alternative methods of inducing solid state reactions. Later sections concentrate on less well-known methods of preparing inorganic solids, such as using high pressures and gas-phase reactions. We also consider some methods used to produce particularly pure solids which are important in the semiconductor industry and the preparation of single crystals.

## 3.2 HIGH TEMPERATURE CERAMIC METHODS

### 3.2.1 DIRECT HEATING OF SOLIDS

#### The Ceramic Method

The simplest and most common way of preparing solids is the **ceramic method**, which consists of heating together two non-volatile solids which react to form the required product. This method is used widely both industrially and in the laboratory, and can be used to synthesize a whole range of materials such as mixed metal oxides, sulfides, nitrides, aluminosilicates and many others — the first high-temperature superconductors were made by a ceramic method. To take a simple example we can consider the formation of zircon, $ZrSiO_4$, which is used in the ceramics industry as the basis of high temperature pigments to colour the glazes on bathroom china. It is made by the direct reaction of zirconia, $ZrO_2$, and silica, $SiO_2$ at 1300°C:

$$ZrO_2(s) + SiO_2(s) = ZrSiO_4(s)$$

The procedure is to take stoichiometric amounts of the binary oxides, grind them in a pestle and mortar to give a uniform small particle size, and then heat in a furnace for several hours in an alumina crucible (Figure 3.1).

Despite its widespread use, the simple ceramic method has several disadvantages. High temperatures are generally required, typically between 500 and 2000°C, and this requires a large input of energy. This is because the coordination numbers in these ionic binary compounds are high, varying from 4 to 12 depending on the size and charge of the ion, and it takes a lot of energy to overcome the lattice energy so that a cation can leave its position in the lattice and diffuse to a different site. In addition, the phase or compound desired may be unstable or decompose at such high temperatures. Reactions such as this can be very slow; increasing the temperature speeds up the reaction as it increases the diffusion rate of the ions. In general, the solids are not raised to their melting temperatures, and so the reaction occurs in the solid state. Solid state reactions can only take place at the interface of the two solids and once the surface layer has reacted, reaction continues as the reactants diffuse from the bulk to the interface. Raising the temperature enables reaction at the interface and diffusion through the solid to go faster than they do at room temperatures; a rule of thumb suggests that a temperature of about two-thirds of the melting temperature of the solids, gives a reasonable reaction time. Even so, diffusion is often the limiting step. Because of this, it is important that the starting materials are ground to give a small particle size, and are very well mixed to maximise the surface contact area and minimise the distance that the reactants have to diffuse.

To achieve a homogeneous mix of small particles it is necessary to be very thorough in grinding the reactants. The number of crystallite faces in direct contact with one another can also be improved by pelletizing the mixed powders in a hydraulic press. Commonly the reaction mixture is removed during the heating process and reground to bring fresh surfaces in contact and so speed up the reaction. Nevertheless, the reaction time is measured in hours; the preparation of the ternary oxide $CuFe_2O_4$ from $CuO$ and $Fe_2O_3$, for example, takes 23 hours. The product is

(a)

(b)

**FIGURE 3.1** The basic apparatus for the ceramic method: (a) pestles and mortars for fine grinding; (b) a selection of porcelain, alumina, and platinum crucibles; and (c) furnace. (*–continued*)

often not homogeneous in composition, this is because as the reaction proceeds a layer of the ternary oxide is produced at the interface of the two crystals, and so ions now need to diffuse through this before they react. It is usual to take the initial product, grind it again, and reheat several times before a phase-pure product is obtained. Trial and error usually has to be used to find out the best reaction conditions, with samples tested by powder X-ray diffraction to determine the phase purity.

Solid state reactions up to 2300 K are usually carried out in furnaces which use resistance heating: the resistance of a metal element results in electrical energy being

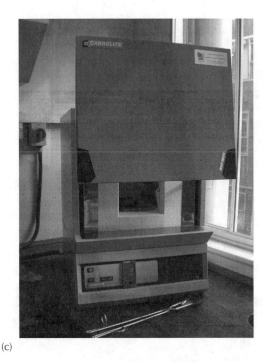

(c)

**Figure 3.1 (continued)**   The basic apparatus for the ceramic method: (c) furnace.

converted to heat, as in an electric fire. This is a common method of heating up to 2300 K, but above this, other methods have to be employed. Higher temperatures in samples can be achieved by directing an electric arc at the reaction mixture (to 3300 K), and for very high temperatures, a carbon dioxide laser (with output in the infrared) can give temperatures up to 4300 K.

Containers must be used for the reactions which can both withstand high temperatures and are also sufficiently inert not to react themselves; suitable crucibles are commonly made of silica (to 1430 K), alumina (to 2200 K), zirconia (to 2300 K) or magnesia (to 2700 K), but metals such as platinum and tantalum, and graphite linings are used for some reactions.

If any of the reactants are volatile or air sensitive, then this simple method of heating in the open atmosphere is no longer appropriate, and a sealed tube method will be needed.

### Sealed Tube Methods

Evacuated tubes are used when the products or reactants are sensitive to air or water or are volatile. An example of the use of this method is the preparation of samarium sulfide. In this case, sulfur has a low boiling temperature (717 K) and an evacuated tube is necessary to prevent it boiling off and being lost from the reaction vessel.

The preparation of samarium sulphide, SmS, is of interest because it contains samarium (a lanthanide element) in an unusual oxidation state +2 instead of the more common state +3. Samarium metal in powder form and powdered sulfur are

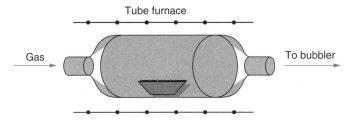

**FIGURE 3.2** Reacting solids under a special gas atmosphere.

mixed together in stoichiometric proportions, and heated to around 1000 K in an evacuated silica tube. (Depending on the temperature of reaction, pyrex or silica are the common choices for these reaction tubes, as they are fairly inert, and can be sealed on to a pyrex vacuum system for easy handling.) The product from the initial heating is then homogenised and heated again, this time to around 2300 K in a tantalum tube (sealed by welding) by passing an electric current through the tube, the resistance of the tantalum providing the heating. The pressures obtained in sealed reaction tubes can be very high, and it is not unknown for tubes to explode however carefully they are made; it is thus very important to take safety precautions, such as surrounding the tube with a protective metal container, and using safety screens.

**Special Atmospheres**

The preparation of some compounds must be carried out under a special atmosphere: an inert gas such as argon may be used to prevent oxidation to a higher oxidation state, an atmosphere of oxygen can be used to encourage the formation of high oxidation states; or conversely a hydrogen atmosphere can be used to produce a low oxidation state. Experiments of this type (Figure 3.2) are usually carried out with the reactant solids in small boat placed in a tube in a horizontal tube furnace. The gas is passed over the reactants for a time to expel all the air from the apparatus, and then flows over the reactants during the heating and cooling cycles, exiting through a bubbler to maintain a positive pressure and prevent the ingress of air by back diffusion.

To obtain products that are more homogeneous as well as better reaction rates, methods involving smaller particles than can be obtained by grinding have been introduced and these methods are described in the next section.

### 3.2.2 REDUCING THE PARTICLE SIZE AND LOWERING THE TEMPERATURE

In a polycrystalline mixture of solid reactants, we might expect the particle size to be of the order of 10 µm; even careful and persistent grinding will only reduce the particle size to around 0.1 µm. Diffusion during a ceramic reaction is therefore taking place across anywhere between 100 and 10,000 unit cells. Various ingenious methods, some physical and some chemical, have been pioneered to bring the components of the reaction either into more intimate contact, or into contact at an atomic level, and so reduce this diffusion path; in doing so, the reactions can often take place at lower temperatures.

## Spray-Drying

The reactants are dissolved in a suitable solvent and sprayed as fine droplets into a hot chamber. The solvent evaporates leaving a mixture of the solids as a fine powder which can then be heated to give the product.

## Freeze-Drying

The reactants are dissolved in a suitable solvent and frozen to liquid nitrogen temperatures (77 K). The solvent is then removed by pumping to leave a fine reactive powder.

## The Co-Precipitation and Precursor Methods

At a simple level, precursors such as nitrates and carbonates can be used as starting materials instead of oxides: they decompose to the oxides on heating at relatively low temperatures, losing gaseous species, and leaving behind fine, more reactive powders.

An even more intimate mixture of starting materials can be made by the **co-precipitation** of solids. A stoichiometric mixture of soluble salts of the metal ions is dissolved and then precipitated as hydroxides, citrates, oxalates, or formates. This mixture is filtered, dried, and then heated to give the final product.

The **precursor method** achieves mixing at the atomic level by forming a solid compound, the precursor, in which the metals of the desired compound are present in the correct stoichiometry. For example if an oxide $MM'_2O_4$ is required, a mixed salt of an oxyacid such as an acetate containing M and M' in the ratio of 1:2 is formed. The precursor is then heated to decompose it to the required product. Homogeneous products are formed at relatively low temperatures. A disadvantage is that it is not always possible to find a suitable precursor, but the preparation of barium titanate gives a good illustration of this method.

Barium titanate, $BaTiO_3$, is a ferroelectric material (see Chapter 9) widely used in capacitors because of its high dielectric constant. It was initially prepared by heating barium carbonate and titanium dioxide at high temperature.

$$BaCO_3(s) + TiO_2(s) = BaTiO_3(s) + CO_2(g)$$

However, for modern electronic circuits, it is important to have a product of controlled grain size and the precursor method is one way to achieve this. (Another way is the sol-gel method which has also been applied to this material.) The precursor used is an oxalate. The first step in the preparation is to prepare an oxo-oxalate of titanium. Excess oxalic acid solution is added to titanium butoxide which initially hydrolyses to give a precipitate which then redissolves in the excess oxalic acid.

$$Ti(OBu)_4(aq) + 4H_2O(l) = Ti(OH)_4(s) + 4BuOH(aq)$$

$$Ti(OH)_4(s) + (COO)_2{}^{2-}(aq) = TiO(COO)_2(aq) + 2OH^-(aq) + H_2O(l)$$

Barium chloride solution is then added and barium titanyl oxalate precipitates.

$$Ba^{2+}(aq) + (COO)_2{}^{2-}(aq) + TiO(COO)_2(aq) = Ba[TiO((COO)_2)_2](s)$$

This precipitate contains barium and titanium in the correct ratio and is easily decomposed by heat, to give the oxide. The temperature used for this final heating is 920 K.

$$Ba[TiO((COO)_2)_2](s) = BaTiO_3(s) + 2CO_2(g) + 2CO(g)$$

The decomposition of oxalates is also used to prepare ferrites $MFe_2O_4$, which are important as magnetic materials (see Chapter 9).

The products of the precursor method are usually crystalline solids, often containing small particles of large surface area. For some applications, such as catalysis and barium titanate capacitors, this is an advantage.

## The Sol-Gel Method

The precipitation methods always have the disadvantage that the stoichiometry of the precipitate(s) may not be exact if one or more ions are left in solution. The sol-gel method overcomes this because the reactants never precipitate out. First, a concentrated solution or colloidal suspension of the reactants, the 'sol', is prepared, which is then concentrated or matured to form the 'gel'. This homogeneous gel is then heat-treated to form the product. The main steps in the sol-gel process are outlined in Figure 3.3.

The first investigation of a sol-gel process for synthesis was made in the mid-19th century. This early investigation studied the preparation of silica glass from a sol made by hydrolysing an alkoxide of silicon. Unfortunately, to prevent the product cracking and forming a fine powder, an aging period of a year or more was required! The sol-gel method was further developed in the 1950s and 1960s after it was realised

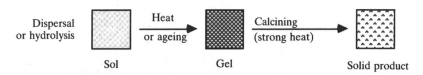

FIGURE 3.3 Steps in the sol-gel synthesis route.

that colloids, which contain small particles (1–1000 nm in diameter), could be highly chemically homogeneous.

A sol is a colloidal suspension of particles in a liquid; for the materials being discussed here, these particles will typically be 1 to 100 nm in diameter. A gel is a semi-rigid solid in which the solvent is contained in a framework of material which is either colloidal (essentially a concentrated sol) or polymeric.

To prepare solids using the sol-gel method, a sol of reactants is first prepared in a suitable liquid. Sol preparation can be either simply the dispersal of an insoluble solid or addition of a precursor which reacts with the solvent to form a colloidal product. A typical example of the first is the dispersal of oxides or hydroxides in water with the pH adjusted so that the solid particles remain in suspension rather than precipitate out. A typical example of the second method is the addition of metal alkoxides to water; the alkoxides are hydrolysed giving the oxide as a colloidal product.

The sol is either then treated or simply left to form a gel over time by dehydrating and/or polymerizing. To obtain the final product, the gel is heated. This heating serves several purposes — it removes the solvent, it decomposes anions such as alkoxides or carbonates to give oxides, it allows rearrangement of the structure of the solid, and it allows crystallisation to occur. Both the time and the temperature needed for reaction in sol-gel processes can reduced from those in the direct ceramic method; in favourable cases, the time from days to hours, and the temperature by several hundred degrees.

Several examples discussed below illustrate the method; two of the examples have been chosen because they have interesting properties and uses, discussed later on in this book. Many other materials have been prepared by the sol-gel method, and other sol-gel preparations have been employed for the materials chosen. Therefore, these examples should be taken as illustrative and not as the main uses of the method.

### Lithium Niobate, $LiNbO_3$

Lithium niobate is a ferroelectric material used as an optical switch. Preparation by the simple ceramic method leads to problems in obtaining the correct stoichiometry, and a mixture of phases often results. Several sol-gel preparations have been described, their advantage being the lower temperature required for the preparation and the greater homogeneity of the product. One such preparation starts with lithium ethoxide ($LiOC_2H_5$ (or LiOEt)) and niobium ethoxide $Nb_2(OEt)_{10}$. Each ethoxide was dissolved in absolute ethanol and the two solutions mixed. The addition of water leads to partial hydrolysis giving hydroxy-alkoxides, for example:

$$Nb_2(OEt)_{10} + 2H_2O = 2Nb(OEt)_4(OH) + 2EtOH$$

The hydroxy-alkoxides condense to form a polymeric gel with metal-oxygen-metal links. Lithium niobate is then formed when the gel is heated. Heating removes any remaining ethanol and any water formed during the condensation. The remaining ethyl groups are pyrolysed (i.e., oxidised to carbon dioxide and water) leaving the oxides.

## Doped Tin Dioxide

Tin dioxide, $SnO_2$, is an oxygen-deficient $n$-type semiconductor (see Chapter 5) whose conductivity is increased by the addition of dopants such as $Sb^{3+}$ ions. One use of doped tin dioxide is to form a transparent electrode. For this, it is deposited on to glass and must have a large surface area with a controlled distribution of dopants. Sol-gel methods have proved suitable for the preparation of such coatings because gels can be sliced finely. In one reported preparation of titanium-doped tin dioxide, the sol was prepared by adding titanium butoxide, $Ti(OC_4H_9)_4$ ($Ti(OBu)_4$), to a solution of tin dichloride dihydrate ($SnCl_2.2H_2O$) in absolute ethanol. Tin(II) salts are easily hydrolysed to give hydroxy complexes and can be oxidised to tin(IV) by oxygen in the air. The sol was left in an open container for five days during which time it formed a gel. The gel was dried by heating at 333 K under reduced pressure for 2 hours. Finally, it was heated to around 600 K for 10 minutes to produce a layer of tin dioxide doped with $Ti^{4+}$ ions.

## Silica for Optical Fibres

Optical fibres need to be free of impurities such as transition metal ions (see Chapter 8) and conventional methods of preparing silica glasses are inadequate. The sol-gel process is one way of forming fibres of sufficient purity (chemical vapour deposition, Section 3.7, is another). These processes use volatile compounds of silicon which are more easily purified, for example by fractional distillation, than silica. It is possible to produce silica fibres using a method similar to that studied in the nineteenth century, but with the gel-drying time reduced from a year to a few days. Liquid silicon alkoxide ($Si(OR)_4$), where R is methyl, ethyl, or propyl, is hydrolysed by mixing with water.

$$Si(OMe)_4 + 4H_2O = Si(OH)_4 + 4MeOH$$

The product $Si(OH)_4$ condenses forming $Si-O-Si$ bonds. Gradually, more and more $SiO_4$ tetrahedra are linked eventually forming $SiO_2$. Early stages of this process are shown:

$$Si(OH)_4 + Si(OH)_4 = (OH)_3Si-O-Si(OH)_3 + H_2O$$
$$(OH)_3Si-O-Si(OH)_3 + 6Si(OH)_4 =$$

$+ 6H_2O$

As the condensed species reach a certain size, they form colloidal particles. The resulting sol is cast into a mould where further cross-linking results in gel formation. Fibres can be pulled as gelation occurs. Condensation continues as the gel ages. During aging, it remains immersed in liquid. The gel is porous, and the alcohol and water produced by hydrolysis and condensation is trapped in the pores. Some is expelled as the gel shrinks during aging, but the rest has to be removed in a drying process. It is at this stage that cracking occurs, but if fibres are drawn with a diameter of less than about 1 cm, then the stresses produced by drying are reduced and cracking is not a problem. For gels of larger cross section, cracking can be reduced by the use of surfactants. Finally, the silica is heated to around 1300 K to increase the density of the glass.

## A Biosensor

An interesting application of the sol-gel method is in manufacturing biosensors. A **biosensor** uses a substance of biological origin, often an enzyme, to detect the presence of a molecule. To be of use as a sensor, the biological substance must be in a form which can easily be carried around, and placed in a solution, in a stream of gas or on a patient's skin. Biological materials such as enzymes are potentially extremely useful as sensors because they can be very specific; for example, where a chemical test may be positive for any oxidising gas, an enzyme may only give a positive test for oxygen. Enzymes are easily denatured, however, by removing them from their aqueous environment or by altering the pH of their environment. One way around this problem is to trap the enzyme with its microenvironment in a silica gel. In one such preparation, a sol of silica was first prepared by acid hydrolysis of tetramethylorthosilicate, $Si(OMe)_4$. In a typical sol-gel synthesis, this compound would be dissolved in methanol, but in this case, methanol was avoided because alcohols denature proteins and so the solvent used was water. The sol was then buffered to be at a suitable pH for the enzyme and the enzyme added. As the sol formed a gel by condensing, the enzyme was trapped in cavities in the gel surrounded by an aqueous medium of the correct pH.

In this last case, the product was left as a gel, but if the product is to be a glass, a powder, or crystalline material, a high temperature calcining step is required after formation of the gel. Thus the sol-gel method improves the homogeneity of the product but reaction times are still long (note the five day gelling time in the tin dioxide synthesis) and high temperatures (1300 K for silica) are still needed.

Use of microwave ovens instead of conventional heating leads to a speeding-up of reaction in favourable cases.

## 3.3 MICROWAVE SYNTHESIS

We are all familiar with use of microwave radiation in cooking food where it increases the speed of reaction. Recently this method has been used to synthesize solid state materials such as mixed oxides. The first solid state reaction experiments were performed in modified domestic ovens, and these are still used, but more specialised (and expensive!) ovens have also been developed to give more control over the conditions. We shall briefly consider how microwaves heat solids and liquids because this gives us insight into which reactions will be good candidates for this method.

In a liquid or solid, the molecules or ions are not free to rotate and so the heating is *not* the result of the absorption of microwaves by molecules undergoing rotational transitions as they would in the gas phase. In a solid or liquid the alternating electric field of the microwave radiation can act in two ways. If charged particles are present that can move freely through the solid or liquid, then these will move under the influence of the field producing an oscillating electric current. Resistance to their movement causes energy to be transferred to the surroundings as heat. This is **conduction heating**. If no particles are present that can move freely, but molecules or units with dipole moments are present, then the electric field acts to align the dipole moments. This effect produces **dielectric heating**, and when it acts on water molecules in food, is generally responsible for the heating/cooking in domestic microwave ovens. The electric field of microwave radiation, like that of all electromagnetic radiation, is oscillating at the frequency of the radiation. The electric dipoles in the solid do not change their alignment instantaneously but with a characteristic time, $\tau$. If the oscillating electric field changes its direction slowly so that the time between changes is much greater than $\tau$ then the dipoles can follow the changes. A small amount of energy is transferred to the surroundings as heat each time the dipole realigns but this is only a small heating effect. If the electric field of the radiation oscillates very rapidly, the dipoles cannot respond fast enough and do not realign. The frequency of microwave radiation is such that the electric field changes sign at a speed that is the same order of magnitude as $\tau$. Under these conditions the dipole realignment lags slightly behind the change of electric field and the solid absorbs microwave radiation. This absorbed energy is converted to heat. The quantities governing this process are the dielectric constant (see Chapter 9), which determines the extent of dipole alignment, and the dielectric loss, which governs how efficiently the absorbed radiation is converted to heat.

To use microwave heating in solid state synthesis, at least one component of the reaction mixture must absorb microwave radiation. The speed of the reaction process is then increased by both increasing the rate of the solid state reaction and by increasing the rate of diffusion, which, as we mentioned earlier, is often the rate-limiting step.

### 3.3.1  The High Temperature Superconductor YBa$_2$Cu$_3$O$_{7-x}$

High temperature superconductors were first prepared using a conventional ceramic method, in this example by baking together yttrium oxide, copper(II) oxide, and barium carbonate. The synthesis however takes 24 hours to complete. A microwave method has been reported that can prepare the superconductor in under two hours. A stoichiometric mixture of copper(II) oxide, CuO, barium nitrate, Ba(NO$_3$)$_2$, and yttrium oxide, Y$_2$O$_3$, was placed in a microwave oven which had been modified to allow safe removal of the nitrogen oxides formed during the reaction. The mixture was treated with 500 watts of microwave radiation for 5 minutes then reground and exposed to microwave radiation at 130 to 500 watts for 15 minutes. Finally, the mixture was ground again and exposed to microwave radiation for an additional 25 minutes. The microwaves in this example couple to the copper(II) oxide. This is one of a range of oxides, mostly non-stoichiometric, which are efficiently heated by microwave radiation. For example, a 5 to 6 g sample of CuO exposed to 500 watts of microwave radiation for half a minute, attained a temperature of 1074 K. Other oxides which strongly absorb microwave radiation include ZnO, $V_2O_5$, $MnO_2$, $PbO_2$, $Co_2O_3$, $Fe_3O_4$, NiO, and $WO_3$. Carbon, $SnCl_2$, and $ZnCl_2$ are also strong absorbers. By contrast, a 5 to 6 g sample of calcium oxide, CaO, exposed to 500 watts microwave radiation only attained a temperature of 356 K after half an hour. Other oxides which do not absorb strongly include $TiO_2$, $CeO_2$, $Fe_2O_3$, $Pb_3O_4$, SnO, $Al_2O_3$, and $La_2O_3$.

Although microwave synthesis dramatically decreases reaction time, it does not solve the problem of chemical inhomogeneity, and temperatures in the reaction are high. It does have some other advantages, however. One advantage is that less problems occur with cracking because the heating is from inside, and not absorbed from outside.

Increased homogeneity can be obtained with the hydrothermal method discussed in Section 3.5.1, which also operates at lower temperatures than conventional methods.

## 3.4 COMBUSTION SYNTHESIS

**Combustion synthesis**, also known as **self-propagating high temperature synthesis**, has been developed as an alternative route to the ceramic method which depends on the diffusion of ions through the reactants, and thus for a uniform product requires repeated heating and grinding for long periods. Combustion synthesis uses highly exothermic ($\Delta H < -170$ kJ mol$^{-1}$) and even explosive reactions to maintain a self-propagating high reaction temperature and has been used to prepare many refractory materials including borides, nitrides, oxides, silicides, intermetallics, and ceramics. The reactants are mixed together, formed into a pellet, and then ignited (laser, electric arc, heating coil) at high temperature. Once ignited the reaction propagates as a wave, the **synthesis wave**, through the pellet, and the reaction must lose less heat than it generates or it will quench; temperatures up to 3000 K are maintained during the fast reactions. Figure 3.4 depicts the synthesis wave moving through a mixture of MgO, Fe, $Fe_2O_3$, and $NaClO_4$, the images were captured by a thermal imaging camera.

Self-ignition can sometimes be achieved by ball milling. The method is now used commercially in Russia, Spain, and Korea because it is fast, economical, and gives high purity products. Thermal ignition of a mixture of $BaO_2$, $Fe_2O_3$ $Cr_2O_3$ and Fe produces a ferrite (Chapter 9) $BaFe_{12-x}Cr_xO_{19}$; in the presence of a magnetic field the reaction proceeds faster and with a higher temperature propagation wave. Examples of other useful products of these reactions are:

- Hydrides (e.g., $MgH_2$) for hydrogen storage
- Borides (e.g., $TiB_2$) for abrasives and cutting tools
- Carbides (e.g., SiC and TiC) for abrasives and cutting tools

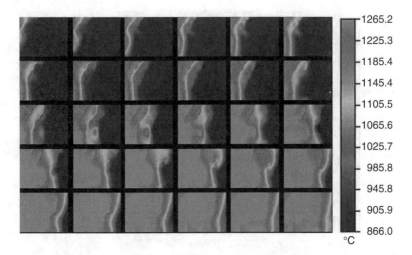

**FIGURE 3.4** Thermal images of the synthesis wave moving through a pellet of MgO, Fe, $Fe_2O_3$, and $NaClO_4$. Each image is of dimension $3 \times 2$ mm. Images were captured at 0.06 s intervals. The first image is top left and the last is bottom right. See colour insert following page 196. (Courtesy of Professor Ivan Parkin, University College, London.)

- Nitrides (e.g., $Si_3N_4$) for high strength, heat resistant ceramics
- Oxides (e.g., cuprates) for high temperature superconductors
- Silicides (e.g., $MoSi_2$) for high temperature heating elements
- Composites (e.g., the thermite reaction, $3Fe_3O_4 + 8Al = 4Al_2O_3 + 9Fe$) have been used to coat the inside of steel pipes with an inner layer of Fe, and a surface layer of $Al_2O_3$, using a centrifugal process.

## 3.5 HIGH PRESSURE METHODS

High pressure methods fall into three main categories: (i) high solvent pressures in an autoclave — the hydrothermal method; (ii) using a high pressure of a reactive gas; and (iii) directly applied hydrostatic pressure on solids. Some solid state applications need single crystals; the hydrothermal method increases homogeneity and lowers operating temperatures, and also grows single crystals.

### 3.5.1 HYDROTHERMAL METHODS

The original **hydrothermal method** involves heating the reactants in a closed vessel, an **autoclave**, with water (Figure 3.5). An autoclave is usually constructed from thick stainless steel to withstand the high pressures, and is fitted with safety valves; it may be lined with non-reactive materials such as teflon. The autoclave is heated, the pressure increases, and the water remains liquid above its normal boiling temperature of 373 K, so-called 'super-heated water'. These conditions, in which the pressure is raised above atmospheric pressure and the temperature is raised above the boiling temperature of water (but not to as high a temperature as used in the

**FIGURE 3.5** Autoclave.

methods described previously), are known as hydrothermal conditions. Hydrothermal conditions exist in nature, and numerous minerals, including naturally occurring zeolites, are formed by this process. Synthetic emeralds are also made under hydrothermal conditions. The term has been extended to other systems with moderately raised pressures and temperatures lower than those typically used in ceramic and sol-gel syntheses. The lower temperatures used are one of the advantages of this method. Others include the preparation of compounds in unusual oxidation states or phases which are stabilised by the raised temperature and pressure. It is useful in metal oxide systems where the oxides are not soluble in water at atmospheric pressure but dissolve in the super-heated water of the hydrothermal set up. Where even these temperatures and pressures are insufficient to dissolve the starting materials, alkali or metal salts can be added whose anions form complexes with the solid and render it soluble. The examples discussed next are chosen to illustrate these advantages and the different variations of the method.

### Quartz

The first industrial process to use hydrothermal methods of synthesis was the production of quartz crystals for use as oscillators in radios. Quartz, $SiO_2$, can be used to generate a high frequency alternating current via the piezoelectric effect (see Chapter 9). The hydrothermal growth of quartz crystals employs a **temperature gradient**. In this variant of hydrothermal processing, the reactant dissolves at a higher

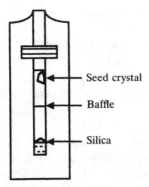

**FIGURE 3.6** Growth of quartz crystals in an autoclave under hydrothermal conditions.

temperature, is transported up the reaction tube by convection, and crystallizes out at a lower temperature. A schematic diagram of the apparatus is presented in Figure 3.6.

One end of the autoclave is charged with silica and an alkaline solution. When the autoclave is heated, the silica is dissolved in the solution and is transported along the autoclave to the cooler part where it crystallizes out. The alkaline solution returns to the hotter region where it can dissolve more silica. In this particular case the autoclave, which is made of steel, can act as the reaction vessel. Because of the corrosive nature of superheated solutions and the possibility of contamination of the product with material from the autoclave walls, it is generally necessary to either line the autoclave with an inert substance, such as teflon, or to perform the reaction in a sealed ampoule in the autoclave.

## Chromium Dioxide

Chromium dioxide ($CrO_2$), which is used on audiotapes because of its magnetic properties (see Chapter 9), contains chromium in the unusual oxidation state of +4. It is prepared by the oxidation of chromium(III) oxide ($Cr_2O_3$), which is the stable oxide of chromium under normal laboratory conditions. Chromium(III) oxide and chromium trioxide (chromium(VI) oxide $CrO_3$) are placed in an autoclave with water and heated to 623 K. Oxygen is produced during the reaction, and because the autoclave is sealed, builds up a high partial pressure. (The pressure in the reaction vessel reaches 440 bar.) This high oxygen partial pressure favours the formation of chromium dioxide through the reactions given next:

$$Cr_2O_3 + CrO_3 = 3CrO_2$$

$$CrO_3 = CrO_2 + \frac{1}{2} O_2$$

## Zeolites

Zeolites are a class of crystalline **aluminosilicates** (see Chapter 7) based on rigid anionic frameworks with well-defined channels and cavities. These cavities contain exchangeable metal cations, and can hold removable and replaceable guest molecules such as water. The primary building units of zeolites are $[SiO_4]^{4-}$ and $[AlO_4]^{5-}$ tetrahedra linked together by **corner sharing**. Silicon-oxygen tetrahedra are electrically neutral when connected together in a three-dimensional network as in quartz. The substitution of Si(IV) by Al(III) in such a structure, however, creates an electrical imbalance, and to preserve overall electrical neutrality, each $[AlO_4]$ tetrahedron needs a balancing positive charge which is provided by exchangeable cations held electrostatically within the zeolite. It is possible for the tetrahedra to link by sharing two, three, or all four corners, forming a huge variety of different structures.

Naturally occurring zeolite minerals are formed through a hydrothermal process geochemically, and it was demonstrated in the 1940s and 1950s that they could be synthesized in the laboratory by simulating these hydrothermal conditions. A general method of preparing zeolites involves mixing an alkali, aluminium hydroxide and silica sol, or an alkali, a soluble aluminate and silica sol. The silica and aluminate condense to form a gel as in the sol-gel method but the gel is then heated in a closed vessel at temperatures close to 373 K. Under these conditions, zeolites instead of other alumino-silicate phases crystallize out. For example, in the synthesis of a typical zeolite, zeolite A $(Na_{12}[(AlO_2)_{12}(SiO_2)_{12}].27H_2O)$ hydrated alumina, $Al_2O_3.3H_2O$, is dissolved in concentrated sodium hydroxide solution. The cooled solution is then mixed with a solution of sodium metasilicate, $Na_2SiO_3.9H_2O$, and a thick white gel forms. The gel is then placed in a closed teflon bottle and heated to about 363 K over 6 hours. The reaction time can be reduced by using microwave radiation for this final step: for example zeolite A has been made by heating for 45 minutes using pulses of 300 watt-power microwave radiation.

Changes in the form of the alumina, in the pH of the solution, the type of base used (organic or inorganic) and in the proportions of alkali, aluminium compound, and silica lead to the production of different zeolites.

## Microporous and Mesoporous Solids

In zeolite synthesis, large cations such as tetramethylammonium $(NMe_4^+)$ and tetrapropylammonium $(N(C_3H_7)_4^+)$ can be used as a **template** around which the aluminosilicate framework crystallizes with large cavities to accommodate the ion. On subsequent heating the cation is pyrolysed, but the structure retains the cavities. Such structures formed around a single molecule template, with pore sizes between 200 and 2000 pm, are known as **microporous**.

Zeolitic structures with pore sizes of 2000 to 10 000 pm are known as **mesoporous** solids, and can be formed by a method known as **liquid crystal templating (LCT)**. The combination of a suitable cationic surfactant together with silicate anions form arrays of rod-like surfactant micelles (Figure 3.7) surrounded by a polymeric siliceous framework. On calcination the mesoporous structure is formed.

If short-chain alkyltrimethylammonium ions, $R(CH_3)_3N^+$, are used as the surfactant, the alkyl chain length determines whether micro- or mesoporous materials are formed. For alkyl chain lengths of $C_6$ or shorter, microporous zeolites such as ZSM-5 are formed; but for longer chains, mesoporous materials are isolated.

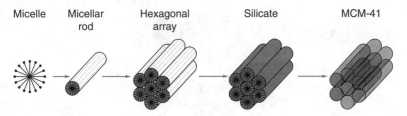

**FIGURE 3.7** The formation of mesoporous MCM-41 by liquid crystal templating.

## Yttrium Aluminium Garnet

The preparation of yttrium aluminium garnet ($Y_3Al_5O_{12}$, YAG) illustrates a variation of the hydrothermal method used if the starting materials have very different solubilities from each other. In this case, yttrium oxide ($Y_2O_3$) was placed in a cooler section of the autoclave and aluminium oxide as sapphire in a hotter section to increase its solubility (Figure 3.8). YAG forms where the two zones meet.

$$3Y_2O_3 + 5Al_2O_3 = 2Y_3Al_5O_{12}$$

### 3.5.2 HIGH PRESSURE GASES

This method is usually used to prepare metal oxides and fluorides in less stable high oxidation states. For instance, the perovskite $SrFeO_3$ containing Fe(IV) can be made from the reaction of $Sr_2Fe_2O_5$ and oxygen at 340 atm. Using fluorine at high pressures (> 4 kbar) has also been used in the synthesis of ternary fluorides containing high oxidation states, for example, $Cs_2CuF_6$ containing Cu(IV).

### 3.5.3 HYDROSTATIC PRESSURES

The application of high pressures and temperatures can induce reactions and phase changes that are not possible under ambient conditions. Applying very high pressures tends to decrease volume and thus improve the packing efficiency; consequently, coordination numbers tend to increase, so for instance Si can be transformed from

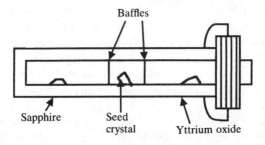

**FIGURE 3.8** Hydrothermal arrangement for synthesis of YAG starting from two reactants of different solubilities.

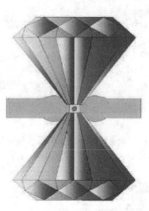

**FIGURE 3.9** High pressure cell. (Courtesy of Cornell High Energy Synchrotron Source).

the four-coordinate diamond structure to the six-coordinate white-tin structure, and NaCl (six-coordinate) can be changed to the CsCl (eight-coordinate) structure.

Various designs of apparatus are available for achieving exceptionally high pressures. Originally a **piston and cylinder** arrangement allowed synthesis at pressures up to 5 GPa (50 000 atm) and 1800 K, but the **belt apparatus** using two opposed tungsten carbide cylinders can reach 15 GPa and 2300 K and is used for making synthetic diamonds. Recently, **diamond anvils** have been used to reach pressures of 20 GPa. Diamond anvils can be arranged either two directly opposed, four tetrahedrally opposed or six in a octahedral configuration (Figure 3.9) but although they achieve greater pressures than other methods, they have the disadvantage that only milligram quantities of material can be processed, and so are more useful for investigating phase transitions than for synthesis.

### 3.5.4 SYNTHETIC DIAMONDS

The drive to produce synthetic diamonds arose during the Second World War; they were urgently needed for the diamond-tipped tools used to manufacture the military hardware and it was feared that the South African supply of natural stones might dry up. GEC started a research programme to mimic the geothermal conditions which produce the natural stones. In 1955 they eventually succeeded in growing crystals up to 1 mm in length by dissolving graphite in a molten metal such as nickel, cobalt, or tantalum in a pyrophyllite vessel, and subjecting the graphite to pressures of 6 GPa and temperatures of 2300 K until diamond crystals form. The molten metal acts as a catalyst, bringing down the working temperature and pressure. By refining the process, gemstone quality stones were eventually made in 1970.

Diamond has the highest thermal conductivity known, about five times that of copper, hence they feel cold to the touch and are nicknamed 'ice'. It has been found that increasing the ratio of $^{12}C$ to $^{13}C$ in synthetic diamond can improve the thermal conductivity by up to 50%. The isotopic enhancement takes place by producing a

CVD diamond film (see Section 3.6) from $^{12}$C-enriched methane. This is crushed and used to make synthetic diamond in the usual way. These electrically insulating diamonds could make highly effective heat-sinks for micro circuits.

### 3.5.5  HIGH TEMPERATURE SUPERCONDUCTORS

High pressure increases the density of $SrCuO_2$ by 8%, and the Cu–O coordination changes from chains to two-dimensional $CuO_2$ sheets. High pressure synthesis has been used to make many high temperature superconductors as this facilitates the formation of the two-dimensional Cu–O sheets necessary for this type of superconductivity (see Chapter 10).

### 3.5.6  HARDER THAN DIAMOND

The search for a substance harder than diamond has recently focussed on buckminsterfullerene, $C_{60}$. Above 8 GPa and 700 K, buckminsterfullerene forms two amorphous phases which are harder than diamond — the samples scratched diamond anvils when squeezed between them! Although harder than diamond, these phases have a lower density and are semiconducting.

## 3.6  CHEMICAL VAPOUR DEPOSITION (CVD)

So far, the preparative methods we have looked at have involved reactants that are in the solid state or in solution. In CVD, powders and microcrystallline compounds are prepared from reactants in the vapour phase, and can be deposited on a substrate (such as a thin sheet of metal or ceramic) to form single-crystal films for devices (see Section 3.7.1). First, the volatile starting materials are heated to form vapours; these are then mixed at a suitable temperature and transported to the substrate for deposition using a carrier gas; finally, the solid product crystallises out. The whole vessel may be heated, which tends to deposit the product all over the walls of the vessel. More commonly, the energy to initiate the reaction is supplied to the substrate. Figure 3.10a illustrates a schematic set-up for chemical-vapour deposition. Typical starting materials are hydrides, halides, and organometallic compounds because these compounds tend to be volatile; if an organometallic precursor is used, the method is often referred to as **MOCVD (metal organic CVD)**.

The advantages of the method are that reaction temperatures are relatively low, the stoichiometry is easily controlled, and dopants can be incorporated.

### 3.6.1  PREPARATION OF SEMICONDUCTORS

The method has been widely used for preparing the III-V semiconductors and silicon, for devices. Various reactions to prepare GaAs involving chlorides and hydrides have been tried, for example the reaction of $AsCl_3$ with Ga in the presence of hydrogen:

$$AsCl_3 + Ga + \tfrac{3}{2} H_2 = GaAs + 3HCl$$

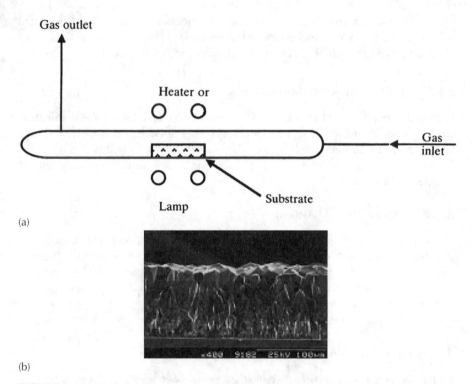

(a)

(b)

**FIGURE 3.10** (a) Chemical-vapour deposition reactor; (b) cross section of a 100 μm-thick CVD diamond film grown by DC arc jet. The columnar nature of the growth is evident, as is the increase in film quality and grain size with growth time. (Courtesy of Dr. Paul May and Prof. Mike Ashfold, Bristol University.)

Silicon for semiconductor devices can be prepared using this method from the reduction of $SiCl_4$:

$$SiCl_4 + 2H_2 = Si + 4HCl$$

### 3.6.2 DIAMOND FILMS

Diamond has many useful properties. As well as being the hardest substance known, it also has the best thermal conductivity (at room temperature), is an electrical insulator, and has the highest transparency in the IR region of the spectrum of any known substance. Diamond films are made by a CVD process at low pressure and temperatures below 1300 K, by heating gas mixtures containing methane (or acetylene) and hydrogen, or by breaking down such mixtures with microwaves; the carbon atoms and carbon-containing radical species are then deposited on a substrate. Early experiments managed to deposit monocrystalline layers on to a seed diamond,

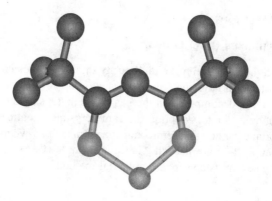

**FIGURE 3.11** β-Diketonate of lithium from 2,2,6,6-tetramethylheptan-3,5-dione. Li, blue; O, pale grey; C, dark grey.

and thus build up larger diamonds layer by layer. Microcrystalline diamond layers have also been deposited on to silicon and metal surfaces. Such layers have been used for non-scratch optical coatings, for coating knives and scalpels so they retain their sharpness, and have the potential to be used as a hard wear-resistant coating for moving parts.

### 3.6.3 LITHIUM NIOBATE

The sol-gel method of preparing lithium niobate used lithium and niobium alkoxides. Alkoxides are often used in CVD methods, but unfortunately for the preparation of lithium niobate, lithium alkoxides are much less volatile than niobium alkoxides and to get the two metals deposited together it is better to use compounds of similar volatility. One way around this problem is to use a more volatile compound of lithium. One reported synthesis uses a β-diketonate of lithium in which lithium is coordinated to 2,2,6,6-tetramethylheptan-3,5-dione ($Me_3CCOCH_2COCMe_3$) (Figure 3.11).

The lithium compound was heated at about 520 K and niobium pentamethoxide at 470 K in a stream of argon containing oxygen. Lithium niobate was deposited on the reaction vessel which was heated to 720 K.

In this example, the volatile precursor compounds were heated to obtain the product. Other energy sources are also used, notably electromagnetic radiation. An example of vapour phase deposition involving photo-decomposition is given in the next section on vapour phase epitaxy.

## 3.7 PREPARING SINGLE CRYSTALS

Single crystals of high purity and defect free, may be needed for many applications, none more so than in the electronics and semiconductor industry. Various methods have been developed for preparing different crystalline forms such as large crystals, films, etc. A few of the important methods are described next.

### 3.7.1 Epitaxy Methods

#### Vapour Phase Epitaxial (VPE) Growth

CVD methods are now used to make high purity thin films in electronics where the deposited layers have to have the correct crystallographic orientation. In epitaxial growth, a precursor is decomposed in the gas phase and a single crystal is built up layer by layer on a substrate, adopting the same crystal structure as the substrate. This is an important method for semiconductor applications where single crystals of controlled composition are needed. Gallium arsenide (GaAs), for example, has been prepared by several vapour phase epitaxial methods.

#### Gallium Arsenide

In one method, arsenic(III) chloride ($AsCl_3$, boiling temperature 376 K) is used to transport gallium vapour to the reaction site where gallium arsenide is deposited in layers. The reaction involved is:

$$2Ga(g) + 2AsCl_3(g) = 2GaAs(s) + 3Cl_2(g)$$

An alternative to gallium vapour is an organometallic compound of gallium. One preparation reacts trimethyl gallium ($Ga(CH_3)_3$) with the highly volatile (although toxic) arsine ($AsH_3$).

$$Ga(CH_3)_3(g) + AsH_3(g) = GaAs(s) + 3CH_4(g)$$

#### Mercury Telluride

Mercury telluride (HgTe) was first made by vapour phase epitaxy in 1984. This preparation illustrates the use of ultraviolet radiation as the energy source for decomposition. Diethyltellurium (($C_2H_5)_2Te$) vapour in a stream of hydrogen carrier gas was passed over heated mercury where it picked up mercury vapour. As the gas passed over the substrate, it was subjected to ultraviolet illumination and mercury telluride was deposited. The substrate was heated to 470 K, which is about 200 K lower than the temperature needed for thermal decomposition. The use of the lower temperature enabled later workers to build up a crystal of alternating mercury telluride and cadmium telluride because diffusion of one layer into the other was slowed by the lower temperature.

Production of single crystals with carefully controlled varying composition is vital to make semiconductor devices. The production of crystals for quantum well lasers illustrates how carefully such syntheses can be controlled.

#### Molecular Beam Epitaxy (MBE)

A molecular beam is a narrow stream of molecules formed by heating a compound in an oven with a hole which is small compared with the mean free path of the gaseous molecules produced. Very thin layers can be built up by directing a beam

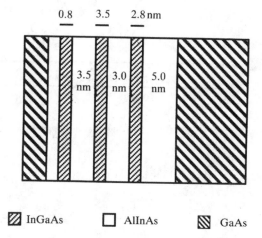

FIGURE 3.12 The layers of an active region of a quantum cascade laser.

of precursor molecules onto the substrate. The system is kept under ultra-high vacuum. Because of the very low pressure, the reactants need not be as volatile as in other vapour deposition methods. An application of this method is the growth of single crystals for quantum cascade lasers (see Chapter 8) where the crystal must contain nanometre thickness layers of $Al_{0.48}In_{0.52}As$ and $Ga_{0.47}In_{0.53}As$. The layers of the laser material are depicted in Figure 3.12.

Beams of aluminium, gallium, arsenic, and indium were directed onto a heated InP crystal. The substrate needs to be heated to allow the atoms deposited from the beams to migrate to their correct lattice position. The relative pressures of the component beams were adjusted for each layer to give the desired compositions.

### 3.7.2 Chemical Vapour Transport

In CVD, solids are formed from gaseous compounds. In **chemical vapour transport**, a solid or solids interact with a volatile compound and a solid product is deposited in a different part of the apparatus. It is used both for preparing compounds and for growing crystals from powders or less pure crystalline material.

#### Magnetite

Chemical vapour transport has been used to grow magnetite crystals using the reaction of magnetite with hydrogen chloride gas:

$$Fe_3O_4(s) + 8HCl(g) = FeCl_2(g) + 2FeCl_3(g) + 4H_2O(g)$$

Powdered magnetite is placed at one end of the reaction vessel and the tube evacuated. HCl gas is introduced and the tube sealed and then heated. The reaction is endothermic ($\Delta H_m^{\ominus}$ positive) and so the equilibrium moves to the right as the temperature is raised. Thus, at the hotter end of the tube, magnetite reacts with the hydrogen chloride gas and transports down the tube as gaseous iron chlorides. At

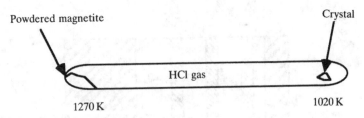

**FIGURE 3.13** Growth of magnetite crystals using chemical vapour transport

the cooler end of the tube, the equilibrium shifts to the left and magnetite is deposited (Figure 3.13).

## Melt Methods

Melt methods depend both on the compound being stable in the liquid phase, and on the availability of a high temperature inert containing vessel.

### The Czochralski Process

The silicon required by the electronics industry for semiconductor devices has to have levels of key impurities, such as phosphorus and boron, of less than 1 atom in $10^{10}$ Si. Silicon is first converted to the highly volatile trichlorosilane, $SiHCl_3$, which is then distilled and decomposed on rods of high purity silicon at 1300 K to give high purity polycrystalline silicon. This is made into large single crystals by the **Czochralski process**. The silicon is melted in an atmosphere of argon, and then a single-crystal rod is used as a seed which is dipped into the melt and then slowly withdrawn, pulling with it an ever-lengthening single crystal in the same orientation as the original seed (Figure 3.14). As well as silicon, this method is also used for preparing other semiconductor materials such as Ge and GaAs, and for ceramics such as perovskites and garnets.

### Using Temperature Gradients

This method is also used to produce large single crystals of silicon which can then be sliced into thin wafers for the semiconductor industry. A polycrystalline rod of silicon is held vertically and a single-crystal seed crystal is attached at one end. A moving heater then heats the area around the seed crystal, and a molten zone is caused to traverse the length of the rod (the **float-zone process**), producing a single crystal as it moves (Figure 3.15). This process also has the advantage of refining the product — so-called **zone-refining** — as impurities tend to stay dissolved in the melt and are thus swept before the forming crystal to the end, where they can be cut off and rejected.

Related methods include the **Bridgman** and **Stockbarger methods** where a temperature gradient is maintained across a melt so that crystallization starts at the cooler end; this can either be achieved using a furnace with a temperature gradient or by pulling the sample through a furnace.

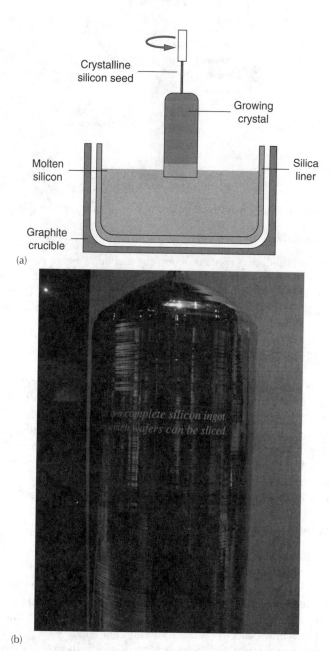

(a)

(b)

**FIGURE 3.14** (a) The Czochralski process for producing a very pure single crystal of silicon; (b) silicon boule. (Photo courtesy of Tonie van Ringelestijn.)

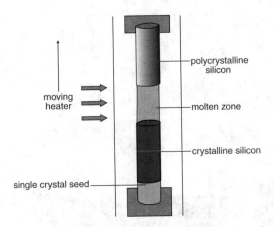

**FIGURE 3.15** The float-zone method for producing very pure crystals of silicon.

## Flame and Plasma Fusion Methods

In the **Verneuil method,** the powdered sample is melted in a high temperature, oxy-hydrogen flame and droplets allowed to fall on to a seed crystal where they crystallize. This is an old method used originally to produce artificial ruby. Using the same method but using a plasma torch to melt the powder, can achieve even higher temperatures.

## Skull Melting

Skull melting is a method used for growing large oxide crystals. A power supply of up to 50 kW, 4 MHz produces radio frequency that is transferred to a coil wrapped around the skull crucible. The crucible is made of copper and consists of water-cooled fingers and base, and the molten material is separated from the copper crucible by a thin layer of sintered material. The container can be evacuated and filled with an appropriate atmosphere. Temperatures of up to 3600 K can be achieved using this method, allowing the growth of large crystals (several centimetres) of refractory oxides such as $ZrO_2$, $ThO_2$, CoO, and $Fe_3O_4$.

### 3.7.3 SOLUTION METHODS

Crystals have traditionally been grown from saturated solutions in a solvent; hot solutions are prepared and then cooled and crystals precipitate from the supersaturated solution. Various techniques can be used to induce crystallization and these include evaporating the solvent either by heating, leaving in the atmosphere or in a desiccator, or under reduced pressure; freezing the solvent; and addition of other components (solvent or salt) to reduce solubility. For solids such as oxides which are very insoluble in water it may be possible to dissolve them in melts of borates, fluorides or even metals, in which case the solvent is generally known as a **flux** as it brings down the melting temperature of the solute. The melt is then cooled slowly until the crystals form, and then the flux poured off or dissolved away. This method has been successfully used for preparing crystalline silicates, quartz, and alumina among many others.

## 3.8 INTERCALATION

The solids produced by the reversible insertion of guest molecules into lattices are known as **intercalation compounds**, and although originally applied to layered solids is now taken to include other solids with similar host-guest interactions. Intercalation compounds have importance as catalysts, as conducting solids and therefore electrode materials, as a means of encapsulating molecules potentially for drug delivery systems, and as a method of synthesizing composite solids.

### 3.8.1 Graphite Intercalation Compounds

Many layered solids form intercalation compounds, where a neutral molecule is inserted between weakly bonded layers. For example when potassium vapour reacts with graphite above the melting temperature of potassium (337 K), it forms a golden compound $KC_8$ in which the potassium ions sit between the graphite layers, and the inter-layer spacing is increased by 200 pm (Figure 3.16). Addition of a small amount of $KO_2$ to the molten potassium results in the formation of a double layer of potassium atoms between the graphite layers and a formula close to $KC_4$.

The potassium donates an electron to the graphite (forming $K^+$) and the conductivity of the graphite increases. Graphite electron-acceptor intercalation compounds have also been made with $NO_3^-$, $CrO_3$, $Br_2$, $FeCl_3$, and $AsF_5$. Some of these compounds have electrical conductivity approaching that of aluminum (see Chapter 6).

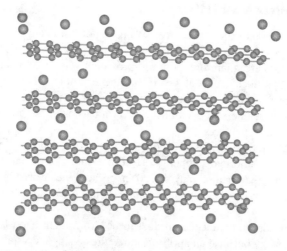

**FIGURE 3.16** The structure of $KC_8$. K, blue; C, grey.

### 3.8.2 TITANIUM DISULFIDE

Layered structures are also found for many oxides and sulfides of transition metals. They can be intercalated with alkali metals (Li, Na, K) to give superconducting solids and conducting solids that are useful for solid state battery materials.

Titanium disulfide has a $CdI_2$ structure (see Chapter 1). The solid is golden-yellow and has a high electrical conductivity along the titanium layers. Forming intercalation compounds with electron donors can increase the conductivity of titanium disulfide, the best example being with lithium, $Li_xTiS_2$. This compound is synthesized in the cathode material of the rechargeable battery described in Chapter 5, and can also be synthesized directly by the lithiation of $TiS_2$ with a solution of butyl lithium:

$$xC_4H_9Li + TiS_2 = Li_xTiS_2 + \tfrac{x}{2} C_8H_{18}$$

### 3.8.3 PILLARED CLAYS

Pillaring is a technique for building strong molecular pillars which prop the layers of a clay sufficiently far apart to create a two-dimensional space between the layers where molecules can interact. In a typical reaction, a clay such as montmorillonite undergoes ion exchange with the large inorganic cation $[Al_{13}O_4(OH)_{24}(H_2O)_{12}]^{7+}$ by slurrying with an alkaline solution of aluminium ions. After washing with water the material is calcined, the hydroxyls and water are lost and alumina pillars created between the silicate layers of the clay, creating a permanently microporous substance.

Many other inorganic pillars have been incorporated into clays; for instance, calcining after ion exchange with zirconium oxychloride can create zirconia pillars.

## 3.9  CHOOSING A METHOD

Our choice of methods is not exhaustive. We have not, for example, covered shock or ultrasonic methods, electrolytic methods, or the preparation of heterogeneous catalysts. Our aim here, therefore, is not to provide a way of choosing the method for a particular product. (Indeed, several of the examples given in this chapter demonstrate that several methods can be suitable for one substance.) Instead, we hope to give a few pointers to deciding whether a particular method is suitable for a particular material.

It is important that you consider the stability of compounds under the reaction conditions and not at normal temperature and pressure. As we have illustrated in this chapter, a particular method may be chosen because the desired product is stable only at raised pressures or because it decomposes if the reaction temperature is too high.

In what form do you want the product to be? You might choose, for example, vapour phase epitaxy because an application requires a single crystal. Alternatively, you might choose a particular method, such as the precursor method or hydrothermal synthesis, because you need a homogeneous product.

How pure must your product be? If you require high purity, you could choose a method that involves a volatile compound as starting material because these are generally easier to purify than solids.

You should consider the availability of reactants required for a particular method. If you are considering a precursor method, is there a suitable precursor with the right stoichiometry? The CVD method needs reactants of similar volatility; do your proposed reactants meet this requirement? In microwave synthesis, does at least one of your starting materials absorb microwaves strongly?

## QUESTIONS

1. The Chevrel phase $CuMo_6S_8$ was prepared by a ceramic method. What would be suitable starting materials and what precaution would you have to take?

2. Which synthetic methods would be suitable for producing the following characteristics?
   a) A thin film of material.
   b) A single crystal.
   c) A single crystal containing layers of different material with the same crystal structure.
   d) A powder of homogeneous composition.

3. What are the advantages and disadvantages of using the sol-gel method to prepare barium titanate for use in a capacitor?

4. A compound $(NH_4)_2Cu(CrO_4)_2.2NH_3$ is known. How could this be used to prepare $CuCr_2O_4$? What would be the advantage of this method over a ceramic method? Suggest which solvent was used to prepare the ammonium compound.

5. $\beta$-TeI is a metastable phase formed at 465 to 470 K. Suggest an appropriate method of preparation.

6. In hydrothermal processes involving alumina $(Al_2O_3)$, such as the synthesis of zeolites, alkali is added to the reaction mixture. Suggest a reason for this addition.

7. The zeolite ZSM-5 is prepared by heating a mixture of silicic acid, $SiO_2.nH_2O$, NaOH, $Al_2SO_4$, water, $n$-propylamine, and tetrapropylammonium bromide in an autoclave for several days at 160°C. The product from this reaction is then heated in air. Why is tetrapropylammonium bromide used in the reaction and what is the effect of the subsequent oxidation reaction?

8. Which of the following oxides would be good candidates for microwave synthesis: $CaTiO_3$, $BaPbO_3$, $ZnFe_2O_4$, $Zr_{1-x}Ca_xO_{2-x}$, $KVO_3$?

9. Crystals of silica can be grown using the chemical vapour transport method with hydrogen fluoride as a carrier gas. The reaction involved is:

$$SiO_2(s) + 4HF(g) = SiF_4(g) + 2H_2O(g)$$

10. In the preparation of lithium niobate by CVD, argon-containing oxygen is used as a carrier gas. In the preparation of mercury telluride, the carrier gas was hydrogen. Suggest reasons for these choices of carrier gas.

# 4 Bonding in Solids and Electronic Properties

## 4.1 INTRODUCTION

Chapter 1 introduced the physical structure of solids — how their atoms are arranged in space. We now turn to a description of the bonding in solids — the electronic structure. Some solids consist of molecules bound together by very weak forces. We shall not be concerned with these because their properties are essentially those of the molecules. Nor shall we be much concerned with 'purely ionic' solids bound by electrostatic forces between ions as discussed in Chapter 1. The solids considered here are those in which all the atoms can be regarded as bound together. We shall be looking at the basic bonding theories of these solids and how the theories account for the very different electrical conductivities of different groups of solids. We will cover both theories based on the free electron model — a view of a solid as an array of ions held together by a sea of electrons — and on molecular orbital theory — a crystal as a giant molecule. Some of the most important of such solids are the semiconductors. Many solid state devices — transistors, photocells, light-emitting diodes (LEDs), solid state lasers, and solar cells — are based on semiconductors. We shall introduce examples of some devices and explain how the properties of semiconductors make them suitable for these applications. We start with the free electron theory of solids and its application to metals and their properties.

## 4.2 BONDING IN SOLIDS — FREE ELECTRON THEORY

Traditionally, bonding in metals has been approached through the idea of free electrons, a sort of electron gas.

The free electron model regards a metal as a box in which electrons are free to roam, unaffected by the atomic nuclei or by each other. The nearest approximation to this model is provided by metals on the far left of the Periodic Table — Group 1 (Na, K, etc.), Group 2 (Mg, Ca, etc.) — and aluminium. These metals are often referred to as simple metals.

The theory assumes that the nuclei stay fixed on their lattice sites surrounded by the inner or core electrons whilst the outer or valence electrons travel freely through the solid. If we ignore the cores then the quantum mechanical description of the outer electrons becomes very simple. Taking just one of these electrons the problem becomes the well-known one of the **particle in a box**. We start by considering an electron in a one-dimensional solid.

The electron is confined to a line of length $a$ (the length of the solid), which we shall call the $x$-axis. Because we are ignoring the cores, there is nothing for the electron to interact with and so it experiences zero potential within the solid. The Schrödinger equation for the electron is

$$\frac{-\hbar^2}{2m_e} \frac{d^2\psi}{dx^2} = (E - V)\psi \tag{4.1}$$

where $\hbar$ is Planck's constant divided by $2\pi$, $m_e$ is the mass of the electron, $V$ is the electrical potential, $\psi$ is the wave function of the electron and $E$ is the energy of an electron with that wave function. When $V = 0$, the solutions to this equation are simple sine or cosine functions, and you can verify this for yourself by substituting

$$\psi = \sin\sqrt{\frac{2m_e E}{\hbar^2}} x$$

into Equation (4.1), as follows:

If

$$\psi = \sin\sqrt{\frac{2m_e E}{\hbar^2}} x$$

then differentiating once gives

$$\frac{d\psi}{dx} = \sqrt{\frac{2m_e E}{\hbar^2}} \cos\sqrt{\frac{2m_e E}{\hbar^2}} x \, .$$

Differentiating twice, you should get

$$\frac{d^2\psi}{dx^2} = -\frac{2m_e E}{\hbar^2} \sin\sqrt{\frac{2m_e E}{\hbar^2}} x \quad \text{which is} \quad -(2m_e \frac{E}{\hbar^2})\psi \, . \quad \text{Thus,} \quad \frac{d^2\psi}{dx^2} = -\frac{2m_e E}{\hbar^2}\psi$$

and multiplying this by $-(\hbar^2 / 2m_e)$ gives (4.1) with $V = 0$.

The electron is not allowed outside the box and to ensure this we put the potential to infinity outside the box. Since the electron cannot have infinite energy, the wave function must be zero outside the box and since it cannot be discontinuous, it must be zero at the boundaries of the box. If we take the sine wave solution, then this is zero at $x = 0$. To be zero at $x = a$ as well, there must be a whole number of half waves in the box. Sine functions have a value of zero at angles of $n\pi$ radians where

$n$ is an integer and so $a\sqrt{2m_e E / \hbar^2} = n\pi$. The energy is thus quantised, $E = \dfrac{n^2 h^2}{8m_e a^2}$,

with quantum number $n$. Because $n$ can take all integral values, this means an infinite number of energy levels exist with larger and larger gaps between each level. Most solids of course are three-dimensional (although we shall meet some later where conductivity is confined to one or two dimensions) and so we need to extend the free electron theory to three dimensions.

For three dimensions, the metal can be taken as a rectangular box $a \times b \times c$. The appropriate wave function is now the product of three sine or cosine functions and the energy is given by

$$E = \frac{h^2}{8m_e}\left(\frac{n_a^2}{a^2} + \frac{n_b^2}{b^2} + \frac{n_c^2}{c^2}\right) \tag{4.2}$$

Each set of quantum numbers $n_a$, $n_b$, $n_c$ will give rise to an energy level. However, in three dimensions, many combinations of $n_a$, $n_b$, and $n_c$ exist which will give the same energy, whereas for the one-dimensional model there were only two levels of each energy ($n$ and $-n$). For example, the following sets of numbers all give $(n_a^2/a^2 + n_b^2/b^2 + n_c^2/c^2) = 108$

| $n_a/a$ | $n_b/b$ | $n_c/c$ |
|---------|---------|---------|
| 6 | 6 | 6 |
| 2 | 2 | 10 |
| 2 | 10 | 2 |
| 10 | 2 | 2 |

and hence the same energy. The number of states with the same energy is known as the **degeneracy**. For small values of the quantum numbers, it is possible to write out all the combinations that will give rise to the same energy. If we are dealing with a crystal of say $10^{20}$ atoms it becomes difficult to work out all the combinations. We can however estimate the degeneracy of any level in a band of this size by introducing a quantity called the wave vector and assuming this is continuous. If we substitute $k_x$, $k_y$, and $k_z$ for $n_a\pi/a$, $n_b\pi/b$ and $n_c\pi/c$, then the energy becomes

$$E = (k_x^2 + k_y^2 + k_z^2)\hbar^2 / 2m_e \tag{4.3}$$

and $k_x$, $k_y$, and $k_z$ can be considered as the components of a vector, **k**; the energy is proportional to the square of the length of this vector. **k** is called the **wave vector** and is related to the momentum of the electron wave as can be seen by comparing the classical expression $E = p^2/2m$, where $p$ is the momentum and $m$ the mass, with the expression above. This gives the electron momentum as $\pm k\hbar$.

All the combinations of quantum numbers giving rise to one particular energy correspond to a wave vector of the same length $|\mathbf{k}|$. Thus, all possible combinations leading to a given energy produce vectors whose ends lie on the surface of a sphere of radius $|\mathbf{k}|$. The total number of wave vectors with energies up to and including that with the given energy is given by the volume of the sphere, that is $4k^3\pi/3$ where $|\mathbf{k}|$ is written as $k$. To convert this to the number of states with energies up to the given energy we have to use the relationships between the components of **k** and the quantum numbers $n_a$, $n_b$, and $n_c$ given previously. This comparison demonstrates that we have to multiply the volume by $abc/\pi^3$. Now we have the number of states with energies *up to* a particular energy, but it is more useful to know the number of states *with* a particular energy. To find this, we define the number of states in a narrow range of $k$ values, d$k$. The number of states up to and including those of wave vector length $k$ + d$k$ is $4/3\pi^2 V(k + dk)^3$ where $V$ (= $abc$) is the volume of the crystal. Therefore, the number with values between $k$ and $k$ + d$k$ is $4/3\pi^2 V ((k + dk)^3 - k^3)$, which when $(k + dk)^3$ is expanded, gives a leading term $4/\pi^2 V k^2 dk$. This quantity is the **density of states**, $N(k)dk$. In terms of the more familiar energy, the density of states $N(E)dE$ is given by

$$\sqrt{(2m_e)^3 E} \times (V/2\pi^2\hbar^3)dE$$

A plot of $N(E)dE$ against $E$ is given in Figure 4.1.

Note that the density of states increases with increasing energy — the higher the energy, the more states there are in the interval d$E$. In metals, the valence electrons fill up the states from the lowest energy up with paired spins. For sodium, for example, each atom contributes one $3s$ electron and the electrons from all the atoms in the crystal occupy the levels in Figure 4.1 until all the electrons are used up. The highest occupied level is called the **Fermi level**.

Now let us see how this theoretical density of states compares to reality. Experimentally the density of states can be determined by X-ray emission spectroscopy. A beam of electrons or high energy X-rays hitting a metal can remove core electrons. In sodium, for example, the $2s$ or $2p$ electrons might be removed. The core energy levels are essentially atomic levels and so electrons have been removed from a discrete well-defined energy level. Electrons from the conduction band can now jump down to the energy level emitting X-rays as they do so. The X-ray energy will depend on the level of the conduction band from which the electron has come. A scan across the emitted X-rays will correspond to a scan across the filled levels. The

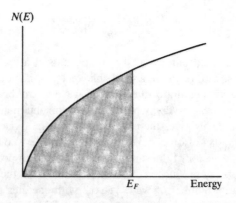

**FIGURE 4.1** A density of states curve based on the free electron model. The levels occupied at 0 K are shaded. Note that later in this book, energy is plotted on the vertical axis. In this figure, energy is plotted along the horizontal axis for comparison with experiment.

intensity of the radiation emitted will depend on the number of electrons with that particular energy, that is the intensity depends on the density of states of the conduction band. Figure 4.2 illustrates some X-ray emission spectra for sodium and aluminium and you can see that the shape of these curves resembles approximately the occupied part of Figure 4.1, so that the free electron model appears to describe these bands quite well.

If we look at the density of states found experimentally for metals with more electrons per atom than the simple metals, however, the fit is not as good. We find that instead of continuing to increase with energy as in Figure 4.1, the density reaches a maximum and then starts to decrease with energy. This is depicted in the $3d$ band of Ni in Figure 4.3.

Extensions of this model in which the atomic nuclei and core electrons are included by representing them by a potential function, $V$, in Equation (4.1) (plane wave methods) can account for the density of states in Figure 4.3 and can be used for semiconductors and insulators as well. We shall however use a different model to describe these solids, one based on the molecular orbital theory of molecules. We describe this in the next section. We end this section by using our simple model to explain the electrical conductivity of metals.

### 4.2.1 ELECTRONIC CONDUCTIVITY

The wave vector, **k**, is the key to understanding electrical conductivity in metals. For this purpose, it is important to note that it is a vector with direction as well as

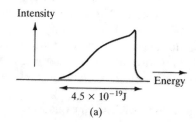

(a)

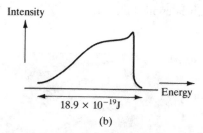

(b)

**FIGURE 4.2** X-ray emission spectra obtained from (a) sodium metal and (b) aluminium metal when conduction electrons fall into the $2p$ level. The slight tail at the high energy end is due to thermal excitation of electrons close to the Fermi level.

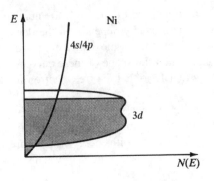

**FIGURE 4.3** The band structure of nickel. Note the shape of the $3d$ band.

magnitude. Thus, there may be many different energy levels with the same value of **k** and hence the same value of the energy, but with different components, $k_x$, $k_y$, $k_z$, giving a different direction to the momentum. In the absence of an electric field, all directions of the wave vector **k** are equally likely and so equal numbers of electrons are moving in all directions (Figure 4.4).

If we now connect our metal to the terminals of a battery producing an electric field, then an electron travelling in the direction of the field will be accelerated and the energies of those levels with a net momentum in this direction will be lowered. Electrons moving in the opposite direction will have their energies raised and so some of these electrons will drop down into levels of lower energy corresponding to momentum in the opposite direction. There will thus be more electrons moving in the direction of the field than in other directions. This net movement of electrons in one direction is an electric current. The net velocity in an electric field is depicted in Figure 4.5.

An important point to note about this explanation is it assumes that empty energy levels are available close in energy to the Fermi level. You will see later that the existence of such levels is crucial in explaining the conductivity of semiconductors.

The model as we have described it so far explains why metals conduct electricity but it does not account for the finite resistance of metals. The current, $i$, flowing through a metal for a given applied electric field, $V$, is given by **Ohm's law** $V = iR$

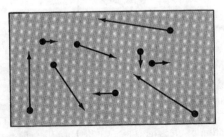

**FIGURE 4.4** Electrons in a metal in the absence of an electric field. They move in all directions, but, overall, no net motion occurs in any direction.

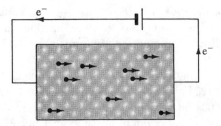

**FIGURE 4.5** The sample of Figure 4.4 in a constant electric field, established by placing the rod between the terminals of a battery. The electrons can move in all directions but now their velocities are modified so that each also has a net movement or drift velocity in the left to right direction.

where $R$ is the resistance. It is characteristic of a metal that $R$ increases with increasing temperature (or putting it another way, the conductance, $\sigma$, decreases with increasing temperature); that is for a given field, the current decreases as the temperature is raised. There is nothing in our theory yet that will impede the flow of electrons. To account for electrical resistance, it is necessary to introduce the ionic cores. If these were arranged periodically on the lattice sites of a perfect crystal, and they were not able to move, then they would not interrupt the flow of electrons. Most crystals contain some imperfections however and these can scatter the electrons. If the component of the electron's momentum in the field direction is reduced, then the current will drop. In addition, even in perfect crystals the ionic cores will be vibrating. A set of crystal vibrations exists in which the ionic cores vibrate together. These vibrations are called **phonons**. An example of a crystal vibrational mode would be one in which each ionic core would be moving out of phase with its neighbours along one axis. (If they moved in phase, the whole crystal would move.) Like vibrations in small molecules, each crystal vibration has its set of quantised vibrational levels. The conduction electrons moving through the crystal are scattered by the vibrating ionic cores and lose some energy to the phonons. As well as reducing the electron flow, this mechanism increases the crystal's vibrational energy. The effect then is to convert electrical energy to thermal energy. This ohmic heating effect is put to good use in, for example, the heating elements of kettles.

## 4.3 BONDING IN SOLIDS — MOLECULAR ORBITAL THEORY

We know that not all solids conduct electricity, and the simple free electron model discussed previously does not explain this. To understand semiconductors and insulators, we turn to another description of solids, molecular orbital theory. In the molecular orbital approach to bonding in solids, we regard solids as a very large collection of atoms bonded together and try to solve the Schrödinger equation for a periodically repeating system. For chemists, this has the advantage that solids are not treated as very different species from small molecules.

However, solving such an equation for a solid is something of a tall order because exact solutions have not yet been found for small molecules and even a small crystal could well contain of the order of $10^{16}$ atoms. An approximation often used for smaller molecules is that combining atomic wave functions can form the molecular wave functions. This **linear combination of atomic orbitals (LCAO)** approach can also be applied to solids.

We shall start by reminding the reader how to combine atomic orbitals for a very simple molecule, $H_2$. For $H_2$, we assume that the molecular orbitals are formed by combining $1s$ orbitals on each of the hydrogen atoms. These can combine in phase to give a bonding orbital or out of phase to give an anti-bonding orbital. The bonding orbital is lower in energy than the $1s$ and the anti-bonding orbital higher in energy. The amount by which the energy is lowered for a bonding orbital depends on the amount of overlap of the $1s$ orbitals on the two hydrogens. If the hydrogen nuclei are pulled further apart, for example, the overlap decreases and so the decrease in energy is less. (If the nuclei are pushed together, the overlap will increase but the electrostatic repulsion of the two nuclei becomes important and counteracts the effect of increased overlap.)

Suppose we form a chain of hydrogen atoms. For $N$ hydrogen atoms, there will be $N$ molecular orbitals. The lowest energy orbital will be that in which all the $1s$ orbitals combine in phase, and the highest energy orbital will be that in which the orbitals combine out of phase. In between are $(N - 2)$ molecular orbitals in which there is some in-phase and some out-of-phase combination. Figure 4.6 is a plot of the energy levels as the length of the chain increases.

Note that as the number of atoms increases, the number of levels increases, but the spread of energies appear to increase only slowly and is levelling off for long chains. Extrapolating to crystal length chains, you can see that there would be a very large number of levels within a comparatively small range of energies. A chain of hydrogen atoms is a very simple and artificial model; as an estimate of the energy separation of the levels, let us take a typical band in an average size metal crystal. A metal crystal might contain $10^{16}$ atoms and the range of energies is only $10^{-19}$ J.

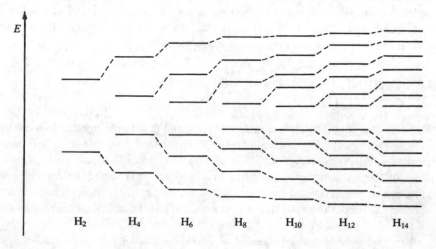

**FIGURE 4.6** Orbital energies for a chain of $N$ hydrogen atoms as $N$ increases.

The average separation between levels would thus be only $10^{-35}$ J. The lowest energy levels in the hydrogen atom are separated by energies of the order of $10^{-18}$ J so that you can see that the energy separation in a crystal is minute. The separation is in fact so small that, as in the free electron model, we can think of the set of levels as a continuous range of energies. Such a continuous range of allowed energies is known as an **energy band**.

In actual calculations on crystals, it is impractical to include all $10^{16}$ atoms and so we use the periodicity of the crystal. We know that the electron density and wave function for each unit cell is identical and so we form combinations of orbitals for the unit cell that reflect the periodicity of the crystal. Such combinations have patterns like the sine waves that we obtained from the particle-in-the-box calculation. For small molecules, the LCAO expression for molecular orbitals is

$$\Psi(i) = \sum_n c_{ni} \phi_n$$

where the sum is over all $n$ atoms, $\Psi(i)$ is the $i$th orbital, and $c_{ni}$ is the coefficient of the orbital on the $n$th atom for the $i$th molecular orbital.

For solids, we replace this with

$$\Psi(\mathbf{k}) = \sum_n e^{i\mathbf{k}.\mathbf{r}_n} a_{n\mathbf{k}} \phi_n$$

where $\mathbf{k}$ is the wave vector, $\mathbf{r}$ is the position vector of a nucleus in the lattice, and $a_{n\mathbf{k}}$ is the coefficient of the orbital on the $n$th atom for a wave vector of $\mathbf{k}$.

The hydrogen chain orbitals were made up from only one sort of atomic orbital — $1s$ — and one energy band was formed. For most of the other atoms in the Periodic Table, it is necessary to consider other atomic orbitals in addition to the $1s$ and we find that the allowed energy levels form a series of energy bands separated by ranges of forbidden energies. The ranges of forbidden energy between the energy bands are known as band gaps. Aluminium, for example, has the atomic configuration $1s^2 2s^2 2p^6 3s^2 3p^1$ and would be expected to form a $1s$ band, a $2s$ band, a $2p$ band, a $3s$ band, and a $3p$ band, all separated by band gaps. In fact the lower energy bands, those formed from the core orbitals $1s$, $2s$, and $2p$, are very narrow and for most purposes can be regarded as a set of localised atomic orbitals. This arises because these orbitals are concentrated very close to the nuclei and so there is little overlap between orbitals on neighbouring nuclei. In small molecules, the greater the overlap the greater the energy difference between bonding and anti-bonding orbitals. Likewise for continuous solids, the greater the overlap, the greater the spread of energies or **bandwidth** of the resulting band. For aluminium, then, $1s$, $2s$, and $2p$ electrons can be taken to be core orbitals and only $3s$ and $3p$ bands considered.

Just as in small molecules, the available electrons are assigned to levels in the energy bands starting with the lowest. Each orbital can take two electrons of opposed spin. Thus, if $N$ atomic orbitals were combined to make the band orbitals, then $2N$ electrons are needed to fill the band. For example, the $3s$ band in a crystal of aluminium containing $N$ atoms can take up to $2N$ electrons whereas the $3p$ band can accommodate up to $6N$ electrons. Because aluminium has only one $3p$ electron per atom, however, there would only be $N$ electrons in the $3p$ band and only $N/2$ levels would be occupied. As in the free electron model, the highest occupied level at 0 K is the Fermi level. Now let us see how this model applies to some real solids and how we can apply the concept of energy bands to understanding some of their properties. First, we revisit the simple metals.

### 4.3.1 SIMPLE METALS

The crystal structures of the simple metals are such that the atoms have high coordination numbers. For example, the Group 1 elements have body-centred cubic structures with each atom surrounded by eight others. This high coordination number increases the number of ways in which the atomic orbitals can overlap. The $ns$ and $np$ bands of the simple metals are very wide, due to the large amount of overlap, and, because the $ns$ and $np$ atomic orbitals are relatively close in energy, the two bands merge. This can be shown even for a small chain of lithium atoms as can be seen in Figure 4.7.

For the simple metals, then, we do not have an $ns$ band and an $np$ band but one continuous band instead, which we shall label $ns/np$. For a crystal of $N$ atoms, this $ns/np$ band contains $4N$ energy levels and can hold up to $8N$ electrons.

The simple metals have far fewer than $8N$ electrons available however; they have only $N$, $2N$, or $3N$. Thus, the band is only partly full. Empty energy levels exist near the Fermi level, and the metals are good electronic conductors.

If we move to the right of the Periodic Table, we find elements that are semi-conductors or insulators in the solid state; we now go on to consider these solids.

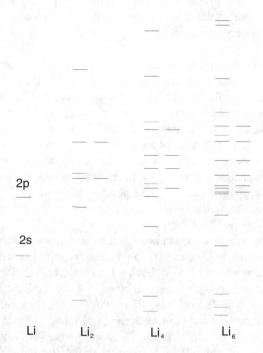

**FIGURE 4.7** $2s$ and $2p$ levels for a Li atom and for chains of 2, 4 and 6 Li atoms.

## 4.4 SEMICONDUCTORS — Si AND Ge

Carbon (as diamond polymorph), silicon, and germanium form structures in which the atoms are tetrahedrally coordinated, unlike simple metals which form structures of high coordination. With these structures, the $ns/np$ bands still overlap but the $ns/np$ band splits into two. Each of the two bands contains $2N$ orbitals and so can accommodate up to $4N$ electrons. You can think of the two bands as analogy of bonding and anti-bonding; the tetrahedral symmetry not giving rise to any non-bonding orbitals. Carbon, silicon, and germanium have electronic configurations $ns^2np^2$ and so they have available $4N$ electrons — just the right number to fill the lower band. This lower band is known as the **valence band**, the electrons in this band essentially bonding the atoms in the solid together.

One question that may have occurred to you is: Why do these elements adopt the tetrahedral structure instead of one of the higher coordination structures? Well, if these elements adopted a structure like that of the simple metals, then the $ns/np$ band would be half full. The electrons in the highest occupied levels would be virtually non-bonding. In the tetrahedral structure, however, all $4N$ electrons would be in bonding levels. This is illustrated in Figure 4.8.

Elements with few valence electrons will thus be expected to adopt high coordination structures and be metallic. Those with larger numbers (4 or more) will be expected to adopt lower coordination structures in which the $ns/np$ band is split and only the lower bonding band is occupied.

Tin (in its most stable room temperature form) and lead, although in the same group as silicon and germanium, are metals. For these elements the atomic $s$–$p$ energy separation is greater and the overlap of $s$ and $p$ orbitals is much less than in silicon and germanium. For tin, the tetrahedral structure would have two $s$–$p$ bands but the band gap is almost zero. Below 291 K, tin undergoes a transition to the diamond structure, but above this temperature, it is more stable for tin to adopt a

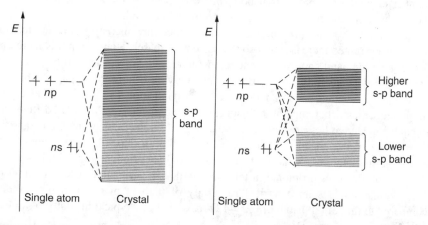

**FIGURE 4.8** Energy bands formed from $ns$ and $np$ atomic orbitals for (a) a body-centred cubic crystal and (b) a crystal of diamond structure, depicting filled levels for $4N$ electrons.

higher coordination structure. The advantage of not having nonbonding levels in the diamond structure is reduced by the small band gap. In lead, the diamond structure would give rise to an $s$ band and a $p$ band instead of $ns/np$ bands because the overlap of $s$ and $p$ orbitals is even smaller. It is more favourable for lead to adopt a cubic close-packed structure and, because lead has only $2N$ electrons to go in the $p$ band, it is metallic. This is an example of the **inert pair effect**, in which the $s$ electrons act as core electrons, on the chemistry of lead. Another example is the formation of divalent ionic compounds containing $Pb^{2+}$ instead of tetravalent covalent compounds like those of silicon and germanium.

Silicon and germanium, therefore, have a completely full valence band; they would be expected to be insulators. However, they belong to a class of materials known as **semiconductors**.

The electronic conductivity, $\sigma$, is given by the expression

$$\sigma = nZe\mu$$

where $n$ is the number of charge carriers per unit volume, $Ze$ is their charge (in the case of an electron this is simply $e$ the electronic charge), and $\mu$, the mobility, is a measure of the velocity in the electric field.

The conductivity of metallic conductors decreases with temperature. As the temperature rises the phonons gain energy; the lattice vibrations have larger amplitudes. The displacement of the ionic cores from their lattice sites is thus greater and the electrons are scattered more, reducing the net current by reducing the mobility, $\mu$, of the electrons.

The conductivity of **intrinsic semiconductors** such as silicon or germanium, however, *increases* with temperature. In these solids, conduction can only occur if electrons are promoted to the higher $s/p$ band, the **conduction band**, because only then will there be a partially full band. The current in semiconductors will depend on $n$, which, in this case, is the number of electrons free to transport charge. The number of electrons able to transport charge is given by the number promoted to the conduction band plus the number of electrons in the valence band that have been freed to move by this promotion. As the temperature increases, the number of electrons promoted increases and so the current increases. How much this increases depends on the band gap. At any one temperature, more electrons will be promoted for a solid with a small band gap than for one with a large band gap so that the solid with the smaller band gap will be a better conductor. The number of electrons promoted varies with temperature in an exponential manner so that small differences in band gap lead to large differences in the number of electrons promoted and hence the number of current carriers. For example, tellurium has a band gap about half that of germanium but, because of the exponential variation, at a given temperature the ratio of electrons promoted in tellurium to that in germanium is of the order of $10^6$. Germanium has an electrical resistivity of 0.46 $\Omega$ m at room temperature compared to that of tellurium of 0.0044 $\Omega$ m, or that of a typical insulator of around $10^{12}$ $\Omega$ m.

The change in resistivity with temperature is used in thermistors, which can be used as thermometers and in fire alarm circuits.

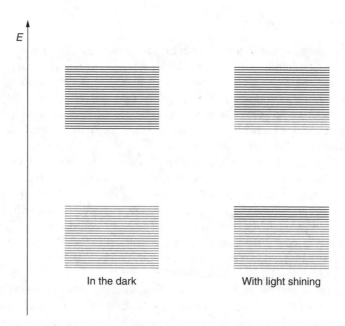

E

In the dark          With light shining

**FIGURE 4.9** Promotion of electrons from the valence band to the conduction band by light.

### 4.4.1 PHOTOCONDUCTIVITY

Forms of energy other than heat energy, for example, light, can also promote electrons. If the photon energy (hv) of light shining on a semiconductor is greater than the energy of the band gap, then valence band electrons will be promoted to the conduction band and conductivity will increase. Promotion of electrons by light is illustrated in Figure 4.9.

Semiconductors with band gap energies corresponding to photons of visible light are **photoconductors,** being essentially non-conducting in the dark but conducting electricity in the light. One use of such photoconductors is in **electrophotography.** In the xerographic process, there is a positively charged plate covered with a film of semiconductor. (In practice, this semiconductor is not silicon but a solid with a more suitable energy band gap such as selenium, the compound $As_2Se_3$ or a conducting polymer.) Light reflected from the white parts of the page to be copied hits the semiconductor film. The parts of the film receiving the light become conducting; an electron is promoted to the conduction band. This electron then cancels the positive charge on the film, the positive hole in the valence band being removed by an electron from the metal backing plate entering the valence band. Now the parts of the film which received light from the original are no longer charged, but the parts underneath the black lines are still positively charged. Tiny negatively charged plastic capsules of ink (toner) are then spread on to the semiconductor film but only stick to the charged bits of the film. A piece of positively charged white paper removes the toner from the semiconductor film and hence acquires an image of the black parts of the original. Finally, the paper is heated to melt the plastic coating and fix the ink.

### 4.4.2 DOPED SEMICONDUCTORS

The properties of semiconductors are extremely sensitive to the presence of impurities at concentrations as low as 1 part in $10^{10}$. For this reason, silicon manufactured for transistors and other devices must be very pure. The deliberate introduction of a very low concentration of certain impurities into the very pure semiconductor, however, alters the properties in a way that has proved invaluable in constructing semiconductor devices. Such semiconductors are known as **doped** or **extrinsic semiconductors**. Consider a crystal of silicon containing boron as an impurity. Boron has one fewer valence electron than silicon. Therefore, for every silicon replaced by boron, there is an electron missing from the valence band (Figure 4.10) (i.e., positive holes occur in the valence band and these enable electrons near the top of the band to conduct electricity). Therefore, the doped solid will be a better conductor than pure silicon. A semiconductor like this doped with an element with fewer valence electrons than the bulk of the material is called a **p-type semiconductor** because its conductivity is related to the number of positive holes (or empty electronic energy levels) produced by the impurity.

Now suppose instead of boron the silicon was doped with an element with more valence electrons than silicon — phosphorus, for example. The doping atoms form a set of energy levels that lie in the band gap between the valence and conduction bands. Because the atoms have more valence electrons than silicon, these energy levels are filled. Therefore, electrons are present close to the bottom of the conduction band and easily promoted into the band. This time the conductivity increases because of extra electrons entering the conduction band. Such semiconductors are called **n-type** — n for negative charge carriers or electrons. Figure 4.10 schematically depicts energy bands in intrinsic, p-type, and n-type semiconductors.

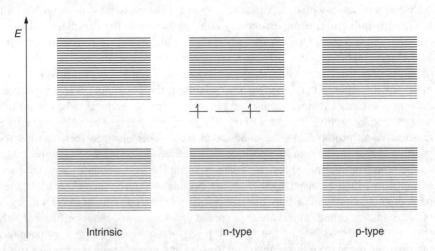

**FIGURE 4.10** Intrinsic, n-type and p-type semiconductors depicting negative charge carriers (electrons in the conduction band) and positive holes.

The $n$- and $p$-type semiconductors in various combinations make up many electronic devices such as rectifiers, transistors, photo-voltaic cells, and LEDs.

### 4.4.3 THE $p$-$n$ JUNCTION — FIELD-EFFECT TRANSISTORS

$p$–$n$ junctions are prepared either by doping different regions of a single crystal with different atoms or by depositing one type of material on top of the other using techniques such as chemical vapour deposition. The use of these junctions stems from what occurs where the two differently doped regions meet. In the region of the crystal where $n$- and $p$-type meet, there is a discontinuity in electron concentration. Although both $n$- and $p$-type are electrically neutral, the $n$-type has a greater concentration of electrons than the $p$-type. To try and equalise the electron concentrations electrons diffuse from $n$- to $p$-type. However, this produces a positive electric charge on the $n$-type and a negative electric charge on the $p$-type. The electric field thus set up encourages electrons to drift back to the $n$-type. Eventually, a state is reached in which the two forces are balanced and the electron concentration varies smoothly across the junction as in Figure 4.11.

The regions located immediately to either side of the junction are known as **depletion regions** because there are fewer current carriers (electrons or empty levels) in these regions. Applying an external electric field across such a junction disturbs the equilibrium and the consequences of this are exploited in LEDs and transistors. In LEDs, which are discussed in Chapter 8, the voltage is applied so that the $n$-type semiconductor is negative relative to the $p$-type. An important feature of most transistors is a voltage applied in the reverse direction, that is, the $n$-type is positive with respect to the $p$-type.

The 1956 Nobel Prize in physics was awarded to Bardeen, Brattain, and Shockley for their work on developing transistors. We consider briefly here an important class

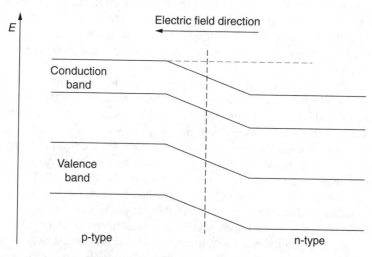

FIGURE 4.11 Bending of energy levels across a $p$–$n$ junction.

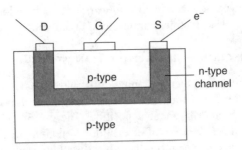

**FIGURE 4.12** Schematic diagram of an FET depicting source, S, drain, D, and gate, G, electrodes.

of transistors, the **field-effect transistors (FET)**. These have many uses including amplifiers in radios etc and gates in computer circuits.

A simple $n$-channel FET consists of a block of heavily doped $p$-type semiconductor with a channel of an $n$-type semiconductor (see Figure 4.12).

Electrodes are attached to the $p$-type block and to the $n$-type channel. The electrodes attached to the $n$-type channel are known as the source (negative electrode which provides electrons) and the drain (positive electrode). The electrode attached to the $p$-type block is known as the gate electrode. A low voltage (typically 6 V) is applied across the source and drain electrodes. To fulfill the transistor's role, as amplifier or switch, a voltage is applied to the gate electrode. Electrons flow into the $p$-type semiconductor but cannot cross the $p$–$n$ junction because the valence band in the $n$-type is full. The electrons, therefore, fill the valence band in the $p$-type. Because of the charge produced, electrons in the $n$-type channel move toward the centre of this channel. The net result is an increased depletion zone and because the $p$-type is more heavily doped, this effect is greater for the $n$-type channel. Depletion zones contain less charge carriers than the bulk semiconductor so the current in the $n$-type channel is greatly reduced. If the voltage on the gate electrode is reduced, then electrons flow out of the $p$-type, the depletion zones shrink, and the current through the $n$-type channel increases. By varying the size of the voltage across the gate electrode, for example, by adding an alternating voltage, the current in the $n$-type channel is turned on or off.

Transistor amplifiers consist of an electronic circuit containing a transistor and other components such as resistors. The signal to be amplified is applied to the gate electrode. The output signal is taken from the drain. Under the conditions imposed by the rest of the circuit, the current through the $n$-type channel increases linearly with the magnitude of the incoming voltage. This current produces a drain voltage that is proportional to the incoming voltage but whose magnitude is greater by a constant factor.

**Metal oxide semiconductor field-effect transistors (MOSFETs)** are field-effect transistors with a thin film of silicon dioxide between the gate electrode and the semiconductor. The charge on the silicon dioxide controls the size of the depletion zone in the $p$-type semiconductor. MOSFETs are easier to mass produce and are used in integrated circuits and microprocessors for computers and in amplifiers for cassette players. Traditionally, transistors have been silicon based but a recent development is field-effect transistors based on organic materials.

## 4.5 BANDS IN COMPOUNDS — GALLIUM ARSENIDE

Gallium arsenide (GaAs) is a rival to silicon in some semiconductor applications, including solar cells, and is used for LEDs and in a solid state laser, as discussed in Chapter 8. It has a diamond-type structure and is similar to silicon except that it is composed of two kinds of atoms. The valence orbitals in Ga and As are the $4s$ and $4p$ and these form two bands each containing $4N$ electrons as in silicon. Because of the different $4s$ and $4p$ atomic orbital energies in Ga and As, however, the lower band will have a greater contribution from As and the conduction band will have a higher contribution from Ga. Thus, GaAs can be considered as having partial ionic character because there is a partial transfer of electrons from Ga to As. The valence band has more arsenic than gallium character and so all the valence electrons end up in orbitals in which the possibility of being near an As nucleus is greater than that of being near a Ga nucleus. The band energy diagram for GaAs is illustrated in Figure 4.13. GaAs is an example of a class of semiconductors known as III/V semiconductors in which an element with one more valence electron than the silicon group is combined with an element with one less valence electron. Many of these compounds are semiconductors (e.g., GaSb, InP, InAs, and InSb). Moving farther along the Periodic Table, there are II/VI semiconductors such as CdTe and ZnS. Toward the top of the Periodic Table and farther out toward the edges (e.g., AlN, AgCl), the solids tend to adopt different structures and become more ionic. For the semiconducting solids, the band gap decreases down a group, for example, GaP > GaAs > GaSb; AlAs > GaAs > InAs.

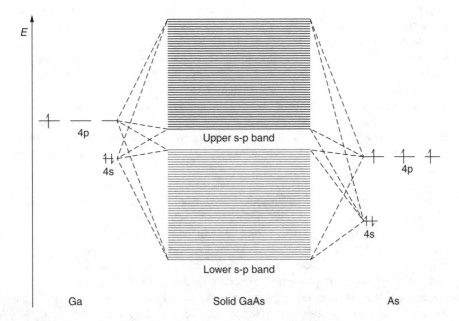

**FIGURE 4.13** Orbital energy level diagram for gallium arsenide.

## 4.6 BANDS IN d-BLOCK COMPOUNDS — TRANSITION METAL MONOXIDES

Monoxides MO with structures based on sodium chloride are formed by the first row transition elements Ti, V, Mn, Fe, Co, and Ni. TiO and VO are metallic conductors, and the others are semiconductors. The O $2p$ orbitals form a filled valence band. The $4s$ orbitals on the metal form another band. What about the $3d$ orbitals?

In the sodium chloride structure, the symmetry enables three of the five $d$ orbitals on different atoms to overlap. Because the atoms are not nearest neighbours, the overlap is not as large as in pure metals and the bands are thus narrow. The other two $d$ orbitals overlap with orbitals on the adjacent oxygens. Thus, two narrow $3d$ bands exist. The lower one, labelled $t_{2g}$, can take up to $6N$ electrons, and the upper one, labelled $e_g$, up to $4N$ electrons. Divalent titanium has two $d$ electrons, therefore, $2N$ electrons fill the $3N$ levels of the lower band. Similarly, divalent vanadium has three $d$ electrons and so the lower band is half full. As in the case of pure metals, a partly filled band leads to metallic conductivity. For FeO, the $t_{2g}$ band would be full, so it is not surprising to find that it is a semiconductor; but MnO with only five electrons per manganese is also a semiconductor. CoO and NiO, which should have partially full $e_g$ levels, are also semiconductors. It is easier to understand these oxides using a localised $d$ electron model.

Going across the first transition series, there is a contraction in the size of the $3d$ orbitals. The $3d$ orbital overlap therefore decreases and the $3d$ band narrows. In a wide band, such as the $s$-$p$ bands of the alkali metals, the electrons are essentially free to move through the crystal keeping away from the nuclei and from each other. In a narrow band by contrast, the electrons are more tightly bound to the nuclei. Interelectron repulsion becomes important, particularly the repulsion between electrons on the same atom. Consider an electron in a partly filled band moving from one nucleus to another. In the alkali metal, the electron would already be in the sphere of influence of surrounding nuclei and would not be greatly repelled by electrons on these nuclei. The $3d$ electron moving from one nucleus to another adds an extra electron near to the second nucleus which already has $3d$ electron density near it. Thus, electron repulsion on the nucleus is increased. For narrow bands, therefore, we have to balance gains in energy on band formation against electron repulsion. For MnO, FeO, CoO, and NiO electron repulsion wins, and it becomes more favourable for the $3d$ electrons to remain in localised orbitals than to be delocalised.

The band gap between the oxygen $2p$ band and the metal $4s$ band is sufficiently wide that the pure oxides would be considered insulators. However, they are almost invariably found to be non-stoichiometric, that is their formulae are not exactly MO, and this leads to semiconducting properties which will be discussed in Chapter 5.

The monoxides are not unique in displaying a variation of properties across the transition series. The dioxides form another series, and $CrO_2$ is discussed later because of its magnetic properties. Several classes of mixed oxides also exhibit a range of electronic properties (e.g., the perovskites $LaTiO_3$, $SrVO_3$, and $LaNiO_3$ are metallic conductors; $LaRhO_3$ is a semiconductor; and $LaMnO_3$ is an insulator). Sulfides also show progression from metal to insulator. In general, compounds with broader $d$ bands will be metallic. Broad bands will tend to occur for elements at the beginning of the transition series and for second and third row metals (e.g., NbO, $WO_2$). Metallic behaviour is also more common amongst lower oxidation state compounds and with less electronegative anions.

## QUESTIONS

1.  In the free electron model, the electron energy is kinetic. Using the formula $E = \frac{1}{2} mv^2$, calculate the velocity of electrons at the Fermi level in sodium metal. The mass of an electron is $9.11 \times 10^{-31}$ kg. Assume the band shown in Figure 4.2a starts at 0 energy.
2.  The density of magnesium metal is 1740 kg m$^{-3}$. A typical crystal has a volume of $10^{-12}$ m$^3$ (corresponding to a cube of side 0.1 cm). How many atoms would such a crystal contain?
3.  An estimate of the total number of occupied states can be obtained by integrating the density of states from 0 to the Fermi level.

$$N = \int_{0}^{E_F} N(E)dE = \int_{0}^{E_F} E^{\frac{1}{2}}(2m_e)^{\frac{3}{2}} V / 2\pi^2 \hbar^3 dE = (2m_e E_F)^{\frac{3}{2}} V / 3\pi^2 \hbar^3$$

Calculate the total number of occupied states for a sodium crystal of volume (a) $10^{-12}$ m$^3$, (b) $10^{-6}$ m$^3$, and (c) $10^{-29}$ m$^3$ (approximately atomic size). Compare your results with the number of electrons available and comment on the different answers to (a), (b), and (c). A crystal of volume $10^{-12}$ m$^3$ contains $2.5 \times 10^{16}$ atoms.
4.  The energy associated with one photon of visible light ranges from 2.4 to $5.0 \times 10^{-19}$ J. The band gap in selenium is $2.9 \times 10^{-19}$ J. Explain why selenium is a good material to use as a photoconductor in applications such as photocopiers.
5.  The band gaps of several semiconductors and insulators are given below. Which substances would be photoconductors over the entire range of visible wavelengths?

| Substance | Si | Ge | CdS |
|---|---|---|---|
| Band gap/$10^{-19}$ J | 1.9 | 1.3 | 3.8 |

6.  Which of the following doped semiconductors will be $p$-type and which will be $n$-type? (a) arsenic in germanium, (b) germanium in silicon, (c) indium in germanium, (d) silicon on antimony sites in indium antimonide (InSb), (e) magnesium on gallium sites in gallium nitride (GaN).
7.  Would you expect carborundum (SiC) to adopt a diamond structure or one of higher coordination? Explain why.

# 5 Defects and Non-Stoichiometry

## 5.1 POINT DEFECTS — AN INTRODUCTION

In a perfect crystal, all atoms would be on their correct lattice positions in the structure. This situation can only exist at the absolute zero of temperature, 0 K. Above 0 K, **defects** occur in the structure. These defects may be **extended defects** such as **dislocations**. The strength of a material depends very much on the presence (or absence) of extended defects, such as dislocations and **grain boundaries**, but the discussion of this type of phenomenon lies very much in the realm of materials science and will not be discussed in this book. Defects can also occur at isolated atomic positions; these are known as **point defects**, and can be due to the presence of a foreign atom at a particular site or to a vacancy where normally one would expect an atom. Point defects can have significant effects on the chemical and physical properties of the solid. The beautiful colours of many gemstones are due to impurity atoms in the crystal structure. Ionic solids are able to conduct electricity by a mechanism which is due to the movement of *ions* through vacant ion sites within the lattice. (This is in contrast to the electronic conductivity that we explored in the previous chapter, which depends on the movement of *electrons*.)

## 5.2 DEFECTS AND THEIR CONCENTRATION

Defects fall into two main categories: **intrinsic defects** which are integral to the crystal in question — they do not change the overall composition and because of this are also known as **stoichiometric defects**; and **extrinsic defects** which are created when a foreign atom is inserted into the lattice.

### 5.2.1 INTRINSIC DEFECTS

Intrinsic defects fall into two categories: **Schottky defects** which consist of vacancies in the lattice, and **Frenkel defects**, where a vacancy is created by an atom or ion moving into an interstitial position.

For a 1:1 solid MX, a Schottky defect consists of a *pair* of vacant sites, a cation vacancy, and an anion vacancy. This is presented in Figure 5.1(a) for an alkali halide type structure: the number of cation vacancies and anion vacancies have to be equal to preserve electrical neutrality. A Schottky defect for an $MX_2$ type structure will consist of the vacancy caused by the $M^{2+}$ ion together with *two* $X^-$ anion vacancies, thereby balancing the electrical charges. Schottky defects are more common in 1:1 stoichiometry and examples of crystals that contain them include rock salt (NaCl), wurtzite (ZnS), and CsCl.

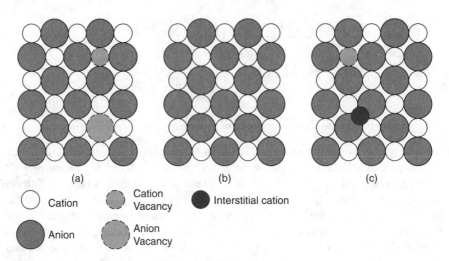

Cation ◯

Cation Vacancy

Interstitial cation ●

Anion

Anion Vacancy

**FIGURE 5.1** Schematic illustration of intrinsic point defects in a crystal of composition MX: (a) Schottky pair, (b) perfect crystal, and (c) Frenkel pair.

A Frenkel defect usually occurs only on one sublattice of a crystal, and consists of an atom or ion moving into an interstitial position thereby creating a vacancy. This is illustrated in Figure 5.1(c) for an alkali–halide-type structure, such as NaCl, where one cation moves out of the lattice and into an interstitial site. This type of behaviour is seen, for instance, in AgCl, where we observe such a *cation Frenkel defect* when $Ag^+$ ions move from their octahedral coordination sites into tetrahedral coordination and this is illustrated in Figure 5.2. The formation of this type of defect is important in the photographic process when they are formed in the light-sensitive AgBr used in photographic emulsions.

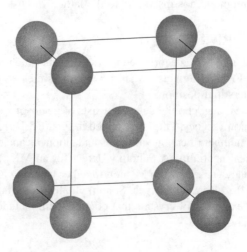

**FIGURE 5.2** The tetrahedral coordination of an interstitial $Ag^+$ ion in AgCl.

It is less common to observe an *anion Frenkel defect* when an anion moves into an interstitial site. This is because anions are commonly *larger* than the cations in the structure and so it is more difficult for them to enter a crowded low-coordination interstitial site.

An important exception to this generalization lies in the formation of anion Frenkel defects in compounds with the fluorite structure, such as $CaF_2$ (other compounds adopting this structure are strontium and lead fluorides, $SrF_2$, $PbF_2$, and thorium, uranium, and zirconium oxides, $ThO_2$, $UO_2$, $ZrO_2$, which are discussed again later in this chapter). One reason for this is that the anions have a lower electrical charge than the cations and so do not find it as difficult to move nearer each other. The other reason lies in the nature of the fluorite structure, which is presented again in Figure 5.3. You may recall (see Chapter 1) that we can think of it as based on a *ccp* array of $Ca^{2+}$ ions with all the tetrahedral holes occupied by the $F^-$ ions. This of course leaves all of the larger octahedral holes unoccupied, giving

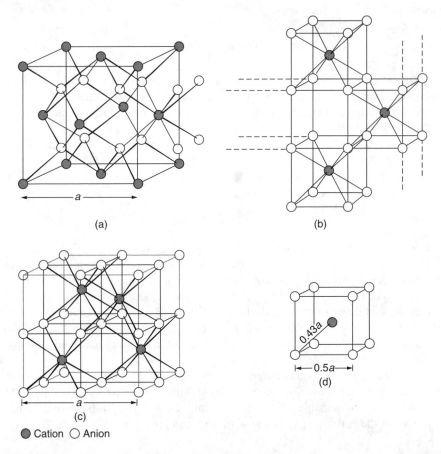

● Cation  ○ Anion

**FIGURE 5.3** The crystal structure of fluorite $MX_2$. (a) Unit cell as a *ccp* array of cations. (b) and (c) The same structure redrawn as a simple cubic array of anions. (d) Cell dimensions.

a very open structure. This becomes clear if we redraw the structure as in Figure 5.3(c), based on a simple cubic array of $F^-$ ions. The unit cell now consists of eight small **octants** with the $Ca^{2+}$ ions occupying every other octant. The two different views are equivalent, but the cell depicted in Figure 5.3(c) shows the possible interstitial sites more clearly.

## 5.2.2 THE CONCENTRATION OF DEFECTS

Energy is required to form a defect. This means that the formation of defects is always an **endothermic process**. It may seem surprising that defects exist in crystals at all, and yet they do even at low temperatures, albeit in very small concentrations. The reason for this is that although it costs energy to form defects, there is a commensurate *gain* in entropy. The enthalpy of formation of the defects is thus balanced by the gain in entropy such that, at equilibrium, the overall change in free energy of the crystal due to the defect formation is zero according to the equation:

$$\Delta G = \Delta H - T\Delta S$$

The interesting point is that thermodynamically we do not expect a crystalline solid to be perfect, contrary, perhaps to our 'commonsense' expectation of symmetry and order! At any particular temperature there will be an equilibrium population of defects in the crystal.

The number of Schottky defects in a crystal of composition MX is given by:

$$n_S \approx N\exp\left(\frac{-\Delta H_S}{2kT}\right) \tag{5.1}$$

where $n_S$ is the number of Schottky defects per unit volume, at $T$ K, in a crystal with $N$ cation and $N$ anion sites per unit volume; $\Delta H_S$ is the enthalpy required to form one defect. It is quite a simple matter to derive these equations for the equilibrium concentration of Schottky defects by considering the change in entropy of a perfect crystal due to the introduction of defects. The change in entropy will be due to the vibrations of atoms around the defects and also to the arrangement of the defects. It is possible to estimate this latter quantity, the **configurational entropy**, using the methods of statistical mechanics.

If the number of Schottky defects is $n_S$ per unit volume at $T$ K, then there will be $n_S$ cation vacancies and $n_S$ anion vacancies in a crystal containing $N$ possible cation sites and $N$ possible anion sites per unit volume. The **Boltzmann formula** tells us that the entropy of such a system is given by:

$$S = k\ln W \tag{5.2}$$

where $W$ is the number of ways of distributing $n_S$ defects over $N$ possible sites at random, and $k$ is the Boltzmann constant $(1.380\ 662 \times 10^{-23}$ J K$^{-1}$ ). Probability theory shows that $W$ is given by

$$W = \frac{N!}{(N-n)!n!} \qquad (5.3)$$

where $N!$ is called 'factorial $N$' and is mathematical shorthand for

$$N \times (N-1) \times (N-2) \times (N-3) \ldots \times 1$$

So, the number of ways we can distribute the cation vacancies will be

$$W_c = \frac{N!}{(N-n_S)!n_S!}$$

and similarly for the anion vacancies

$$W_a = \frac{N!}{(N-n_S)!n_S!}$$

The total number of ways of distributing these defects, $W$, is given by the product of $W_c$ and $W_a$:

$$W = W_c W_a$$

and the *change* in entropy due to introducing the defects into a perfect crystal is thus:

$$\Delta S = k\ln W = k\ln \left( \frac{N!}{(N-n_S)!n_S!} \right)^2$$

$$= 2k\ln \left( \frac{N!}{(N-n_S)!n_S!} \right)$$

We can simplify this expression using **Stirling's approximation** that

$$\ln N! \approx N\ln N - N$$

and the expression becomes (after manipulation)

$$\Delta S = 2k\ \{N\ln N - (N-n_S)\ln(N-n_S) - n_S\ln n_S\}$$

If the enthalpy change for the formation of a single defect is $\Delta H_S$ and we assume that the enthalpy change for the formation of $n_S$ defects is $n_S \Delta H_S$, then the Gibbs free energy change is given by:

$$\Delta G = n_S \Delta H_S - 2kT\{N\ln N - (N - n_S)\ln(N - n_S) - n_S \ln n_S\}$$

At equilibrium, at constant $T$, the Gibbs free energy of the system must be a minimum with respect to changes in the number of defects $n_S$; thus

$$\left(\frac{d\Delta G}{dn_S}\right) = 0$$

So,

$$\Delta H_S - 2kT \frac{d}{dn_S}[N\ln N - (N - n_S)\ln(N - n_S) - n_S \ln n_S] = 0$$

$N\ln N$ is a constant and hence its differential is zero; the differential of $\ln x$ is $\frac{1}{x}$ and of $(x\ln x)$ is $(1 + \ln x)$. On differentiating, we obtain:

$$\Delta H_S - 2kT[\ln(N - n_S) + 1 - \ln n_S - 1] = 0$$

thus,

$$\Delta H_S = 2kT \ln\left[\frac{(N - n_S)}{n_S}\right]$$

and

$$n_S = (N - n_S) \exp\left(\frac{-\Delta H_S}{2kT}\right)$$

as $N \gg n_S$ we can approximate $(N - n_S)$ by $N$, finally giving:

$$n_S \approx N\exp\left(\frac{-\Delta H_S}{2kT}\right) \tag{5.3}$$

If we express this equation in molar quantities, it becomes:

$$n_S \approx N\exp\left(\frac{-\Delta H_S}{2RT}\right) \tag{5.4}$$

where now $\Delta H_S$ is the enthalpy required to form one mole of Schottky defects and $R$ is the gas constant, 8.314 J K$^{-1}$ mol$^{-1}$. The units of $\Delta H_S$ are J mol$^{-1}$.

By a similar analysis, we find that the number of Frenkel defects present in a crystal MX is given by the expression:

$$n_F = (NN_i)^{1/2} \exp\left(\frac{-\Delta H_F}{2kT}\right) \tag{5.5}$$

where $n_F$ is the number of Frenkel defects per unit volume, $N$ is the number of lattice sites and $N_i$ the number of interstitial sites available. $\Delta H_F$ is the enthalpy of formation of one Frenkel defect. If $\Delta H_F$ is the enthalpy of formation of one mole of Frenkel defects the expression becomes:

$$n_F = (NN_i)^{1/2} \exp\left(\frac{-\Delta H_F}{2RT}\right) \tag{5.6}$$

Table 5.1 lists some enthalpy-of-formation values for Schottky and Frenkel defects in various crystals.

Using the information in Table 5.1 and Equation (5.1), we can now get an idea of how many defects are present in a crystal. Assume that $\Delta H_S$ has a middle-of-the-range value of $5 \times 10^{-19}$ J. Substituting in Equation (5.1) we find that the proportion

**TABLE 5.1**
**The formation enthalpy of Schottky and Frenkel defects in selected compounds**

|  | Compound | $\Delta H/10^{-19}$ J | $\Delta H/eV^a$ |
|---|---|---|---|
| Schottky defects | MgO | 10.57 | 6.60 |
|  | CaO | 9.77 | 6.10 |
|  | LiF | 3.75 | 2.34 |
|  | LiCl | 3.40 | 2.12 |
|  | LiBr | 2.88 | 1.80 |
|  | LiI | 2.08 | 1.30 |
|  | NaCl | 3.69 | 2.30 |
|  | KCl | 3.62 | 2.26 |
| Frenkel defects | $UO_2$ | 5.45 | 3.40 |
|  | $ZrO_2$ | 6.57 | 4.10 |
|  | $CaF_2$ | 4.49 | 2.80 |
|  | $SrF_2$ | 1.12 | 0.70 |
|  | AgCl | 2.56 | 1.60 |
|  | AgBr | 1.92 | 1.20 |
|  | $\beta$-AgI | 1.12 | 0.70 |

[a]The literature often quotes values in eV, so these are included for comparison.
1 eV = $1.60219 \times 10^{-19}$ J.

of vacant sites $\frac{n_S}{N}$ at 300 K is $6.12 \times 10^{-27}$; at 1000 K this rises to $1.37 \times 10^{-8}$. This shows what a low concentration of Schottky defects is present at room temperature. Even when the temperature is raised to 1000 K, we still find only of the order of one or two vacancies per hundred million sites!

Whether Schottky or Frenkel defects are found in a crystal depends in the main on the value of $\Delta H$, the defect with the lower $\Delta H$ value predominating. In some crystals it is possible for *both* types of defect to be present.

We will see in a later section that in order to change the properties of crystals, particularly their ionic conductivity, we may wish to introduce more defects into the crystal. It is important, therefore, at this stage to consider how this might be done.

First, we have seen from the previous calculation that raising the temperature introduces more defects. We would have expected this to happen because defect formation is an endothermic process and **Le Chatelier's principle** tells us that increasing the temperature of an endothermic reaction will favour the products — in this case defects. Second, if it were possible to decrease the enthalpy of formation of a defect, $\Delta H_S$ or $\Delta H_F$, this would also increase the proportion of defects present. A simple calculation as we did before, again using Equation (5.1), but now with a lower value for $\Delta H_S$, for instance, $1 \times 10^{-19}$ J allows us to see this. Table 5.2 compares the results. This has had a dramatic effect on the numbers of defects! At 1000 K, there are now approximately 3 defects for every 100 sites. It is difficult to see how the value of $\Delta H$ could be manipulated within a crystal, but we do find crystals where the value of $\Delta H$ is lower than usual due to the nature of the structure, and this can be exploited. This is true for one of the systems that we shall look at in detail later, $\alpha$-AgI. Third, if we introduce impurities selectively into a crystal, we can increase the defect population.

### 5.2.3 EXTRINSIC DEFECTS

We can introduce vacancies into a crystal by **doping** it with a selected impurity. For instance, if we add $CaCl_2$ to a NaCl crystal, each $Ca^{2+}$ ion replaces *two* $Na^+$ ions in order to preserve electrical neutrality, and so *one* cation vacancy is created. Such created vacancies are known as **extrinsic**. An important example that you will meet later in the chapter is that of **zirconia, $ZrO_2$**. This structure can be stabilised by doping with CaO, where the $Ca^{2+}$ ions replace the Zr(IV) atoms in the lattice. The charge compensation here is achieved by the production of anion vacancies on the oxide sublattice.

**TABLE 5.2**
**Values of $n_s/N$**

| $T$ (K) | $\Delta H_S = 5 \times 10^{-19}$ J | $\Delta H_S = 1 \times 10^{-19}$ J |
|---|---|---|
| 300 | $6.12 \times 10^{-27}$ | $5.72 \times 10^{-6}$ |
| 1000 | $1.37 \times 10^{-8}$ | $2.67 \times 10^{-2}$ |

## 5.3 IONIC CONDUCTIVITY IN SOLIDS

One of the most important aspects of point defects is that they make it possible for atoms or ions to move through the structure. If a crystal structure were perfect, it would be difficult to envisage how the movement of atoms, either **diffusion** through the lattice or **ionic conductivity** (ion transport under the influence of an external electric field) could take place. Setting up equations to describe either diffusion or conductivity in solids is a very similar process, and so we have chosen to concentrate here on conductivity, because many of the examples later in the chapter are of solid electrolytes.

Two possible mechanisms for the movement of ions through a lattice are sketched in Figure 5.4. In Figure 5.4(a), an ion hops or jumps from its normal position on the lattice to a neighbouring equivalent but vacant site. This is called the **vacancy mechanism**. (It can equally well be described as the movement of a vacancy instead of the movement of the ion.) Figure 5.4(b) depicts the **interstitial mechanism**, where an interstitial ion jumps or hops to an adjacent equivalent site. These simple pictures of movement in an ionic lattice are known as the **hopping model**, and ignore more complicated cooperative motions.

Ionic conductivity, $\sigma$, is defined in the same way as electronic conductivity:

$$\sigma = nZe\mu \qquad (5.7)$$

where $n$ is the number of charge carriers per unit volume, $Ze$ is their charge (expressed as a multiple of the charge on an electron, $e = 1.602\ 189 \times 10^{-19}$ C), and $\mu$ is their **mobility**, which is a measure of the drift velocity in a constant electric field. Table 5.3 lists the sort of conductivity values one might expect to find for different materials. As we might expect, ionic crystals, although they *can* conduct, are poor conductors compared with metals. This is a direct reflection of the difficulty the charge carrier (in this case an ion, although sometimes an electron) has in moving through the crystal lattice.

Equation (5.7) is a general equation defining conductivity in all conducting materials. To understand why some ionic solids conduct better than others it is useful to look at the definition more closely in terms of the hopping model that we have

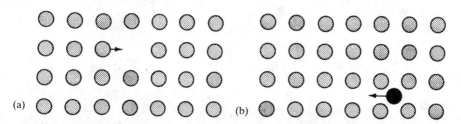

**FIGURE 5.4** Schematic representation of ionic motion by (a) a vacancy mechanism and (b) an interstitial mechanism.

**TABLE 5.3**
**Typical values of electrical conductivity**

| | Material | Conductivity/S m$^{-1}$ |
|---|---|---|
| Ionic conductors | Ionic crystals | $< 10^{-16} - 10^{-2}$ |
| | Solid electrolytes | $10^{-1} - 10^{3}$ |
| | Strong (liquid) electrolytes | $10^{-1} - 10^{3}$ |
| Electronic conductors | Metals | $10^{3} - 10^{7}$ |
| | Semiconductors | $10^{-3} - 10^{4}$ |
| | Insulators | $< 10^{-10}$ |

set up. First of all, we have said that an electric current is carried in an ionic solid by the defects. In the cases of crystals where the ionic conductivity is carried by the vacancy or interstitial mechanism, $n$ the concentration of charge carriers will be closely related to the concentration of defects in the crystal, $n_S$ or $n_F$. $\mu$ will thus refer to the mobility of these defects in such cases.

Let us look more closely at the mobility of the defects. Take the case of NaCl, which contains Schottky defects. The Na$^+$ ions are the ones which move because they are the smaller; however, even these meet quite a lot of resistance, as a glance at Figure 5.5 will show. We have used dotted lines to illustrate two possible routes that the Na$^+$ could take from the centre of the unit cell to an adjacent vacant site. The direct route

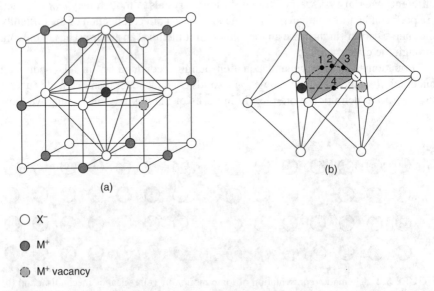

(a)

(b)

○ X$^-$

● M$^+$

◐ M$^+$ vacancy

**FIGURE 5.5** Sodium chloride type structure, depicting (a) coordination octahedron of central cation and (b) coordination octahedra of central cation and adjacent vacancy.

(labelled 4) is clearly going to be very unlikely because it leads directly between two Cl⁻ ions, which in a close-packed structure such as this, are going to be very close together. The other pathway first passes through a one of the triangular faces of the octahedron (point 1), then through one of the tetrahedral holes (point 2), and finally through another triangular face (point 3), before finally arriving at the vacant octahedral site. The coordination of the $Na^+$ ion will change from $6{\to}3{\to}4{\to}3{\to}6$ as it jumps from one site to the other. While there is clearly going to be an energy barrier to this happening, it will not be as large as for the direct path where the $Na^+$ ion becomes 2-coordinate. Generally, we would expect the ion to follow the lowest energy path available. A schematic diagram of the energy changes involved in such a pathway is depicted in Figure 5.6. Notice that the energy of the ion is the same at the beginning and end of the jump; the energy required to make the jump, $E_a$, is known as the **activation energy** for the jump. This means that the temperature dependence of the mobility of the ions can be expressed by an **Arrhenius equation**:

$$\mu \propto \exp\left(\frac{-E_a}{kT}\right) \tag{5.8}$$

or

$$\mu = \mu_0 \exp\left(\frac{-E_a}{kT}\right) \tag{5.9}$$

where $\mu_0$ is a proportionality constant known as a pre-exponential factor. $\mu_0$ depends on several factors: the number of times per second that the ion attempts the move, $v$, called the **attempt frequency** (this is a frequency of vibration of the lattice of the order of $10^{12}$–$10^{13}$ Hz); the distance moved by the ion; and the size of the external field. If the external field is small (up to about 300 V cm⁻¹), a temperature dependence of $1/T$ is introduced into the pre-exponential factor.

If we combine all this information in Equation (5.7), we arrive at an expression for the variation of ionic conductivity with temperature that has the form:

$$\sigma = \frac{\sigma_0}{T}\exp\left(\frac{-E_a}{T}\right) \tag{5.10}$$

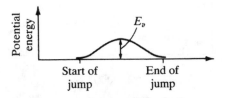

**FIGURE 5.6** Schematic representation of the change in energy during motion of an ion along the lowest energy path.

The term $\sigma_0$ now contains $n$ and $Ze$, as well as the information on attempt frequency and jump distance. This expression accounts for the fact that ionic conductivity *increases* with temperature. If we now take logs of Equation (5.10), we obtain:

$$\ln\sigma T = \ln\sigma_0 - \left(\frac{E_a}{T}\right)$$

Plotting $\ln\sigma T$ against $\frac{1}{T}$ should produce a straight line with a slope of $-E_a$. The expression in Equation (5.10) is sometimes plotted empirically as

$$\sigma = \sigma_0 \exp\left(\frac{-E_a}{T}\right)$$

because plotting either $\ln\sigma T$ or $\ln\sigma$ makes little difference to the slope; both types of plot are found in the literature. The results of doing this for several compounds are presented in Figure 5.7.

For the moment, ignore the AgI line, which is discussed in the next section. The other lines on the plot are straight lines apart from the one for LiI, where we can clearly see two lines of differing slope. Therefore, it looks as though the model we have set up does describe the behaviour of many systems. But how about LiI? In

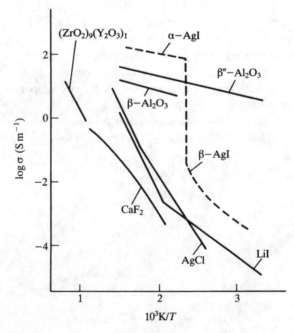

**FIGURE 5.7** The conductivities of selected solid electrolytes over a range of temperatures.

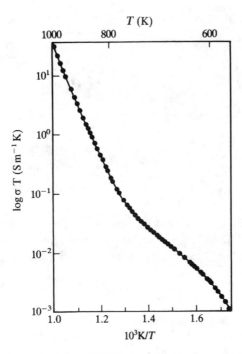

**FIGURE 5.8** The ionic conductivity of NaCl plotted against reciprocal temperature.

fact, other crystals also show this kink in the plot (some experimental data is presented for NaCl in Figure 5.8, where it can also be seen). Is it possible to explain this using the equations that we have just set up? The answer is yes.

The explanation for the two slopes in the plot lies in the fact that even a very pure crystal of NaCl contains some impurities, and the line corresponding to low temperatures (on the right of the plot) is due to the **extrinsic vacancies**. At low temperatures, the concentration of **intrinsic vacancies** is so small that it can be ignored because it is dominated by the defects created by the impurity. For a particular amount of impurity, the number of vacancies present will be essentially *constant*. $\mu$ in this **extrinsic region** thus depends only on the cation mobility due to these extrinsic defects, whose temperature dependence is given by Equation (5.9):

$$\mu = \mu_0 \exp \left( \frac{-E_a}{kT} \right) \tag{5.9}$$

However, at the higher temperatures on the left-hand side of the graph, the concentration of intrinsic defects has increased to such an extent that it now is similar to, or greater than, the concentration of extrinsic defects. The concentration of intrinsic defects, unlike that of the extrinsic defects will *not* be constant; indeed, it will vary according to Equation (5.1), as discussed earlier:

$$n_S \approx N \exp\left(\frac{-\Delta H_S}{2kT}\right) \tag{5.1}$$

Therefore, the conductivity in this **intrinsic region** on the left-hand side of the plot is given by:

$$\sigma = \frac{\sigma'}{T}\exp\left(\frac{-E_a}{kT}\right)\exp\left(\frac{-\Delta H_S}{2kT}\right) \tag{5.11}$$

A plot of $\ln\sigma T$ vs. $\frac{1}{T}$ in this case will give a greater value for the activation energy, $E_S$, because it will actually depends on two terms, the activation energy for the cation jump, $E_a$, and the enthalpy of formation of a Schottky defect:

$$E_S = E_a + \frac{1}{2}\Delta H_S \tag{5.12}$$

Similarly, for a system with Frenkel defects:

$$E_F = E_a + \frac{1}{2}\Delta H_F \tag{5.13}$$

From plots such as these, we find that the activation energies lie in the range of 0.05 to 1.1 eV, which is lower than the enthalpies of defect formation. As we have seen, raising the temperature increases the number of defects, and so increases the conductivity of a solid. Better than increasing the temperature to increase conductivity is to find materials that have low activation energies, less than about 0.2 eV. We find such materials in the top right-hand corner of Figure 5.7.

## 5.4  SOLID ELECTROLYTES

Much of the recent research in solid state chemistry is related to the ionic conductivity properties of solids, and new electrochemical cells and devices are being developed that contain solid, instead of liquid, electrolytes. Solid-state batteries are potentially useful because they can perform over a wide temperature range, they have a long shelf life, it is possible to make them very small, and they are spill-proof. We use batteries all the time — to start cars, in toys, watches, cardiac pacemakers, and so on. Increasingly we need lightweight, small but powerful batteries for a variety of uses such as computer memory chips, laptop computers, and mobile phones. Once a **primary battery** has discharged, the reaction cannot be reversed and it has to be thrown away, so there is also interest in solid electrolytes in the production of **secondary** or **storage batteries**, which are reversible because once the chemical reaction has taken place the reactant concentrations can be restored by reversing the cell reaction using an external source of electricity. If storage batteries can be made to give sufficient power and they are not too heavy, they become useful as alternative fuel sources (e.g., to power cars). Where possible, batteries need to be made out of nontoxic materials. All these goals give plenty of scope for research. Chemical sensors, electrochromic devices, and fuel cells that depend on the conducting properties of solids, are also being developed and we look at examples of all these later in the chapter.

## 5.4.1 FAST-ION CONDUCTORS: SILVER ION CONDUCTORS

Some ionic solids have been discovered that have a much higher conductivity than is typical for such compounds and these are known as **fast-ion conductors**. One of the earliest to be noticed, in 1913 by Tubandt and Lorenz, was a high temperature phase of silver iodide.

### α-AgI

Below 146°C, two phases of AgI exist: γ-AgI, which has the zinc blende structure, and β-AgI with the wurtzite structure. Both are based on a close-packed array of iodide ions with half of the tetrahedral holes filled. However, above 146°C a new phase, α-AgI, is observed where the iodide ions now have a body-centred cubic lattice. If you look back to Figure 5.7, you can see that a dramatic increase in conductivity is observed for this phase: the conductivity of α-AgI is very high, 131 S m$^{-1}$, a factor of $10^4$ higher than that of β- or γ-AgI, comparable with the conductivity of the best conducting liquid electrolytes. How can we explain this startling phenomenon?

The explanation lies in the crystal structure of α-AgI. The structure is based on a body-centred cubic array of I$^-$ ions as illustrated in Figure 5.9(a). Each I$^-$ ion in the array is surrounded by eight equidistant I$^-$ ions. To see where the Ag$^+$ ions fit into the structure, we need to look at the *bcc* structure in a little more detail. Figure 5.9(b) presents the same unit cell but with the next-nearest neighbours added in; these are the six at the body-centres of the surrounding unit cells, and they are only 15% farther away than the eight immediate neighbours. This means that the atom marked A is effectively surrounded by *14* other identical atoms, although not in a completely regular way: these 14 atoms lie at the vertices of a **rhombic dodecahedron** (Figure 5.9(c)). It can also be convenient to describe the structure in terms of the space-filling **truncated octahedron** depicted in Figure 5.9(d), which has six square faces and eight hexagonal faces corresponding to the two sets of neighbours. This is called the **domain** of an atom. Each vertex of the domain lies at the centre of an interstice (like the tetrahedral and octahedral holes found in the close-packed structures), which in this structure is a distorted tetrahedron. Figure 5.9(e) shows two such adjacent distorted tetrahedral holes, and you can see clearly from this the 'tetrahedral' site lying on the face of the unit cell. Where the two 'tetrahedra' join, they share a common triangular face — a trigonal site in the centre of this face is also marked in Figure 5.9(e). A third type of site in the centre of each face and of each edge of the unit cell can also be defined: these sites have distorted octahedral coordination. The structure thus possesses

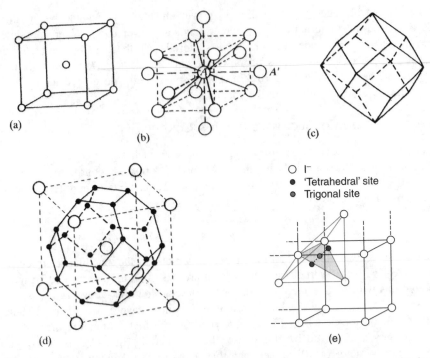

**FIGURE 5.9** Building up the structure α-AgI. (a) The body-centered cubic array of I⁻ ions. (b) *bcc* array extended to next-nearest neighbours. (c) Rhombic dodecahedron. (d) *bcc* array with enclosed truncated octahedron; the 24 vertices (black dots) lie at the centres of the distorted tetrahedra. (Because these 24 sites lie on the unit cell faces they are each shared between two unit cells, making 12 such sites per unit cell on average.) (e) The positions of two tetrahedral sites and a trigonal site between them.

the unusual feature that a variety of positions can be adopted by the Ag⁺ ions, and Figure 5.10 sketches many of them. Each AgI unit cell possesses two I⁻ ions ([8 × ¹/₈] at the corners and 1 at the body-centre) and so positions for only two Ag⁺ ions need be found. From the possible sites that we have described, 6 are distorted octahedral, 12 are 'tetrahedral', and 24 are trigonal — a choice of 42 possible sites — therefore, the Ag⁺ ions have a huge choice of positions open to them! Structure determinations indicate that the Ag⁺ are statistically distributed among the twelve tetrahedral sites all of which have the same energy. Therefore, counting these sites only, we find that *five* spare sites are available per Ag⁺ ion. We can visualise the Ag⁺ ions moving from tetrahedral site to tetrahedral site by jumping through the vacant trigonal site, following the paths marked with solid lines in Figure 5.10, continually creating and destroying Frenkel defects and able to move easily through the lattice. The paths marked with thin dotted lines in Figure 5.10 require higher energy, as they pass through the vacant 'octahedral' sites which are more crowded. The jump that we have described only changes the

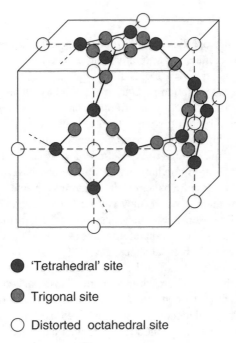

● 'Tetrahedral' site

◕ Trigonal site

○ Distorted  octahedral site

**FIGURE 5.10** Possible cation sites in the *bcc* structure of α-AgI. The thick solid and dashed lines mark possible diffusion paths.

coordination number from $4 \rightarrow 3 \rightarrow 4$  and experimentally the activation energy is found to be very low, 0.05 eV. This easy movement of the $Ag^+$ ions through the lattice has often been described as a **molten sublattice** of $Ag^+$ ions.

The very high conductivity of α-AgI appears to arise because of a conjunction of favourable factors, and we can list the features that have contributed to this:

- the charge on the ions is low, the mobile $Ag^+$ ions are monovalent;
- a large number of vacant sites are available into which the cations can move;
- the structure has an open framework with pathways that the ions can move through;
- the coordination around the ions is also low, so that when they jump from site to site, the coordination changes only changes by a little, affording a route through the lattice with a low activation energy;
- the anions are rather **polarizable**; this means that the electron cloud surrounding an anion is easily distorted, making the passage of a cation past an anion rather easier.

These are properties that are important when looking for other fast-ion conductors.

## RbAg₄I₅

The special electrical properties of $\alpha$-AgI inevitably led to a search for other solids exhibiting high ionic conductivity preferably at temperatures lower than 146°C. The partial replacement of Ag by Rb, forms the compound $RbAg_4I_5$. This compound has an ionic conductivity at room temperature of 25 S m$^{-1}$, with an activation energy of only 0.07 eV. The crystal structure is different from that of $\alpha$-AgI, but similarly the $Rb^+$ and $I^-$ ions form a rigid array while the $Ag^+$ ions are randomly distributed over a network of tetrahedral sites through which they can move.

If a conducting ionic solid is to be useful as a solid electrolyte in a battery, not only must it possess a high conductivity, but also must have *negligible electronic conductivity*. This is to stop the battery short-circuiting: the electrons must only pass through the external circuit, where they can be harnessed for work. $RbAg_4I_5$ has been used as the solid electrolyte in batteries with electrodes made of Ag and $RbI_3$. Such cells operate over a wide temperature range ($-55$ to $+200$°C), have a long shelf life, and can withstand mechanical shock.

A table of ionic conductors that behave in a similar way to $\alpha$-AgI is given in Table 5.4. Some of these structures are based on a close-packed array of anions and this is noted in the table; the conducting mechanism in these compounds is similar to that in $\alpha$-AgI. The chalcogenide structures, such as silver sulfide and selenide, tend to demonstrate electronic conductivity as well as ionic, although this can be quite useful in an *electrode material* as opposed to an electrolyte.

### 5.4.2 FAST-ION CONDUCTORS: OXYGEN ION CONDUCTORS

#### Stabilized Zirconias

The fluorite structure, as we know from Figure 5.3, has plenty of empty space which can enable an $F^-$ ion to move into an interstitial site. If the activation energy for this process is low enough, we might expect compounds with this structure to show ionic conductivity. Indeed, $PbF_2$ has a low ionic conductivity at room temperature, but this increases smoothly with temperature to a limiting value of ~500 S m$^{-1}$ at 500°C. Uranium, thorium, and cerium readily form oxides with the fluorite structure, $UO_2$, $ThO_2$, and $CeO_2$, respectively, but zirconium is different. The cubic (fluorite) form of $ZrO_2$ is only formed at high temperature or when doped with another element.

## TABLE 5.4
## $\alpha$-AgI-related ionic conductors

| Anion structure | bcc | ccp | hcp | other |
|---|---|---|---|---|
| | $\alpha$-AgI | $\alpha$-CuI | $\beta$-CuBr | $RbAg_4I_5$ |
| | $\alpha$-CuBr | $\alpha$-Ag$_2$Te | | |
| | $\alpha$-Ag$_2$S | $\alpha$-Cu$_2$Se | | |
| | $\alpha$-Ag$_2$Se | $\alpha$-Ag$_2$HgI$_4$ | | |

At the turn of the last century, Nernst found that mixed oxides of yttria, $Y_2O_3$, and zirconia, $ZrO_2$, glowed white hot if an electric current was passed, and attributed this to the conduction of oxide ions. Nernst used this doped zirconia for the filaments in his 'glower' electric lights which started to replace candles, and oil and gas lamps, making his fortune before the arrival of tungsten filament bulbs. $ZrO_2$ is found as the mineral baddeleyite, which has a monoclinic structure. Above 1000°C this changes to a tetragonal form and at high temperatures $ZrO_2$ adopts the cubic fluorite structure. The cubic form can be stabilized at room temperature by the addition of other oxides such as yttria, $Y_2O_3$ (known as **yttria-stabilized zirconia, YSZ**), and CaO. CaO forms a new phase with $ZrO_2$ after heating at about 1600°C; the phase diagram is depicted in Figure 5.11 and shows that at about 15 mole% CaO the new phase, known as **calcia-stabilized zirconia**, appears alone and persists until about 28 mole% CaO. Above 28 mole%, another new phase appears, finally appearing pure as $CaZrO_3$. If the $Ca^{2+}$ ions sit on $Zr^{4+}$ sites, compensating vacancies are created

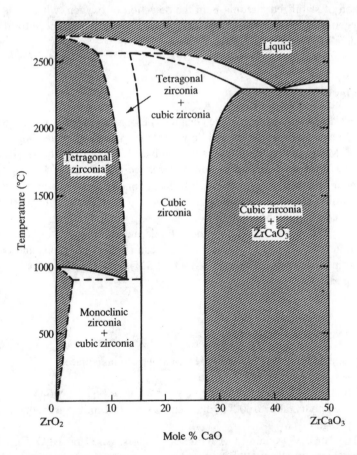

**FIGURE 5.11** Phase diagram of the pseudobinary $CaO-ZrO_2$ system. The cubic calcia-stabilized zirconia phase occupies the central band in the diagram and is stable to about 2400°C.

in the $O^{2-}$ sublattice. Thus, for every $Ca^{2+}$ ion taken into the structure *one* anion vacancy is created. Consequently, these materials are exceptionally good fast-ion conductors of $O^{2-}$ anions (Figure 5.7).

Conductivity maximizes at relatively low concentrations of dopant, but not at the 50% one might expect; this is because there is an elastic interaction between the substituted ion and the vacancy created. The best conductivity seems to be achieved when the crystal lattice is distorted as little as possible (i.e., when the dopant ion is similar in size to the cation it is replacing). Consequently, two of the best oxygen–ion conductors are zirconia doped with scandia, $Sc_2O_3$, and ceria doped with gadolinia, $Gd_2O_3$ (CGO); these are, however, rather expensive. Many other materials are also made with this type of structure, which are based on oxides such as $CeO_2$, $ThO_2$, and $HfO_2$ as well as $ZrO_2$, and doped with rare earth or alkaline earth oxides: these are collectively known as stabilized zirconias and are widely used in electrochemical systems.

YSZ is the usual material for use in **solid oxide fuel cells**. Another interesting application of stabilized zirconia is in the detection of oxygen, where it is used in both **oxygen meters** and **oxygen sensors**, which are based on a specialized electro-chemical cell (Section 5.4.4).

### Perovskites

The perovskite structure, $ABX_3$ (see Chapter 1, Figure 1.44) has two different metal sites which could be substituted with lower valence metal cations leading to oxygen vacancies. Materials based on lanthanum gallate, $LaGaO_3$, have been successfully doped with strontium and magnesium to produce $La_{1-x}Sr_xGa_{1-y}Mg_yO_{3-\delta}$ (LSGM) with similar conductivities to the stabilized zirconias, but at lower temperatures. Using them in oxide-ion conducting devices would have the advantage of bringing down the operating temperature.

For some applications, such as the cathode materials in solid oxide fuel cells (see Section 5.4.4), a material is needed that can conduct both ions and electrons. The strontium-doped perovskites $LaMnO_3$ (LSM), and $LaCrO_3$ (LSC) have both these properties.

### Other Oxygen Ion Conductors

Development of other oxide conductors continues to take place.

- The LAMOX family of oxide conductors, based on $La_2Mo_2O_9$, has high conductivity above 600°C, but tend to be susceptible to reduction by hydrogen.
- The BIMEVOX family of oxide conductors, based on $Bi_2O_3$, has high conductivity above 600°C.
- The apatite structures, $La_{10-x}M_6O_{26+y}$ (M = Si or Ge), conduct well at very high temperatures.

## 5.4.3 FAST-ION CONDUCTORS: SODIUM ION CONDUCTORS

### β-alumina

β-alumina is the name given to a series of compounds that demonstrate fast-ion conducting properties. The parent compound is sodium β-alumina, $Na_2O.11Al_2O_3$ ($NaAl_{11}O_{17}$), and is found as a by-product from the glass industry. (The compound was originally thought to be a polymorph of $Al_2O_3$, and was named as such — it was only later found to contain sodium ions! However, the original name has stuck.) The general formula for the series is $M_2O.nX_2O_3$, where $n$ can range from 5 to 11: M is a monovalent cation such as (alkali metal)$^+$, $Cu^+$, $Ag^+$, or $NH_4^+$, and X is a trivalent cation $Al^{3+}$, $Ga^{3+}$, or $Fe^{3+}$. The real composition of β-alumina actually varies quite considerably from the ideal formula and the materials are always found to be rich in $Na^+$ and $O^{2-}$ ions to a greater or lesser extent.

Interest in these compounds started in 1966 when research at the Ford Motor Company demonstrated that the $Na^+$ ions were very mobile both at room temperature and above. The high conductivity of the $Na^+$ ions in β-alumina is due to the crystal structure. This can be thought of as close-packed layers of oxide ions, but in every fifth layer three-quarters of the oxygens are missing (Figure 5.12). The four close-packed layers contain the $Al^{3+}$ ions in both octahedral and tetrahedral holes. (They are known as the 'spinel blocks' because of their similarity to the crystal structure of the mineral spinel [$MgAl_2O_4$], which is discussed in Chapter 1 and illustrated in Figure 1.43.) The groups of four close-packed oxide layers are held apart by a rigid $Al-O-Al$ linkage; this O atom constituting the fifth oxide layer which contains only a quarter of the number of oxygens of each of the other layers. The $Na^+$ ions

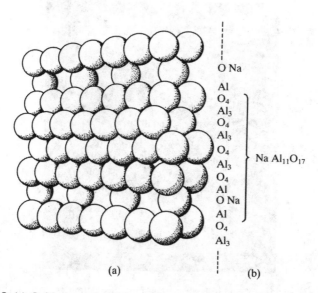

(a)     (b)

**FIGURE 5.12** (a) Oxide layers in β-alumina. (b) The ratio of atoms in each layer of the structure.

are found in these fifth oxide layers, which are mirror planes in the structure. The overall stoichiometry of the structure is illustrated layer by layer in Figure 5.12(b). The sequence of layers is

$$B(ABCA) \ C \ (ACBA) \ B \ldots$$

where the brackets enclose the four close-packed layers and the intermediate symbols refer to the fifth oxide layer.

The crystal structure of β-alumina is depicted in Figure 5.13. The $Na^+$ ions move around easily as plenty of vacancies exist, so there is a choice of sites. Conduction in the β-aluminas only occurs within the planes containing the oxygen vacancies; these are known as the **conduction planes**. The alkali metal cations cannot penetrate the dense 'spinel blocks', but can move easily from site to site within the plane. β-alumina is not found in the stoichiometric form — it is usually $Na_2O$ rich, and the sodium-rich compounds have a much higher conductivity than stoichiometric β-alumina. The extra sodium ions have to be compensated by a counter-defect in order to keep the overall charge on the compound at zero. There is more than one possibility for this, but in practice it is found that extra oxide ions provide the compensation and the overall formula can be written as $(Na_2O)_{1+x}.11Al_2O_3$. The extra sodium and oxide ions both occupy the fifth oxide layer; the $O^{2-}$ ions are locked into position by an $Al^{3+}$ moving out from the spinel block, and the $Na^+$ ions become part of the mobile pool of ions. The $Na^+$ ions are so fluid that the ionic conductivity in β-alumina at 300°C is close to that of typical liquid electrolytes at ambient temperature.

β-aluminas are used as electrolytes in the manufacture of electrochemical cells particularly for power supplies.

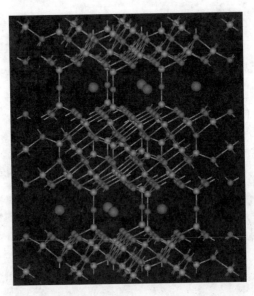

**FIGURE 5.13** Structure of stoichiometric β-alumina. See colour insert following this page. Na, purple; Al, pink; O, red.

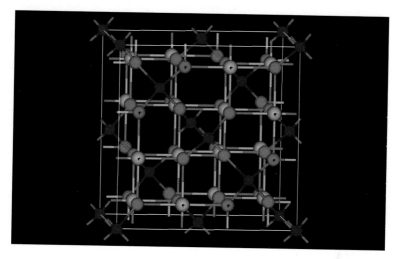

**FIGURE 1.43** The spinel structure, $CuAl_2O_4$ ($AB_2O_4$). Cu, blue spheres; Al, pink spheres; O, red spheres.

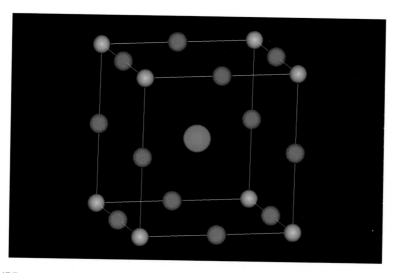

**FIGURE 1.44** The perovskite structure of compounds $ABX_3$, such as $CaTiO_3$. Ca, green sphere; Ti, silver spheres; O, red spheres.

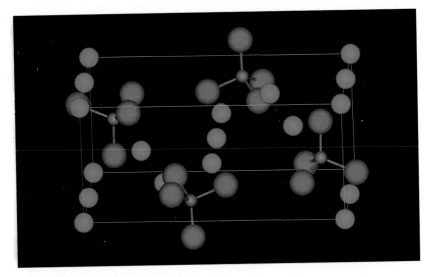

**FIGURE 1.53** The unit cell of olivine. Key: Mg,Fe, green;  Si, grey; O, red.

**FIGURE 1.54** The structure of an amphibole. Key: Mg, green;  Si, grey; O,  red;  Na, purple.

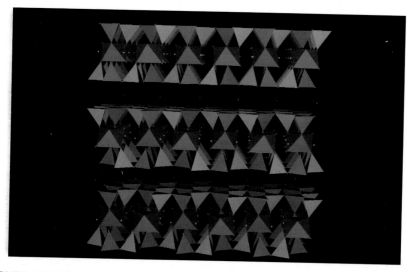

**FIGURE 1.55** The structure of biotite. Key: Mg, green; Si, grey; O, red; Na, purple; Al, pink; Fe, blue.

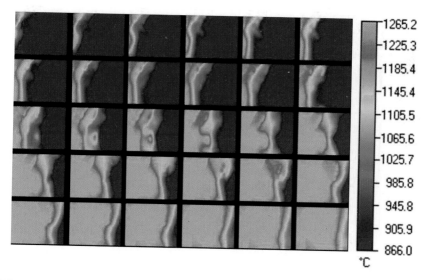

**FIGURE 3.4** Thermal images of the synthesis wave moving through a pellet of MgO, Fe, $Fe_2O_3$, and $NaClO_4$. Each image is of dimension $3 \times 2$ mm. Images were captured at 0.06 s intervals. The first image is top left and the last is bottom right. (Courtesy of Professor Ivan Parkin, University College, London.)

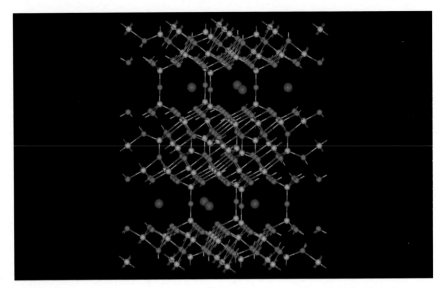

**FIGURE 5.13** Structure of stoichiometric β-alumina. Na, purple; Al, pink; O, red.

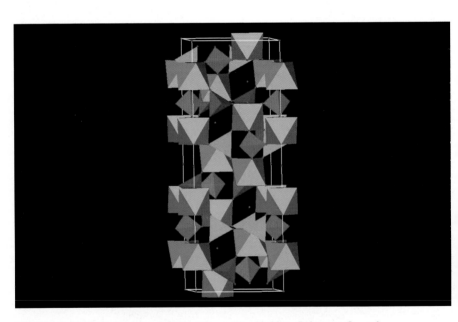

**FIGURE 5.14** The structure of NZP. Na, purple; Si, blue; P, brown; O, red.

## NASICON

In 1968, Hagman and Kierkegaard found an exciting new material, sodium zirconium phosphate ($NaZr_2(PO_4)_3$), known as **NZP** (Figure 5.14). The structure consists of corner-linked $ZrO_6$ octahedra joined by $PO_4$ tetrahedra, each of which corner-shares to four of the octahedra. This creates a three-dimensional system of channels through the structure containing two types of vacant site: Type I, a single distorted octahedral site (occupied by $Na^+$ ions in NZP), and three larger Type II sites (vacant in NZP). This has proved to be an incredibly versatile and stable structure with the general formula $A_xM_2((Si,P)O_4)_3$, which is adopted by hundreds of compounds. The balancing cations, A, are usually alkali or alkaline earth metals, and the structural metal(s), M, is a transition metal such as Ti, Zr, Nb, Cr, or Fe. The phosphorus can be substituted by silicon. The most famous member of this family is known as **NASICON** (from **Na S**uper**I**onic **Con**ductor). This has proved to be a very good $Na^+$ fast ion conductor with a conductivity of 20 S $m^{-1}$ at 300°C. It has the formula $Na_3Zr_2(PO_4)(SiO_4)_2$, and three out of the four vacant sites are occupied by $Na^+$, allowing a correlated motion as the ions diffuse through the channels.

### 5.4.4 APPLICATIONS

The development of solid state conducting solids that are on a par with liquid electrolytes has revolutionized the design of batteries and other **solid state ionic devices (SSIs)** in recent years, and this section explains the operating principles behind some of these devices. Figure 5.15 is a simple schematic diagram which we can use to explain the operation of several different types of electrochemical device.

**FIGURE 5.14** The structure of NZP. See colour insert on previous page. Na, purple; Si, blue; P, brown; O, red.

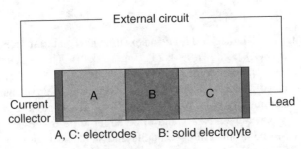

**FIGURE 5.15** Diagram of a general electrochemical device using a solid electrolyte.

The solid electrolyte, B (which is able to conduct ions but *not* electrons) is sandwiched between two electrodes, A and C, which are connected by an external circuit around which electrons can flow.

## Batteries

A battery is an electrochemical cell that produces an electric current at a constant voltage as a result of a chemical reaction. The ions taking part in the reaction pass through an **electrolyte** and are then either oxidised or reduced at an **electrode**. The electrode at which oxidation takes place is called the **anode**, and reduction takes place at the **cathode**. To be useful as a solid electrolyte in a battery, the conducting solid must have a high ionic conductivity, but be an electronic insulator, so that it can separate the two reactants of the device (Figure 5.15) allowing only ions, and not electrons, to travel through the solid — or the device would short-circuit. The electrons are released at the positively charged electrode and travel round the external circuit where they can be harnessed for useful work. The electromotive force (emf), or voltage produced by the cell under standard open circuit conditions, is related to the standard Gibb's free energy change for the reaction by the following equation:

$$\Delta G^{\ominus} = -nE^{\ominus}F \qquad (5.14)$$

where $n$ is the number of electrons transferred in the reaction, $E^{\ominus}$ is the standard emf of the cell (the voltage delivered under standard, zero-current conditions), and F is the Faraday constant (96 485 C mol$^{-1}$ or 96 485 J V$^{-1}$). The energy stored in a battery is a factor of the energy generated by the cell reaction and the amount of materials used; it is usually expressed in watt-hours, Wh (current × voltage × discharge time).

The reason that solid state batteries are potentially useful is that they can perform over a wide temperature range, they have a long shelf life, and it is possible to manufacture them so that they are extremely small. Lightweight rechargeable batteries can be now be made to give sufficient power to maintain mobile phones for several days and laptop computers for several hours. They are used for backup power supplies and may eventually become useful as alternative fuel sources to power cars.

*Lithium Batteries*

Because of the need for small, lightweight batteries, the energy stored in a battery is not always the most useful merit indicator. More important can be the **energy density** (watt-hours divided by battery volume in litres) or **specific energy** (watt-hours divided by battery weight in kilograms). Light elements, such as lithium, are obvious candidates for such battery materials.

Although LiI has a fairly low ionic conductivity (Figure 5.7), its use in heart pacemaker batteries for implantation was first proposed by Wilson Greatbatch and went into production in the early 1970s (Figure 5.16). These batteries only need to provide a low current, but can be made very small, they last a long time, they produce no gas so they can be hermetically sealed, and, above all, they are very reliable. LiI is the solid electrolyte that separates the anode (lithium) from the cathode (iodine embedded in a conducting polymer, poly-2-vinyl-pyridine), which based on Figure 5.15, we can depict as:

$$\begin{matrix} A & B & C \\ \text{Li} & // \text{ LiI } // & I_2 \text{ and polymer} \end{matrix}$$

The electrode reactions are:

$$\textbf{\textit{Anode A}}: 2Li(s) = 2Li^+(s) + 2e^-$$
$$\textbf{\textit{Cathode C}}: I_2(s) + 2e^- = 2I^-(s)$$

Because LiI contains intrinsic Schottky defects, the small $Li^+$ cations are able to pass through the solid electrolyte, while the released electrons go round an external circuit.

Making reliable rechargeable lithium batteries has proved to be a very difficult problem because the lithium redeposits in a finely divided state which is very reactive and can even catch fire. However, the technology of lithium-containing batteries has developed immensely in recent years. In the 1990s, the Sony Corporation in Japan developed rechargeable **lithium-ion batteries**, which now recycle reliably many times. Their lightness makes them popular for use in mobile phones and in laptop computers (Figure 5.17). The driving reaction of these cells is that of Li with a transition metal oxide, such as $CoO_2$, to form an intercalation compound $Li_xCoO_2$. The anode is made of lithium embedded in graphitic carbon, which also makes an intercalation compound, typically $C_6Li$, but one which easily releases $Li^+$, that then travels through a conducting polymer electrolyte (see Chapter 6) to the $CoO_2$ cathode. The $Li^+$ ions 'rock' between the two intercalation compounds; no lithium metal is ever present, eliminating many of the hazards associated with lithium batteries. Various cathode materials, such as $CoO_2$, $NiO_2$, $TiS_2$, and $MnO_2$, can be used.

$$\begin{matrix} A & B & C \\ \text{Li/C} & // \text{ polymer gel } // & CoO_2 \text{ and polymer} \end{matrix}$$

$$\textbf{\textit{Anode A}}: Li_xC_6(s) = xLi^+ + 6C + xe^-$$
$$\textbf{\textit{Cathode C}}: xLi^+ + CoO_2(s) + xe^- = Li_xCoO_2(s)$$

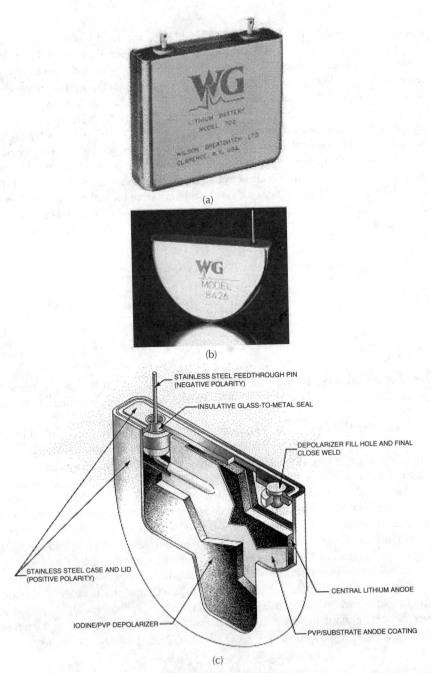

(a)

(b)

STAINLESS STEEL FEEDTHROUGH PIN
(NEGATIVE POLARITY)

INSULATIVE GLASS-TO-METAL SEAL

DEPOLARIZER FILL HOLE AND FINAL
CLOSE WELD

STAINLESS STEEL CASE AND LID
(POSITIVE POLARITY)

CENTRAL LITHIUM ANODE

IODINE/PVP DEPOLARIZER

PVP/SUBSTRATE ANODE COATING

(c)

**FIGURE 5.16** (a) An early LiI Greatbatch heartpacemaker battery (1973) approximate size
5 × 8 cms. (b) Modern versions are much smaller. (c) construction of modern LiI battery.
(Courtesy of Wilson Greatbatch Technologies, Inc.)

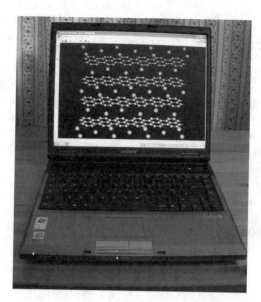

**FIGURE 5.17** Laptop computer powered by lightweight lithium-ion rechargeable battery.

Once the cell has been discharged, it can be regenerated by passing an electric current which breaks down the complex.

*Sodium Batteries*

$Na^+$ conduction has been put to good use in a secondary battery that operates at high temperatures — the **sodium sulfur battery**. This system uses either NASICON or β-alumina as the electrolyte. From Equation (5.14), to obtain a large voltage from a cell, we must have a cell reaction with a large negative Gibbs free energy change, such as a reaction between an alkali metal and a halogen might give. In terms of specific energy, such a reaction can yield about 110 Wh kg$^{-1}$ of material because it incorporates a highly energetic reaction between light substances. Such reactive materials have to be separated by an electrolyte impermeable to electrons but which can be crossed by ions. The electrolyte separates molten sodium from molten sulfur, and at the sulfur/electrolyte interface a complex reaction forming polysulfides of sodium takes place. The heat required to maintain the cell at the operating temperature of 300°C is supplied by the actual cell reaction. The cell reaction is given here for one such reaction (Figure 5.15):

A　　　　　B　　　　　C
Na(l)//　β-alumina //S(l) and graphite


*Anode A*: $2Na(l) = 2Na^+ + 2e^-$
*Cathode C*: $2Na^+ + 5S(l) + 2e^- = Na_2S_5(l)$
*Overall reaction*: $2Na(l) + 5S(l) = Na_2S_5(l)$

Later, low polysulfides are formed, and the discharge is terminated at a composition of about $Na_2S_3$. Despite the complexity of the reactions, applying a current from an external source can reverse the electrode process. For many years, there was intense interest in the development of this lightweight, high energy system for powering electric cars, but interest has now largely waned due to the stringent safety features needed (they contain highly reactive and corrosive chemicals at 300°C), together with reliability problems.

Currently interest has now been directed toward a similar high temperature system, the **ZEBRA Battery**, which also uses β-alumina as a $Na^+$ ion conductor. The sulfur electrode is replaced by nickel chloride or by a mixture of ferrous and nickel chlorides. Contact between the $NiCl_2$ electrode and the solid electrolyte is poor as they are both solids, and current flow is improved by adding a second liquid electrolyte (molten $NaAlCl_4$) between this electrode and the β-alumina. The overall cell reaction is now:

$$2Na + NiCl_2 = Ni + 2NaCl$$

The high specific energy for this cell of >100 Wh $kg^{-1}$ give electric vehicles powered by these batteries a range of up to 250 km, sufficient for the daily use in a city. These batteries are fully rechargeable, safe, and have been found to need no maintenance over 100 000 km; they are thus considered an attractive proposition for the electric vehicles of the future.

## Fuel Cells

Fuel cells work on the same principles as a battery (Figure 5.18) — the difference is that instead of the 'fuel' for the reaction being contained in the electrode materials as

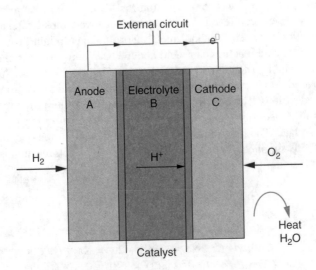

**FIGURE 5.18** Schematic diagram of a fuel cell.

in a battery, it is now fed in externally to the electrodes. This has the huge advantage that the cell can operate continuously as long as the fuel is available, unlike a battery which either has to be thrown away once the battery material is exhausted (primary) or recharged by plugging in to an electricity supply for several hours (secondary/rechargeable) to reverse the cell reaction. In a fuel cell, the fuels used are usually hydrogen and oxygen (air), which react together electrochemically to produce water, electricity, and heat. $H_2$ is fed to the anode where it is oxidized to $H^+$ ions and electrons. The electrons travel round the external circuit, and the $H^+$ ions travel through the electrolyte to the cathode, where they react with a supply of $O_2$. Another other great advantage of fuel cells, is that the by-products of water and heat are both potentially useful and nonpolluting. The fact that hydrogen and oxygen combine at low temperatures compared with normal combustion, means that side-reactions producing polluting $NO_x$s are avoided. Fuel cells operate with an efficiency of 50% or more. This compares with 15 to 20% for internal combustion engines and 30% for diesel engines.

Fuel cells are currently being intensively developed as they have the potential to provide power in a relatively nonpolluting fashion. Legislation in the United States requires that a percentage of all new vehicles should emit no hydrocarbons or oxides of nitrogen (so-called *zero emission vehicles*). The current internal combustion engine cannot meet such stringent demands and so alternatives have to be found. The main contenders are electric cars which run on either batteries or fuel cells, or a combination of the two. Current developments now include not only fuel-cell-driven buses and cars, but also power sources for homes and factories. Micro-fuel cells for mobile phones and laptops have been developed.

Notwithstanding all the advantages of fuel cells, complications and drawbacks to their use also exist. The reduction of oxygen at the cathode is rather slow at low temperatures. To increase the rate of reaction, an expensive Pt catalyst is incorporated into the carbon electrodes. If the electrolyte allows the passage of $H^+$ ions, the cell reaction can be written:

**A**                **B**                **C**

$H_2(g)//Pt/C$ electrode//   hydrogen electrolyte   //Pt/C electrode//$O_2(g)$

*Anode A*: $H_2 = 2H^+ + 2e^-$
*Cathode C*: $\frac{1}{2} O_2 + 2H^+ + 2e^- = H_2O$

The theoretical emf for this cell, calculated from the Gibbs function for the decomposition of water, is $E^{\ominus} = 1.229$ V at 298 K, but this decreases with temperature to about 1 V at 500 K. A compromise in cell design is therefore always needed between the voltage generated and an operating temperature high enough to maintain a fast reaction.

The oxygen supply comes simply from the air, but a major drawback is the supply, transportation and storage of the hydrogen. The current infrastructure of fuel stations is for the supply of liquid fuel: petrol (gasoline) and diesel. Hydrogen, a gas at room temperature, is the least dense of all the elements and has to be compressed or liquefied in order to store it in a manageable volume. Compressed hydrogen needs a strong tank that can withstand the high pressures involved; these cylinders are very heavy. Liquefied hydrogen requires cooling and must be kept very cold (boiling point is −253°C), so the tanks must be extremely well insulated. Research is under way to develop materials, such as carbon nanotubes, which will store large amounts of hydrogen and release it when needed. Although hydrogen is probably less dangerous than a liquid hydrocarbon fuel, public perception about its safety needs to be overcome: many people have seen old footage of the hydrogen-filled Hindenberg airship exploding on landing.

Hydrogen also has to be generated; very pure hydrogen, required by some fuel cells, is produced electrolytically, which is an expensive process unless cheap sources of electricity can be found, such as solar energy or hydroelectric power. The alternative is to use a **reforming reaction**, where hydrogen-rich sources, such as methane or methanol, are reacted with steam to produce hydrogen and carbon dioxide. To avoid the transport and storage problems, hydrogen can also be produced *in situ* using a reformer and fed directly into the fuel cell. The reforming reaction has its own problems, in that the catalyst is poisoned by sulfur in the fuel. In addition, small quantities of carbon monoxide are produced along with the hydrogen in the reforming reaction, and this also poisons the Pt catalyst in the fuel cell thereby reducing its efficiency.

Fuel cells have been around for quite a long time; it was in 1839 that William Grove, a Welsh physicist, made the first working fuel cell, but not until 1959 that Tom Bacon at Cambridge University produced a stack of 40 alkaline fuel cells that produced 5 kW of power. Around the same time, Willard Grubb and Leonard Niedrach at General Electric developed a conducting membrane fuel cell with a Pt catalyst on a Ti gauze which was used in the Gemini earth-orbit space programme. By 1965, Pratt & Whitney had improved (longer life) the alkaline fuel cells (AFC) for use in the Apollo missions where they provided both power and drinking water for the astronauts. In 1983, Ballard Power (Canada) was established, and in 1993, unveiled the first fuel cell buses which went into service in Vancouver and Chicago. 2004 saw the first fuel-cell powered buses in London (Figure 5.19).

**FIGURE 5.19** A zero-emission Mercedes Citaro bus operating in London, powered by a Ballard® fuel-cell engine. (Courtesy of Ballard Power Systems.)

Several types of fuel cell are currently under development, using different electrolyte systems: phosphoric acid (PAFC), alkaline, molten carbonate (MCFC), regenerative, zinc-air, protonic ceramic, (PCFC), proton exchange membrane (PEM), direct methanol (DMFC), and solid oxide (SOFC). The last four contain solid electrolytes.

*Solid Oxide Fuel Cells (SOFCs)*

SOFCs employ a ceramic oxide (ceria- or yttria-doped zirconia, $Y_2O_3/ZrO_2$) electrolyte which becomes $O^{2-}$ conducting at very high temperatures (800–1000°C); this system has the disadvantage of needing energy to heat the cell, but the advantage that at this temperature, reforming and $H^+$ production can take place internally without the need for expensive Pt catalysts. Because of the long time needed to get the cell up to its operating temperature, it is not useful for powering vehicles, but it can be used for power generation in buildings and industry (Figure 5.20). The cell reaction is:

$$\begin{array}{ccc} \text{A} & \text{B} & \text{C} \end{array}$$
$$H_2(g) \ //\text{electrode}// \ \text{solid oxide electrolyte} \ //\text{electrode}// \ O_2(g)$$

*Anode A*: $H_2 + O^{2-} = H_2O + 2e^-$
*Cathode C*: $^1/_2 O_2 + 2e^- = O^{2-}$

It is vital that the solid oxide electrolyte can withstand the extreme conditions of hydrogen at the anode at 800°C or above. Under these conditions, many oxides would be reduced, liberating electrons and thus leading to unwanted electronic conductivity.

The cathode materials used have to conduct both oxide ions and electrons satisfactorily, but, in addition, for compatibility, they must have similar thermal expansion coefficients as the electrolyte. The strontium-doped perovskite, LSM (see Section 5.4.2), is one of the materials of choice.

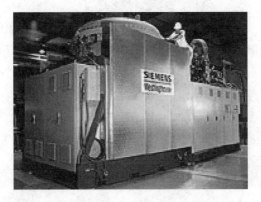

**FIGURE 5.20** Industrial-sized 220 kW solid oxide fuel cell made by Siemens. (Courtesy of Siemens PG CTET: Stationary Fuel Cells, Europe.)

*Proton Exchange Membrane Fuel Cells (PEM)*

There is currently great interest in the PEM cells which operate at much lower temperatures (80°C). Here the electrolyte is a conducting polymer membrane (see Chapter 6), usually Nafion, which is a sulfonated fluoropolymer made by Dupont, strengthened by Gore-Tex™. The strongly acidic $-SO_2OH$ group allows the passage of $H^+$ ions, but not of atoms or electrons. Output is typically 1 V at 80°C; with a current flow of 0.5 A cm$^{-2}$, this drops to 0.5 V because of ohmic losses. A membrane of 1 m$^2$ provides about 1 kW. To produce the correct power output, a number of cells are placed together to form a **stack**. Most car manufacturers now have prototype zero emission vehicles (Figure 5.21), and the first hydrogen fuel-cell consumer vehicle — a Nissan 4x4 pick-up truck — went on sale in the USA in 2005.

250 kilowatt fuel cells are made which provides enough heat and electricity to power industry, and indeed for establishments needing high reliability and backup power supplies, such as banks and hospitals (Figure 5.20). Smaller systems (7 kW), about the size of a refrigerator, are produced that can provide all the power necessary for a house and the heat produced can also be harnessed to provide hot water. Fuel cells are still used to power the space shuttle. Because of a European Union initiative, DaimlerChrysler built between 20 and 30 fuel cell buses for use in eight European cities, three of which were destined for London (2004). The buses run on compressed hydrogen stored in tanks in the roof with a power of 200 kW. The cylinders are recharged from a hydrogen filling station set up by British Petroleum for the project.

**FIGURE 5.21** Honda Motor Company's hydrogen-fuelled FCX fuel-cell car with a range of 170 miles and a top speed of 93 mph. (Courtesy of Ballard Power Systems.)

Each bus has a range of up to 300 km, a top speed of 80 km per hour, and carries 70 passengers.

On a smaller scale, microfuel cells might soon replace batteries in electrical equipment such as mobile phones and palmtop computers. These cells can be replenished easily and quickly by adding more fuel in the form of a methanol capsule.

Although fuel cells offer exciting possibilities for the future, the current reality is that the electricity produced by them is still expensive — anything up to eight times that produced in a traditional generating plant with gas-fired turbines. Fuel cells also need to improve in reliability with better catalysts which are not poisoned by emissions from the reformer. There is no doubt, however, that they are strong contenders for the technology of the future, offering as they do the possibility to move toward a greener 'hydrogen economy', thus eliminating greenhouse gas emissions.

### Sensors

Calcia-stabilized zirconia is used in the detection of oxygen, in both **oxygen meters** and **oxygen sensors**. Figure 5.22 depicts a slab of calcia-stabilized zirconia acting as the solid electrolyte, B, which separates two regions containing oxygen at different pressures. Gas pressures tend to equalize if they can and so, if $p' > p''$, oxygen ions, which are able to pass through the stabilized zirconia, tend to pass through the solid from the right-hand side to the left. This tendency produces a potential difference (because the ions are charged), indicating that oxygen is present (in the sensor) and measurement of this potential gives a measure of the oxygen pressure difference (in the oxygen meter).

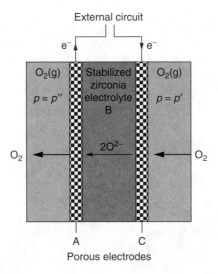

**FIGURE 5.22** Schematic representation of an oxygen meter.

Oxygen gas is reduced to $O^{2-}$ at the right-hand electrode (C). The oxide ions are able to pass through the doped zirconia and are oxidised to oxygen gas at the left-hand electrode (A). The equation for the cell reaction is:

**Anode A:** $2O^{2-} \rightarrow O_2(p'') + 4e^-$
**Cathode C:** $O_2(p') + 4e^- \rightarrow 2O^{2-}$
**Overall:** $O_2(p') \rightarrow O_2(p'')$

Under standard conditions, we can relate the change in Gibb's free energy for the preceding reaction to the standard emf of the cell:

$$\Delta G^{\ominus} = -nE^{\ominus}F \tag{5.14}$$

The **Nernst equation** allows calculation of the cell *emf* under nonstandard conditions, $E$. If the cell reaction is given by a general equation:

$$aA + bB + \ldots + ne = xX + yY + \ldots$$

then,

$$E = E^{\ominus} - \frac{2.303RT}{nF} \log \left\{ \frac{a_X^x a_Y^y \ldots}{a_A^a a_B^b \ldots} \right\} \tag{5.15}$$

where the quantities $a_X$, etc. are the **activities** of the reactants and products. Applying the Nernst equation to the cell reaction in the oxygen meter, we obtain:

$$E = E^{\ominus} - \frac{2.303RT}{4F} \log \frac{p''}{p'}$$

In this particular case, $E^{\ominus}$ is actually zero because under standard conditions, the pressure of the oxygen on each side would be 1 atm, and there would be no net potential difference. The pressure of the oxygen on one side of the cell (for instance, $p''$) is set to be a known reference pressure, usually either pure oxygen at 1 atm or atmospheric oxygen pressure (~0.21 atm). Making these two changes, we obtain:

$$E = \frac{2.303RT}{4F} \log \frac{p'}{P_{ref}}$$

All the quantities in this equation are now known or can be measured, affording a direct measure of the unknown oxygen pressure $p'$. For this cell to operate, there must be no *electronic* conduction through the electrolyte.

Oxygen meters find industrial uses in the detection of oxygen in waste gases from chimneys or exhaust pipes, in investigation into the operation of furnaces, and in measuring the oxygen content in molten metals during the production process. This same principle has been employed in developing sensors for other gases by using different electrolytes. Examples include $H_2$, $F_2$, $Cl_2$, $CO_2$, $SO_x$, and $NO_x$.

## Electrochromic devices

Electrochromic devices work in the opposite sense to an electrochemical cell: instead of harnessing the chemical reaction between the electrodes to give an electric current, an electric current is applied *to* the cell, causing the movement of ions through the electrolyte and creating a coloured compound in one of the electrodes. When an electric current is applied to the following device (Figure 5.23):

$$\begin{matrix} \mathbf{A} & & \mathbf{B} & & \mathbf{C} \\ LiCoO_2 & // & LiNbO_3 & // & WO_3 \end{matrix}$$

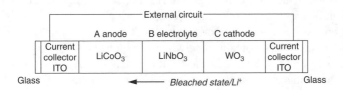

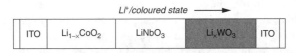

(a)

(b)

**FIGURE 5.23** (a) Diagram of an electrochromic device; (b) electrochromic office windows. (Courtesy of Pilkington plc.)

$Li^+$ ions flow from the anode, through the colourless electrolyte to form $Li_xWO_3$ at the cathode, changing it from almost colourless to deep blue. This gives the potential for interesting 'smart' devices, such as windows, which can be switched from clear to coloured to control temperature in buildings and cars.

## 5.5  PHOTOGRAPHY

A photographic emulsion consists of very small crystallites of AgBr (or AgBr-AgI) dispersed in gelatin. This is usually supported on paper or thin plastic to form photographic film. The crystallites are usually small triangular or hexagonal platelets, known as **grains**. They are grown very carefully *in situ* with as few structural defects as possible, and they range in size from 0.05 to $2 \times 10^{-6}$ m. During the photographic process, light falling on the AgBr produces Ag atoms in some of the grains; these eventually form the dark parts of the negative. The grains which are affected by the light, contain the so-called **latent image**. It is important for the grains to be free from structural defects such as dislocations and grain boundaries, because these interfere with the deposition of the Ag atoms on the surface of the grains. However, the formation of the latent image is dependent on the *presence* of point defects.

AgBr and AgI both have the NaCl or rock salt crystal structure. However, unlike the alkali halides, which contain mainly Schottky defects, AgBr has been shown to contain mostly Frenkel defects in the form of interstitial $Ag^+$ ions. For a grain to possess a latent image, a cluster of as few as *four* Ag atoms forms on the surface. The formation of the clusters of Ag atoms is a complex process that is still not fully understood. However, it is thought to take place in several stages. The first stage is when light strikes one of the AgBr crystals, and an electron is promoted from the valence band to the conduction band. The band gap of AgBr is 2.7 eV, so the light absorbed is visible from the extreme blue end of the spectrum. This electron eventually neutralises one of the interstitial silver ions ($Ag_i^+$):

$$Ag_i^+ + e^- = Ag$$

In the next stages, this Ag atom speck has to grow into a cluster of atoms on the surface of the crystal. A possible sequence of events for this is:

$$Ag + e^- = Ag^-$$

$$Ag^- + Ag_i^+ = Ag_2$$

$$Ag_2 + Ag_i^+ = Ag_3^+$$

$$Ag_3^+ + e^- = Ag_3$$

$$Ag_3 + e^- = Ag_3^-$$

$$Ag_3^- + Ag_i^+ = Ag_4 + \ldots \ldots \ldots$$

Notice that only the odd numbered clusters appear to interact with the electrons.

In reality, the process is even more complex than this because emulsions made from pure AgBr are not sensitive enough, and so they also contain **sensitizers**, such as sulfur or organic dyes, which absorb light of longer wavelength than AgBr and so extend the spectral range. The sensitizers form traps for the photoelectrons on the surfaces of the grains; these electrons then transfer from an excited energy level of the sensitizer to the conduction band of AgBr.

The film containing the latent image is then treated with various chemicals to produce a lasting negative. First of all, it is developed: a reducing agent, such as an alkaline solution of hydroquinone, is used to reduce the AgBr crystals to Ag. The clusters of Ag atoms act as a catalyst to this reduction process, and all the grains with a latent image are reduced to Ag. The process is rate controlled, so the grains that have not reacted with the light are unaffected by the developer (unless the film is developed for a very long time, when eventually they will be reduced and a fogged picture results). The final stage in producing a negative is to dissolve out the remaining light-sensitive AgBr. This is done using 'hypo' – sodium thiosulfate $(Na_2S_2O_3)$, which forms a water soluble complex with $Ag^+$ ions.

## 5.6 COLOUR CENTRES

During early research in Germany, it was noticed that if crystals of the alkali halides were exposed to X-rays, they became brightly coloured. It was thought that the colour was associated with a defect known then as a **Farbenzentre (colour centre)**, now abbreviated as **F-centre**. Since then, it has been found that many forms of high energy radiation (UV, X-rays, neutrons) will cause F-centres to form. The colour produced by the F-centres is always characteristic of the host crystal, so, for instance, NaCl becomes deep yellowy-orange, KCl becomes violet, and KBr becomes blue-green.

Subsequently, it was found that F-centres can also be produced by heating a crystal in the vapour of an alkali metal: this gives a clue to the nature of these defects. The excess alkali metal atoms diffuse into the crystal and settle on cation sites; at the same time, an equivalent number of anion site vacancies are created, and ionisation gives an alkali metal cation with an electron trapped at the anion vacancy (Figure 5.24). In fact, it does not even matter which alkali-metal is used; if NaCl is heated with potassium, the colour of the F-centre does not change because

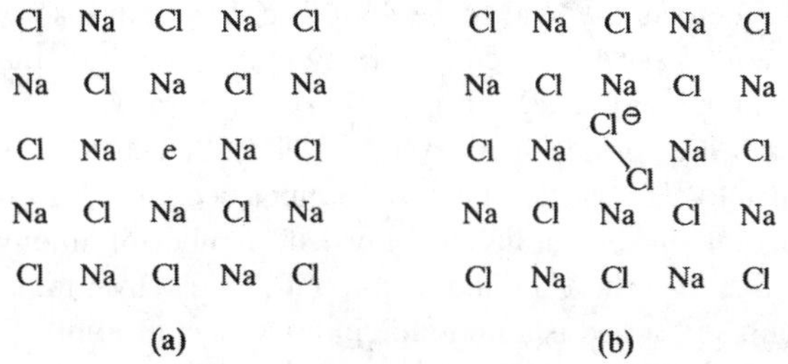

| Cl | Na | Cl | Na | Cl |     | Cl | Na | Cl | Na | Cl |
|----|----|----|----|----|-----|----|----|----|----|----|
| Na | Cl | Na | Cl | Na |     | Na | Cl | Na | Cl | Na |
| Cl | Na | e  | Na | Cl |     | Cl | Na |    | Na | Cl |
| Na | Cl | Na | Cl | Na |     | Na | Cl | Na | Cl | Na |
| Cl | Na | Cl | Na | Cl |     | Cl | Na | Cl | Na | Cl |

(a)                                (b)

**FIGURE 5.24** (a) The F-centre, an electron trapped on an anion vacancy; (b) H-centre.

it is characteristic of the electron trapped at the anion vacancy in the host halide. Work with Electron Spin Resonance spectroscopy, ESR, has confirmed that F-centres are indeed unpaired electrons trapped at vacant lattice (anion) sites.

The trapped electron provides a classic example of an 'electron in a box'. A series of energy levels are available for the electron, and the energy required to transfer from one level to another falls in the visible part of the electromagnetic spectrum, hence the colour of the F-centre. There is an interesting natural example of this phenomenon: The mineral fluorite ($CaF_2$) is found in Derbyshire, United Kingdom where it is known as 'Blue John', and its beautiful blue-purple colouration is due to the presence of F-centres.

Many other colour centres have now been characterized in alkali halide crystals. The **H-centre** is formed by heating, for instance, NaCl in $Cl_2$ gas. In this case, a $[Cl_2]^-$ ion is formed and occupies a single anion site (Figure 5.24(b)). F-centres and H-centres are perfectly complementary — if they meet, they cancel one another out!

Another interesting natural example of colour centres lies in the colour of smoky quartz and amethyst. These semi-precious stones are basically crystals of silica, $SiO_2$, with some impurities present. In the case of smoky quartz, the silica contains a little aluminium impurity. The $Al^{3+}$ substitutes for the $Si^{4+}$ in the lattice, and the electrical neutrality is maintained by $H^+$ present in the same amount as $Al^{3+}$. The colour centre arises when ionising radiation interacts with an $[AlO_4]^{5-}$ group, liberating an electron which is then trapped by $H^+$:

$$[AlO_4]^{5-} + H^+ = [AlO_4]^{4-} + H$$

The $[AlO_4]^{4-}$ group is now electron-deficient and can be considered as having a 'hole' trapped at its centre. This group is the colour centre, absorbing light and producing the smoky colour. In crystals of amethyst, the impurity present is $Fe^{3+}$. On irradiation, $[FeO_4]^{4-}$ colour centres are produced which absorb light to give the characteristic purple coloration.

## 5.7 NON-STOICHIOMETRIC COMPOUNDS

### 5.7.1 INTRODUCTION

Previous sections of this chapter have shown that it is possible to *introduce* defects into a perfect crystal by adding an impurity. Such an addition causes point defects of one sort or another to form, but they no longer occur in complementary pairs. Impurity-induced defects are said to be **extrinsic**. We have also noted that when assessing what defects have been created in a crystal, it is important to remember that the overall charge on the crystal must always be zero.

Colour centres are formed if a crystal of NaCl is heated in sodium vapour; sodium is taken into the crystal, and the formula becomes $Na_{1+x}Cl$. The sodium atoms occupy cation sites, creating an equivalent number of anion vacancies; they subsequently ionize to form a sodium cation with an electron trapped at the anion vacancy. The solid so formed is a **non-stoichiometric compound** because the ratio of the atomic components is no longer the simple integer that we have come to expect for well-characterized compounds. A careful analysis of many substances, particularly inorganic solids, demonstrates that it is common for the atomic ratios to be non-integral. Uranium dioxide, for instance, can range in composition from $UO_{1.65}$ to $UO_{2.25}$, certainly not the perfect $UO_2$ that we might expect! Many other examples exist, some of which we discuss in some detail.

What kind of compounds are likely to be non-stoichiometric? 'Normal' covalent compounds are assumed to have a fixed composition where the atoms are usually held together by strong covalent bonds formed by the pairing of two electrons. Breaking these bonds usually takes quite a lot of energy, and so under normal circumstances, a particular compound does not show a wide range of composition; this is true for most molecular organic compounds, for instance. Ionic compounds also are usually stoichiometric because to remove or add ions requires a considerable amount of energy. We have seen, however, that it is possible to make ionic crystals non-stoichiometric by doping them with an impurity, as with the example of Na added to NaCl. Another mechanism also exists, whereby ionic crystals can become non-stoichiometric: if the crystal contains an element with a variable valency, then a change in the number of ions of that element can be compensated by changes in ion charge; this maintains the charge balance but alters the stoichiometry. Elements with a variable valency mostly occur in the transition elements, the lanthanides and the actinides.

In summary, non-stoichiometric compounds can have formulae that do not have simple integer ratios of atoms; they also usually exhibit a range of composition. They can be made by introducing impurities into a system, but are frequently a consequence of the ability of the metal to exhibit variable valency. Table 5.5 lists a few non-stoichiometric compounds together with their composition ranges.

## TABLE 5.5
### Approximate composition ranges for some non-stoichiometric compounds

| Compound | | Composition range[a] |
|---|---|---|
| $TiO_x$ | [≈TiO] | $0.65 < x < 1.25$ |
| | [≈$TiO_2$] | $1.998 < x < 2.000$ |
| $VO_x$ | [≈VO] | $0.79 < x < 1.29$ |
| $Mn_xO$ | [≈MnO] | $0.848 < x < 1.000$ |
| $Fe_xO$ | [FeO] | $0.833 < x < 0.957$ |
| $Co_xO$ | [≈CoO] | $0.988 < x < 1.000$ |
| $Ni_xO$ | [≈NiO] | $0.999 < x < 1.000$ |
| $CeO_x$ | [≈$Ce_2O_3$] | $1.50 < x < 1.52$ |
| $ZrO_x$ | [≈$ZrO_2$] | $1.700 < x < 2.004$ |
| $UO_x$ | [≈$UO_2$] | $1.65 < x < 2.25$ |
| $Li_xV_2O_5$ | | $0.2 < x < 0.33$ |
| $Li_xWO_3$ | | $0 < x < 0.50$ |
| $TiS_x$ | [≈TiS] | $0.971 < x < 1.064$ |
| $Nb_xS$ | [≈NbS] | $0.92 < x < 1.00$ |
| $Y_xSe$ | [≈YSe] | $1.00 < x < 1.33$ |
| $V_xTe_2$ | [≈$VTe_2$] | $1.03 < x < 1.14$ |

[a] Note that all composition ranges are temperature dependent and the figures here are intended only as a guide.

Until recently, defects both in stoichiometric and non-stoichiometric crystals were treated entirely from the point of view that point defects are randomly distributed. However, isolated point defects are not scattered at random in non-stoichiometric compounds but are often dispersed throughout the structure in some kind of regular pattern. The following sections attempt to explore the relationship between stoichiometry and structure.

It is difficult to determine the structure of compounds containing defects, and it is only very recently that much of our knowledge has been formed. X-ray diffraction is the usual method for the determination of the structure of a crystal; however, this method yields an *average* structure for a crystal. For pure, relatively defect-free structures, this is a good representation but for non-stoichiometric and defect structures it avoids precisely the information that you want to know. For this kind of structure determination, a technique that is sensitive to *local* structure is needed, and such techniques are very scarce. Structures are often elucidated from a variety of sources of evidence: X-ray and neutron diffraction, density measurements, spectroscopy (when applicable), and high resolution electron microscopy (HREM); magnetic measurements have also proved useful in the case of FeO. HREM has probably done the most to clarify the understanding of defect structures because it is capable under favourable circumstances of giving information on an atomic scale by 'direct lattice imaging'.

Non-stoichiometric compounds are of potential use to industry because their electronic, optical, magnetic, and mechanical properties can be modified by changing the proportions of the atomic constituents. This is widely exploited and researched by the electronics and other industries. Currently, the best known example of non-stoichiometry is probably that of oxygen vacancies in the high temperature superconductors such as YBCO (1-2-3) ($YBa_2Cu_3O_{7-x}$). The structure of these is discussed in detail in Chapter 10.

## 5.7.2 NON-STOICHIOMETRY IN WUSTITE

Ferrous oxide is known as **wustite (FeO)**, and it has the NaCl (rock salt) crystal structure. Accurate chemical analysis demonstrates that it is non-stoichiometric: it is always deficient in iron. The FeO phase diagram (Figure 5.25) illustrates that the compositional range of wustite increases with temperature and that stoichiometric FeO is *not* included in the range of stability. Below 570°C, wustite disproportionates to $\alpha$-iron and $Fe_3O_4$.

An iron deficiency could be accommodated by a defect structure in two ways: either *iron vacancies*, giving the formula $Fe_{1-x}O$, or alternatively, there could be an *excess of oxygen in interstitial positions*, with the formula $FeO_{1+x}$. A comparison of the theoretical and measured densities of the crystal distinguishes between the alternatives. The easiest method of measuring the density of a crystal is the flotation method. Liquids of differing densities which dissolve in each other, are mixed together until a mixture is found that will just suspend the crystal so that it neither floats nor sinks. The density of that liquid mixture must then be the same as that of the crystal, and it can be found by weighing an accurately measured volume.

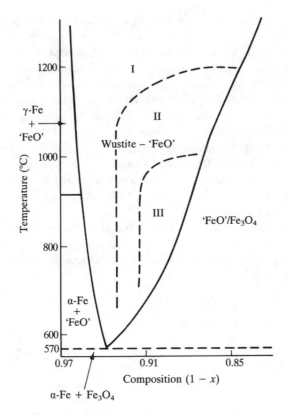

**FIGURE 5.25** Phase diagram of the FeO system. I, II, and III together comprise the wustite region.

The theoretical density of a crystal can be obtained from the volume of the unit cell and the mass of the unit cell contents. The results of an X-ray diffraction structure determination gives both of these data, as the unit cell dimensions are accurately measured and the type and number of formula units in the unit cell are also determined. An example of this type of calculation for FeO follows:

A particular crystal of FeO was found to have a unit cell dimension of 430.1 pm, a measured density of 5.728 kg m$^{-3}$, and an iron to oxygen ratio of 0.945. The unit cell volume (which is a cube) is thus $(430.1 \text{ pm})^3 = 7.956 \times 10^7 \text{ (pm)}^3 = 7.956 \times 10^{-29} \text{ m}^3$.

There are four formula units of FeO in a perfect unit cell with the rock salt structure. The mass of these four units can be calculated from their relative atomic masses: Fe, 55.85 and O, 16.00. One mole of FeO weighs $(55.85 + 16.00)$ g $= 0.07185$ kg; four moles weigh $(4 \times 0.07185)$ kg; and four formula units weigh $(4 \times 0.07185)/N_A$ kg $= 4.773 \times 10^{-25}$ kg, where **Avogadro's number**, $N_A = 6.022 \times 10^{23}$ mol$^{-1}$.

The sample under consideration has an Fe:O ratio of 0.945. Assume, in the first instance, that it has *iron vacancies*: The unit cell contents in this case will be

(4 × 0.945) = 3.78Fe and 4O. The mass of the contents will be [(3.78 × 55.85) + (4 × 16.00)]/($N_A$ × $10^3$) kg. Dividing by the volume of the unit cell, we obtain a value of 5.742 × $10^3$ kg m$^{-3}$ for the density. If instead the sample possesses *interstitial oxygens*, the ratio of oxygens to iron in the unit cell will be given by 1/0.945 = 1.058. The unit cell in this case will contain 4Fe and (4 × 1.058) = 4.232 O. The mass of this unit cell is given by: [(4 × 55.85) + (4.232 × 16.00)]/($N_A$ × $10^3$) kg, giving a density of 6.076 × $10^3$ kg m$^{-3}$. Comparing the two sets of calculations with the experimentally measured density of 5.728 × $10^3$ kg m$^{-3}$, it is clear that this sample contains *iron vacancies* and that the formula should be written as $Fe_{0.945}O$. A table of densities is drawn up for FeO in Table 5.6.

It is found to be characteristic of most non-stoichiometric compounds that *their unit cell size varies smoothly with composition but the symmetry is unchanged*. This is known as **Vegard's Law**.

In summary, non-stoichiometric compounds are found to exist over a range of composition, and throughout that range the unit cell length varies smoothly with no change of symmetry. It is possible to determine whether the non-stoichiometry is accommodated by vacancy or interstitial defects using density measurements.

## Electronic Defects in FeO

The discussion of the defects in FeO has so far been only structural. Now we turn our attention to the balancing of the charges within the crystal. In principle the compensation for the iron deficiency can be made either by oxidation of some Fe(II) ions *or* by reduction of some oxide anions. It is energetically more favourable to oxidise Fe(II). For each $Fe^{2+}$ vacancy, *two* $Fe^{2+}$ cations must be oxidised to $Fe^{3+}$. In the overwhelming majority of cases, defect creation involves changes in the cation oxidation state. In the case of metal excess in simple compounds, we would usually expect to find that neighbouring cation(s) would be reduced.

In a later section we will look at some general cases of non-stoichiometry in simple oxides, but before we do that we will complete the FeO story with a look at its detailed structure.

### TABLE 5.6
**Experimental and theoretical densities ($10^3$ kg m$^{-3}$) for FeO**

| | | | | Theoretical density | |
| O:Fe ratio | Fe:O ratio | Lattice parameter/pm | Observed density | Interstitial O | Fe vacancies |
|---|---|---|---|---|---|
| 1.058 | 0.945 | 430.1 | 5.728 | 6.076 | 5.742 |
| 1.075 | 0.930 | 429.2 | 5.658 | 6.136 | 5.706 |
| 1.087 | 0.920 | 428.5 | 5.624 | 6.181 | 5.687 |
| 1.099 | 0.910 | 428.2 | 5.613 | 6.210 | 5.652 |

## The Structure of FeO

FeO has the NaCl structure with $Fe^{2+}$ ions in octahedral sites. The iron deficiency manifests itself as cation vacancies, and the electronic compensation made for this is that for every $Fe^{2+}$ ion vacancy there are two neighbouring $Fe^{3+}$ ions, and this is confirmed by Mössbauer spectroscopy. If the rock salt structure were preserved, the $Fe^{2+}$, $Fe^{3+}$, and the cation vacancies would be distributed over the octahedral sites in the *ccp* $O^{2-}$ array. However, $Fe^{3+}$ is a high-spin $d^5$ ion and has no crystal field stabilization in either octahedral or tetrahedral sites, and therefore no preference. Structural studies (X-ray, neutron and magnetic) have demonstrated that some of the $Fe^{3+}$ ions are in *tetrahedral* sites. If a tetrahedral site is occupied by an $Fe^{3+}$ ion, the immediate surrounding four $Fe^{2+}$ octahedral sites must be vacant as they are too close to be occupied at the same time, and this type of defect is found for low values of $x$. At higher values of $x$, the structure contains various types of **defect clusters**, which are distributed throughout the crystal. A defect cluster is a region of the crystal where the defects form an *ordered structure*. One possibility is known as the **Koch–Cohen cluster** (Figure 5.26). This has a standard NaCl-type unit cell at the centre, but with four interstitial, tetrahedrally coordinated, $Fe^{3+}$ ions in the tetrahedral holes; the 13 immediately surrounding octahedral $Fe^{2+}$ sites must be vacant. Surrounding this central unit cell, the other octahedral iron sites in the cluster are occupied, but they may contain either $Fe^{2+}$ or $Fe^{3+}$ ions. We therefore designate them simply as $Fe_{oct}$. The front and back planes have been cut away from the diagram in Figure 5.26 to make the central section more visible.

It is instructive to consider the composition of a cluster such as the one shown in detail in Figure 5.26. Clusters such as this are often referred to by the ratio of cation vacancies to interstitial $Fe^{3+}$ in tetrahedral holes, in this case 13:4. The complete cluster, allowing for the back and front faces which are not illustrated here, contains 8 NaCl-type unit cells, and thus 32 oxide ions. A stoichiometric array with no defects would also contain 32 $Fe^{2+}$ cations on the octahedral sites. Taking into account the 13 octahedral vacancies, there must be 19 $Fe_{oct}$ ions and the 4 interstitial $Fe_{tet}^{3+}$ ions making *twenty-three* iron cations in all. The overall formula for the cluster is thus, $Fe_{23}O_{32}$, almost $Fe_3O_4$. It bears a strong resemblance to the structure of $Fe_3O_4$, the next-highest oxide of iron.

Having determined the atomic contents of the cluster we now turn our attention to the charges. There are 32 oxide ions, so to balance them, the Fe cations overall must have 64 positive charges: 12 are accounted for by the 4 $Fe^{3+}$ ions in tetrahedral positions, leaving 52 to find from the remaining 19 $Fe_{oct}$ cations. This accounting is difficult to do by inspection. Suppose that $x Fe^{2+}$ ions and $y Fe^{3+}$ ions are present in octahedral sites, we know that

$$x + y = 19$$

We also know that their total charges must equal 52, so:

$$2x + 3y = 52$$

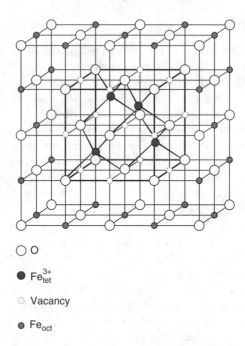

O  O

● $Fe^{3+}_{tet}$

◌  Vacancy

● $Fe_{oct}$

**FIGURE 5.26** The Koch–Cohen cluster illustrated with the back and front planes cut away for clarity. The central section with four tetrahedrally coordinated $Fe^{3+}$ ions is picked out in bold.

giving two simultaneous equations. Solving gives $y = 14$ and $x = 5$. The octahedral sites surrounding the central (bold) unit cell are thus occupied by 5 $Fe^{2+}$ ions and by 14 $Fe^{3+}$ ions. By injecting such clusters throughout the FeO structure, the non-stoichiometric structure is built up. The exact formula of the compound (the value of $x$ in $Fe_{1-x}O$) will depend on the average separation of the randomly injected clusters. Neutron scattering experiments indicate that as the concentration of defects increases, the clusters order into a regular repeating pattern with its own unit cell of lower symmetry; the new structure is referred to as a **superstructure** or **super-lattice** of the parent. In the oxygen-rich limit when the whole structure is composed of these clusters, there would be a new ordered structure of formula $Fe_{23}O_{32}$, based on the structure of the parent defect cluster.

### 5.7.3 URANIUM DIOXIDE

Above 1127°C, a single oxygen-rich non-stoichiometric phase of $UO_2$ is found with formula $UO_{2+x}$, ranging from $UO_2$ to $UO_{2.25}$. Unlike FeO, where a metal-deficient oxide was achieved through cation vacancies, in this example the metal-deficiency arises from interstitial anions.

$UO_{2.25}$ corresponds to $U_4O_9$, which is a well-characterized oxide of uranium known at low temperature. $UO_2$ has the fluorite structure. The unit cell is depicted in Figure 5.27(a) and contains four formula units of $UO_2$. (There are four uranium ions contained within the cell boundaries; the eight oxide ions come from: $(8 \times \frac{1}{8}) = 1$ at the corners;

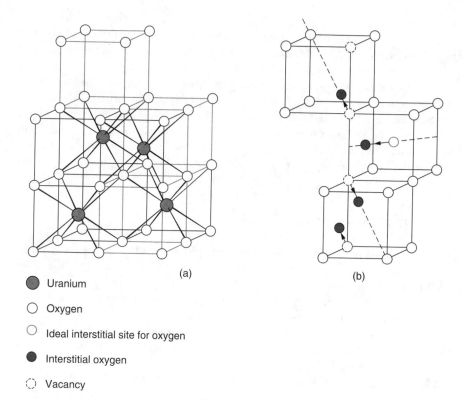

● Uranium

○ Oxygen

○ Ideal interstitial site for oxygen

● Interstitial oxygen

⊙ Vacancy

**FIGURE 5.27** (a) The fluorite structure of $UO_2$ with a unit cell marked in bold. (b) Interstitial defect cluster in $UO_{2+x}$. Uranium positions (not shown) are in the centre of every other cube.

$(6 \times 1/2) = 3$ at the face centres; $(12 \times 1/4) = 3$ at the cell edges; and 1 at the cell body centre.)

As more oxygen is taken into $UO_2$, the extra oxide ions go into interstitial positions. The most obvious site available is in the middle of one of the octants (the vacant octahedral holes) where there is no metal atom. This site however, is not ideal for an extra oxide ion, as not only is it crowded, but it is also surrounded by eight ions of the same charge. Neutron diffraction demonstrates that an interstitial oxide anion does not sit exactly in the centre of an octant but is displaced sideways; this has the effect of moving two other oxide ions from their lattice positions by a very small amount, leaving two vacant lattice positions. This is illustrated in Figure 5.27(b) where three vacant octants are picked out, and the positions of one additional interstitial oxide and the two displaced oxides with their vacancies are depicted. This defect cluster can be considered as two vacancies: one interstitial of one kind, $O'$, and two of another, $O''$, and it is called the 2:1:2 **Willis cluster**. The movement of the ions from 'ideal' positions is depicted by small arrows: The movement of the interstitial oxide $O'$ from the centre of an octant is along the direction of a diagonal of one of the cube faces (the *110* direction), whereas the movement of the oxide ions $O''$ on lattice positions is along cube diagonals (the *111* direction). The atomic composition of a fluorite unit cell (Figure 5.27(a)) when modified by this defect cluster is $U_4O_9$; the net oxygen gain is just the single new interstitial in the central octant of Figure 5.27(b). A more commonly found defect cluster contains two vacancies, two interstitial oxides, O′, and two interstitial oxides, O″, leading to a 2:2:2 Willis cluster (not illustrated here). In the oxygen-rich limit for the $UO_{2+x}$ non-stoichiometric structure ($UO_{2.25}$), these are ordered throughout the structure, and the structure has a very large unit cell based on $4 \times 4 \times 4$ fluorite unit cells (with a volume of 64 times that of the fluorite unit cell). We can think of $UO_{2+x}$ as containing **microdomains** of the $U_4O_9$ structure within that of $UO_2$. The electronic compensation for the extra interstitial oxide ions will most likely be the oxidation of neighbouring U(IV) atoms to either U(V) or U(VI).

### 5.7.4 THE TITANIUM MONOXIDE STRUCTURE

Titanium and oxygen form non-stoichiometric phases which exist over a range of composition centered about the stoichiometric 1:1 value, from $TiO_{0.65}$ to $TiO_{1.25}$. We shall look at what happens in the upper range from $TiO_{1.00}$ to $TiO_{1.25}$.

At the stoichiometric composition of $TiO_{1.00}$ the crystal structure can be thought of as a NaCl-type structure with vacancies in both the metal and the oxygen sub-lattices: one-sixth of the titaniums and one-sixth of the oxygens are missing. Above 900°C, these vacancies are randomly distributed, but below this temperature, they are ordered as shown in Figure 5.28.

In Figure 5.28(a) we show a layer through an NaCl-type structure. Every *third* vertical diagonal plane has been picked out by a dashed line. In the $TiO_{1.00}$ structure, every other atom along those dashed lines is missing. This is illustrated in Figure 5.28(b). If we consider that in these first two diagrams we are looking at the structure along the y-axis, and that this layer is the top of the unit cells, b = 0 or 1, then the layer below this and parallel to it will be the central horizontal plane of the unit cell at b = $^1/_2$. This is drawn in Figure 5.28(c), and again we notice that every other atom along every third diagonal plane is missing. This is true throughout the structure. In the figure, the unit cell of a perfect NaCl type structure is marked on (a), whereas the boundaries of the new unit cell, taking the ordered defects into account, are marked on (b) and (c). The new unit cell of the superlattice is **monoclinic** (see Chapter 1) because the angle in the xz plane ($\beta$) is not equal to 90°. This structure is unusual in that it appears to be stoichiometric, but, in fact, contains defect vacancies on both the anion and cation sublattices.

As discussed in Chapter 4, unusually for a transition metal monoxide, $TiO_{1.00}$ demonstrates metallic conductivity. The existence of the vacant sites within the TiO structure is thought to permit sufficient contraction of the lattice that the 3d orbitals on titanium overlap, thus broadening the conduction band and allowing electronic conduction.

When titanium monoxide has the limiting formula $TiO_{1.25}$, it has a different defect structure, still based on the NaCl structure, but with *all* the oxygens present, and one in every five titaniums missing (Figure 5.29). The pattern of the titanium vacancies is shown in Figure 5.30, which is a layer of the type in Figure 5.28 but

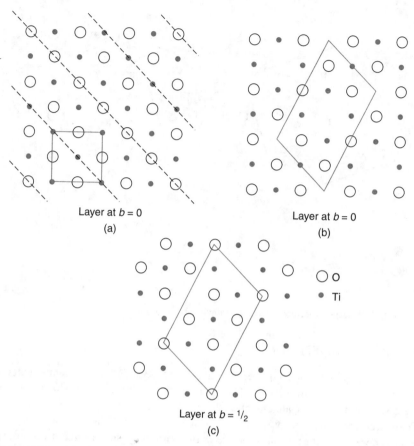

**FIGURE 5.28** Layers parallel to the horizontal planes of Figure 1.31. (a) The hypothetical TiO structure of the NaCl type shown in Figure 1.31; the line of intersection of every third vertical diagonal plane is marked by a dashed line. (b) The same plane in the observed structure of TiO; every alternate atom is removed along the diagonal lines in (a). (c) The plane directly beneath the layer in (b); again, every alternate atom is removed along the cuts made by the planes with the intersection lines that are shown in (a). In (b) and (c), the cross section of a monoclinic unit cell is indicated.

with the oxygens omitted; only titaniums are marked. If you draw lines through the titaniums, every fifth one is missing. The ordering of the defects has again produced a superlattice. Where samples of titanium oxide have formulae which lie between the two limits discussed here, $TiO_{1.00}$ and $TiO_{1.25}$, the structure seems to consist of portions of the $TiO_{1.00}$ and $TiO_{1.25}$ structures intergrown. Note that although most texts refer to the structure as $TiO_{1.25}$, as we have, on the definitions that we have used previously when discussing 'FeO', it is more correctly written as $Ti_{0.8}O$ ($Ti_{1-x}O$) because this indicates that the structure contains titanium vacancies rather than interstitial oxygens.

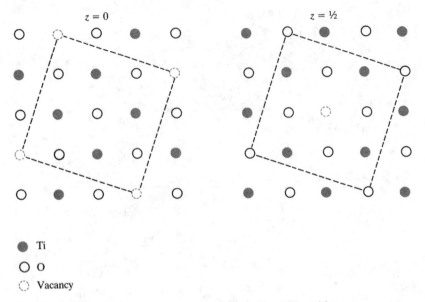

- ● Ti
- ○ O
- ○ Vacancy

**FIGURE 5.29** The structure of $TiO_{1.25}$, showing both O and Ti positions.

## 5.8 PLANAR DEFECTS

The introduction to this chapter mentions that crystals often contain **extended defects** as well as point defects. The simplest *linear* defect is a **dislocation** where there is a fault in the arrangement of the atoms in a line through the crystal lattice. There are many different types of *planar* defects, most of which we are not able to discuss here either for reasons of space or of complexity, such as **grain boundaries**, which are of more relevance to materials scientists, and **chemical twinning**, which can contain unit cells mirrored about the twin plane through the crystal. However,

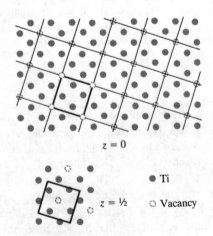

**FIGURE 5.30** Successive Ti layers in the structure of $TiO_{1.25}$.

we shall look briefly at two forms of planar defect which are relevant to the kind of structural discussions in this book:

- **crystallographic shear planes** where the oxygen vacancies effectively collect together in a plane which runs through the crystal
- **intergrowth structures** where two different but related structures alternate throughout the crystal

## 5.8.1 CRYSTALLOGRAPHIC SHEAR PLANES

Non-stoichiometric compounds are found for the higher oxides of tungsten, molybdenum, and titanium — $WO_{3-x}$, $MoO_{3-x}$, and $TiO_{2-x}$, respectively. The reaction of these systems to the presence of point defects is entirely different from what has been discussed previously. In fact, the point defects are eliminated by a process known as **crystallographic shear (CS)**.

In these systems, a series of closely related compounds with very similar formulae and structure exists. The formulae of these compounds all obey a general formula, which for the molybdenum and tungsten oxides can be $Mo_nO_{3n-1}$, $Mo_nO_{3n-2}$, $W_nO_{3n-1}$, and $W_nO_{3n-2}$ and for titanium dioxide is $Ti_nO_{2n-1}$; $n$ can vary taking values of 4 and above. The resulting series of oxides is known as an **homologous series** (like the alkanes in organic chemistry). The first seven members of the molybdenum trioxide series is: $Mo_4O_{11}$, $Mo_5O_{14}$, $Mo_6O_{17}$, $Mo_7O_{20}$, $Mo_8O_{23}$, $Mo_9O_{26}$, $Mo_{10}O_{29}$, and $Mo_{11}O_{32}$.

In these compounds, we find regions of corner-linked octahedra separated from each other by thin regions of a different structure known as the crystallographic shear (CS) planes. The different members of a homologous series are determined by the fixed spacing between the CS planes. The structure of a shear plane is quite difficult to understand, and these structures are usually depicted by the linking of octahedra as described in Chapter 1.

$WO_3$ has several polymorphs, but above 900°C the $WO_3$ structure is that of $ReO_3$, which is illustrated in Figure 5.31 (ignore the bold squares for the time being). (The structures of the other polymorphs are distortions of the $ReO_3$ structure.) $ReO_3$ is made up of [$ReO_6$] octahedra that are linked together via their corners; each corner of an octahedron is shared with another. Figure 5.31(a) shows part of one layer of linked octahedra in the structure. Notice that within the layer, any octahedron is linked to four others; it is also linked via its upper and lower corners, to octahedra in the layers above and below. Part of the $ReO_3$ structure is drawn in Figure 5.32, and you can see that every oxygen atom is shared between two metal atoms. As six oxygens surround each Re, the overall formula is $ReO_3$.

The non-stoichiometry in $WO_{3-x}$ is achieved by some of the octahedra in this structure changing from corner-sharing to edge-sharing. Look back now to the octahedra marked in bold in Figure 5.31(a). The edge-sharing corresponds to shearing the structure so that the chains of bold octahedra are displaced to the positions in Figure 5.31(b). This shearing occurs at regular intervals in the structure and is interspersed with slabs of the 'ReO$_3$' structure (corner linked [$WO_6$] octahedra). It creates groups of four octahedra which share edges. The direction of maximum

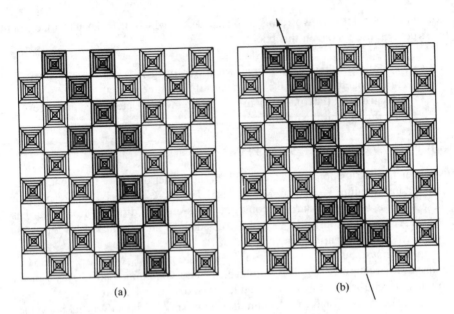

(a)                                    (b)

**FIGURE 5.31** Formation of shear structure.

density of the edge-sharing groups is called the crystallographic shear plane and is indicated by an arrow in (b).

To see how crystallographic shear alters the stoichiometry of $WO_3$, we need to find the stoichiometry of one of the groups of four octahedra that are linked together by sharing edges; part of the structure depicting one of these groups is presented in Figure 5.33(a). The four octahedra consist of 4 W atoms and 18 O atoms (Figure 5.33(b)); 14 of the oxygen atoms are linked out to other octahedra (these bonds are indicated), so are each shared by two W atoms, while the remaining 4 oxygens are only involved in the edge-sharing within the group. The overall stoichiometry is given by [4 W + (14 × ½)O+ 4 O], giving $W_4O_{11}$.

Clearly, if groups of four octahedra with stoichiometry $W_4O_{11}$ are interspersed throughout a perfect $WO_3$ structure, then the amount of oxygen in the structure is reduced and we can write the formula as $WO_{3-x}$. The effect of introducing the groups of four in an ordered way can be quantified. If the structure sheared in such a way that the entire structure was composed of these groups, the formula would become $W_4O_{11}$. If one $[WO_6]$ octahedron is used for each group of four, then the overall formula becomes $[W_4O_{11} + WO_3] = W_5O_{14}$. Clearly, we can extend this process to any number of $[WO_6]$ octahedra regularly interspersed between the groups:

$W_4O_{11} + 2WO_3 = W_6O_{17};$         $W_4O_{11} + 3WO_3 = W_7O_{20};$

$W_4O_{11} + 4WO_3 = W_8O_{23};$         $W_4O_{11} + 5WO_3 = W_9O_{26};$

$W_4O_{11} + 6WO_3 = W_{10}O_{29};$        $W_4O_{11} + 7WO_3 = W_{11}O_{32}.$

The basic formula of the group of four, $W_4O_{11}$ can be written as $W_nO_{3n-1}$ where $n = 4$. This formula also holds for all the other formulae that are listed. Therefore,

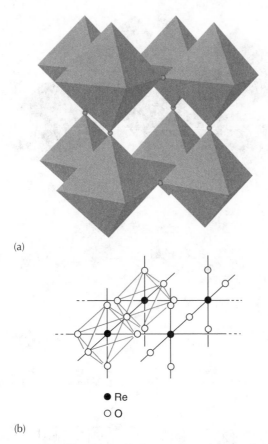

(a)

(b)

● Re
○ O

**FIGURE 5.32** (a) and (b): Part of the ReO$_3$ structure, showing the linking of octahedra through the corners.

we have produced the general formula for the homologous series simply by intro-ducing set ratios of the edge-sharing groups in among the [WO$_6$] octahedra.

The shear planes are found to repeat throughout a particular structure in a regular and ordered fashion, so any particular sample of WO$_{3-x}$ will have a specific formula corresponding to one of those listed previously. The different members of the homologous series are determined by the fixed spacing between the CS planes. An example of one of the structures is illustrated in Figure 5.34. A unit cell has been marked so that the ratio of [WO$_6$] octahedra to the groups of four is clear. Within the marked unit cell there is one group of four and seven octahedra giving the overall formula W$_4$O$_{11}$ + 7WO$_3$ = W$_{11}$O$_{32}$.

Members of the Mo$_n$O$_{3n-1}$ series have the same structure as their W$_n$O$_{3n-1}$ analog, even though unreduced MoO$_3$ does not have the ReO$_3$ structure, but a layer structure.

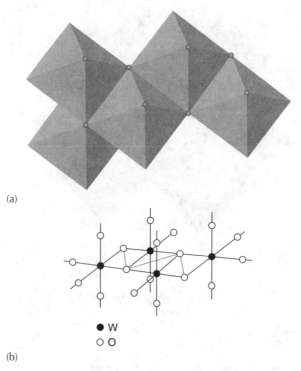

(a)

(b)

**FIGURE 5.33** (a) and (b): Group of four $[WO_6]$ octahedra sharing edges formed by the creation of shear planes in $W_nO_{3n-1}$.

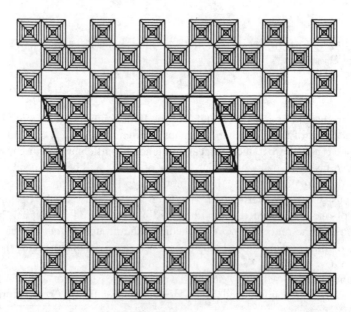

**FIGURE 5.34** A member of the $W_nO_{3n-1}$ homologous series with the projection of a unit cell marked.

If the structures shear in such a way that groups of six octahedra share edges regularly throughout the structure, then homologous series with the general formula $M_nO_{3n-2}$ are formed.

The homologous series for oxygen deficient $TiO_2$ is given by the formula $Ti_nO_{2n-1}$. In this case, the octahedra along the CS planes are joined to each other by sharing *faces*, whereas in the unreduced parts of the $TiO_2$ structure the octahedra share edges as in rutile.


## 5.8.2 PLANAR INTERGROWTHS

Many systems show examples of intergrowth where a solid contains regions of more than one structure with clear solid–solid interfaces between the regions. Epitaxy (see Chapter 3) is an example of such a phenomenon with great technological interest for the production of circuits and 'smart' devices. The zeolites ZSM-5 and ZSM-11 can form intergrowths (see Chapter 7), as can the barium ferrites (see Chapter 9). We will only look at one example here that of intergrowths in the **tungsten bronzes**. The term *bronze* is applied to metallic oxides that have a deep colour, metallic lustre and are either metallic conductors or semiconductors. The sodium-tungsten bronzes, $Na_xWO_3$, have colours that range from yellow to red and deep purple depending on the value of $x$.

We have already seen that $WO_3$ has the rhenium oxide ($ReO_3$) structure, with $[WO_6]$ octahedra joined through the corners. This is illustrated in Figure 5.31(a) and Figure 5.32. The structure contains a three-dimensional network of channels throughout the structure and it has been found that alkali metals can be incorporated into the structure in these channels. The resultant crystal structure depends on the proportion of alkali metal in the particular compound. The structures are based on three main types; there are *cubic* phases where the alkali metal occupies the centre of the unit cell (similar to perovskite; see Chapters 1 and 10) and *tetragonal* and *hexagonal* phases. The basic structures of two of these are illustrated in Figure 5.35. The electronic conductivity properties of the bronzes are due to the fact that charge compensation has to be made for the presence of $M^+$ ions in the structure. This is achieved by the change in oxidation state of some of the tungsten atoms from VI to V (such processes are discussed in more detail in the final section [Section 5.10] of this chapter).

The hexagonal bronze structure illustrated is formed when potassium reacts with $WO_3$ (K needs a bigger site) and the composition lies in the range $K_{0.19}WO_3$ to $K_{0.33}WO_3$. If the proportion of potassium in the compound is less than this, the structure is found to consist of $WO_3$ intergrown with the hexagonal structure in a regular fashion. The layers of hexagonal structure can be either one or two tunnels wide, as depicted in Figure 5.36. Similar structures are observed for tungsten bronzes containing metals other than potassium such as Rb, Cs, Ba, Sn, and Pb. A high resolution electron micrograph of the barium tungsten bronze clearly illustrates its preferred single tunnel structure (Figure 5.37). In other samples, the separation of the tunnels increases as the concentration of barium decreases.

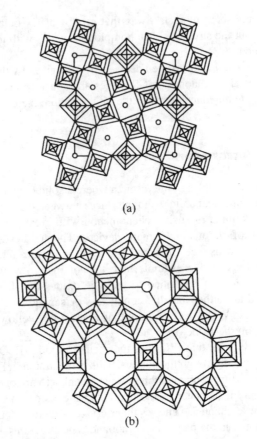

(a)

(b)

**FIGURE 5.35** (a) The tetragonal tungsten bronze structure; (b) the hexagonal tungsten bronze structure. The shaded squares represent $WO_6$ octahedra, which are linked to form pentagonal, square and hexagonal tunnels. These are able to contain a variable population of metal atoms, shown as open circles.

## 5.9 THREE-DIMENSIONAL DEFECTS

### 5.9.1 BLOCK STRUCTURES

In oxygen-deficient $Nb_2O_5$, and in mixed oxides of Nb and Ti, and Nb and W, the crystallographic shear planes occur in two sets at right angles to each other. The intervening regions of perfect structure now change from infinite sheets to infinite *columns* or *blocks*. These structures are known as **double shear** or **block structures** and are characterized by the cross-sectional size of the blocks. The block size is expressed as the number of octahedra sharing vertices. In addition to having phases built of blocks of one size, the complexity can be increased by having blocks of two or even three different sizes arranged in an ordered fashion. The block size (or sizes)

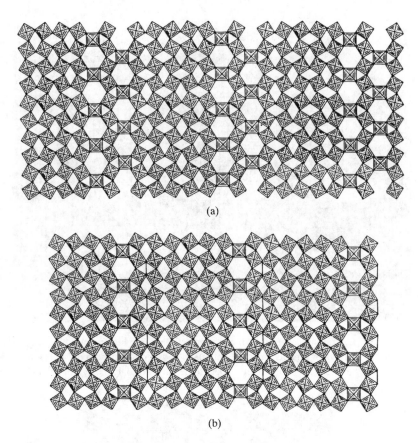

(a)

(b)

**FIGURE 5.36** The idealized structures of two intergrowth tungsten bronze phases: (a) containing double rows of hexagonal tunnels, and (b) containing single rows of tunnels. The tungsten trioxide matrix is shown as shaded squares. The hexagonal tunnels are empty, although in the known intergrowth tungsten bronzes, the tunnels contain variable amounts of metal atoms.

determines the overall stoichiometry of the solid. An example of a crystal with two different block sizes is shown in Figure 5.38.

### 5.9.2 Pentagonal Columns

Three-dimensional faults also occur in the so-called **PC** or **pentagonal column structures**. These structures contain the basic repeating unit depicted in Figure 5.39(a), which consists of a pentagonal ring of five $[MO_6]$ octahedra. When these are stacked on top of each another, a pentagonal column is formed that contains chains of alternating M and O atoms. These pentagonal columns can fit inside a $ReO_3$ type of structure in an ordered way, and depending on the spacing, a homologous series is formed. One example is given in Figure 5.39(b) for the

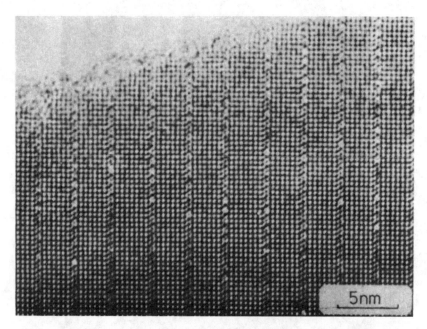

**FIGURE 5.37** Electron micrograph of the intergrowth tungsten bronze phase $Ba_xWO_3$, illustrating the single rows of tunnels clearly. Each black spot on the image represents a tungsten atom, and many of the hexagonal tunnels appear to be empty or only partly filled with barium.

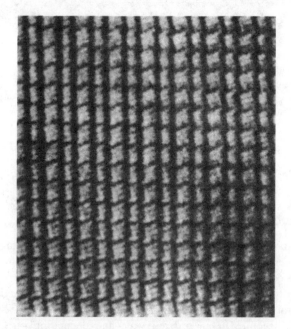

**FIGURE 5.38** High resolution electron micrograph of the $W_4Nb_{26}O_{77}$ structure, depicting strings of (4 × 4) and (3 × 4) blocks. CS planes between the blocks have a darker contrast. (Photograph courtesy of Dr. J.L. Hutchison.)

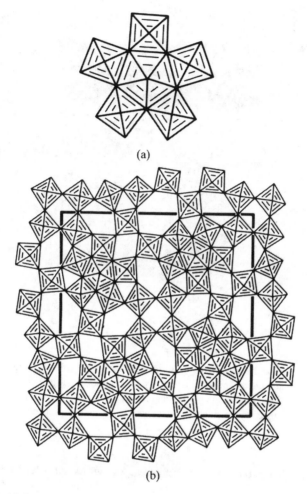

**FIGURE 5.39** (a) A pentagonal column; (b) the structure of $Mo_5O_{14}$.

compound $Mo_5O_{14}$. This type of structure is also found in the tetragonal tungsten bronzes.

### 5.9.3 INFINITELY ADAPTIVE STRUCTURES

In Section 5.9.1, we saw that the mixed oxides of niobium and tungsten could have a range of different compositions made by fitting together rectangular columns or blocks. The closely related $Ta_2O_5$–$WO_3$ system does something even more unusual. A large number of compounds form, but they are built up by fitting together pentagonal columns (PCs). The idealized structures of two of these compounds are depicted in Figure 5.40; the structures have a *wavelike* skeleton of PCs. As the composition varies, so the wavelength of the backbone changes, giving rise to a

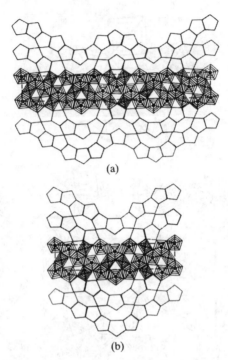

(a)

(b)

**FIGURE 5.40** The idealized structures of (a) $Ta_{22}W_4O_{67}$ and (b) $Ta_{30}W_2O_{81}$. These phases are built from pentagonal columns depicted as shaded pentagons; the octahedra are depicted as shaded squares. The wavelength of the chains of pentagonal columns varies with composition in such a way that any given anion to cation ration can be accommodated by an ordered structure.

huge number of possible ordered structures, known as **infinitely adaptive compounds**.

## 5.10 ELECTRONIC PROPERTIES OF NON-STOICHIOMETRIC OXIDES

Earlier, we considered the structure of non-stoichiometric FeO in some detail. If we apply the same principle to other binary oxides, we can define four types of compound:

*Metal excess (reduced metal)*
   Type A: anion vacancies present; formula $MO_{1-x}$
   Type B: interstitial cations; formula $M_{1+x}O$

*Metal deficiency (oxidised metal)*
   Type C: interstitial anions; formula $MO_{1+x}$
   Type D: cation vacancies; formula $M_{1-x}O$

## TABLE 5.7
### Types of non-stoichiometric oxides (MO)

| Metal excess Reduced metal M | | Metal deficiency Oxidized metal M | |
|---|---|---|---|
| A anion vacancies $MO_{1-x}$ | B interstitial cations $M_{1+x}O$ | C interstitial anions $MO_{1+x}$ | D cation vacancies $M_{1-x}O$ |
| TiO, VO, (ZrS) | CdO, ZnO | | TiO, VO, MnO, FeO, CoO, NiO |

The four types are summarized in Table 5.7, and Figure 5.41 illustrates some of the structural possibilities for simple oxides with A, B, C, and D type non-stoichiometry, assuming an NaCl-type structure.

We looked in some detail earlier at the structure of FeO; this falls into the **type D** category, with metal deficiency and cation vacancies resulting in oxidized metal. (Other compounds falling into this category are MnO, CoO, and NiO.)

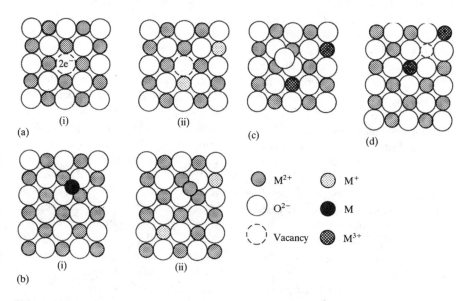

**FIGURE 5.41** Structural possibilities for binary oxides. (a) Type A oxides: metal excess/anion vacancies. (i) This shows the two electrons that maintain charge neutrality, localized at the vacancy. (ii) The electrons are associated with the normal cations making them into $M^+$. (b) Type B oxides: metal excess/interstitials. (i) This shows an interstitial atom, whereas in (ii) the atom has ionized to $M^{2+}$, and the two liberated electrons are now associated with two normal cations, reducing them to $M^+$. (c) Type C oxides: metal deficiency/interstitial anions. The charge compensation for an interstitial anion is by way of two $M^{3+}$ ions. (d) Type D oxides: metal deficiency/cation vacancies. The cation vacancy is compensated by two $M^{3+}$ cations.

**Type A** oxides compensate for metal excess with **anion vacancies**. To maintain the overall neutrality of the crystal, two electrons have to be introduced for each anion vacancy. These can be trapped at a vacant anion site, as depicted in (a)(i). However, it is an extremely energetic process to introduce electrons into the crystal and so we are more likely to find them associated with the metal cations as shown in (a)(ii), which we can describe as reducing those cations from $M^{2+}$ to $M^+$. $VO_{1-x}$ is an example of this type of system.

**Type B** oxides have a metal excess which is incorporated into the lattice in **interstitial** positions. This is shown in (b)(i) as an interstitial atom, but it is more likely that the situation in (b)(ii) will hold, where the interstitial atom has ionised and the two electrons so released are now associated with two neighbouring ions, reducing them from $M^{2+}$ to $M^+$. Cadmium oxide, CdO, has this type of structure. Oxygen is lost when zinc(II) oxide is heated, forming $Zn_{1+x}O$, oxide vacancies form and to compensate, $Zn^{2+}$ ions migrate to interstitial positions and are reduced to $Zn^+$ ions or Zn atoms. Electron transfer can take place between the $Zn^{2+}$ and $Zn^+/Zn$ resulting in the yellow coloration seen when ZnO is heated.

**Type C** oxides compensate for the lack of metal with **interstitial anions**. The charge balance is maintained by the creation of two $M^{3+}$ ions for each interstitial anion (c), each of which we can think of as $M^{2+}$ associated with a positive hole.

Before considering the conductivity of these non-stoichiometric oxides it is probably helpful to recap what we know about the structure and properties of the stoichiometric binary oxides of the first-row transition elements. A summary of the properties of binary oxides is given in Table 5.8.

We discussed the conductivity of the stoichiometric oxides in Chapter 4, and their conductivity is dependent on two competing effects: (i) the $d$ orbitals overlap to give a band — the bigger the overlap the greater the band width — and electrons in the band are delocalized over the whole structure, and (ii) interelectronic repulsion tends to keep electrons localized on individual atoms. TiO and VO behave as metallic conductors and must therefore have good overlap of the $d$ orbitals producing a $d$ electron band. This overlap arises partly because Ti and V are early in the transition series (the $d$ orbitals have not suffered the contraction, due to increased nuclear charge, seen later in the series) and partly because of the crystal stuctures, TiO has one-sixth of the titaniums and oxygens are missing from an NaCl–type structure), which allow contraction of the structures and thus better $d$ orbital overlap.

Further along the series, we saw that stoichiometric MnO, FeO, CoO, and NiO are insulators. This situation is not easily described by band theory because the $d$ orbitals are now too contracted to overlap much, typical band widths are 1 eV, and the overlap is not sufficient to overcome the localizing influence of interelectronic repulsions. (It is this localization of the $d$ electrons on the atoms that gives rise to the magnetic properties of these compounds that are discussed in Chapter 9.)

Going back now to the non-stoichiometric oxides, in the excess metal monoxides of type A and type B, we saw that extra electrons have to compensate for the excess metal in the structure. Figure 5.41 illustrates that these could be associated either with an anion vacancy or alternatively they could be associated with metal cations within the structure. Although we have described this association as reducing neighbouring cations, this association can be quite weak, and these electrons can be free

**TABLE 5.8**
**Properties of the first-row transition element monoxides**

| Element | Ca | Sc | Ti | V | Cr | Mn | Fe | Co | Ni | Cu | Zn |
|---|---|---|---|---|---|---|---|---|---|---|---|
| Structure of stoichiometric oxide MO | NaCl structure | Does not exist | Defect NaCl structure, $1/6$ vacancies | Defect NaCl | Does not exist | NaCl structure | NaCl structure[a] | NaCl structure | NaCl structure | PtS structure | Wurtzite structure (NaCl at high pressure) |
| Defect structure | | | $Ti_{1-\delta}O$ Ti vacancies (intergrowths of $TiO_{1.00}$ and $TiO_{1.25}$ structures) | $V_{1-\delta}O$ V vacancies and tetrahedral V interstitials in defect clusters | | $Mn_{1-\delta}O$ Mn vacancies | $Fe_{1-\delta}O$ Fe vacancies and tetrahedral Fe interstitials in defect clusters | $Co_{1-\delta}O$ Co vacancies | $Ni_{1-\delta}O$ Ni vacancies | | $Zn_{1+\delta}O$ Interstitial Zn |
| Conductivity of stoichiometric compound | | | Metallic | Metallic < 120 K | | Insulator | Insulator | Insulator | Insulator | | Insulator |
| Conductivity of non-stoichiometric compound | | | Metallic | Metallic | | $p$-type hopping semiconductor | $p$-type | $p$-type | $p$-type | | $n$-type |
| Magnetism (see Chapter 9) | | | Diamagnetic | Diamagnetic | | Paramagnetic $\mu = 5.5\ \mu_B$ (antiferromagnetic when cooled, $T_N = 122$ K) | Paramagnetic (antiferromagnetic when cooled, $T_N = 198$ K) | Antiferromagnetic ($T_N = 292$ K) | Antiferromagnetic $T_N = 530$ K | | |

[a] Exactly stoichiometric FeO is never found.

to move through the lattice; they are not necessarily strongly bound to particular atoms. Thermal energy is often sufficient to make these electrons move and so conductivity *increases* with temperature. We associate semiconductivity with such behaviour (metallic conductivity *decreases* with temperature).

Chapter 4 discussed semiconductivity in terms of band theory. An intrinsic semiconductor has an empty conduction band lying close above the filled valence band. Electrons can be promoted into this conduction band by heating, leaving positive holes in the valence band; the current is carried by both the electrons in the conduction band and by the positive holes in the valence band. Semiconductors, such as silicon, can also be doped with impurities to enhance their conductivity. For instance, if a small amount of phosphorus is incorporated into the lattice the extra electrons form impurity levels near the empty conduction band and are easily excited into it. The current is now carried by the electrons in the conduction band and the semiconductor is known as **n-type** (**n** for negative). Correspondingly, doping with Ga increases the conductivity by creating positive holes in the valence band and such semiconductors are called **p-type** (**p** for positive).

Compounds of type A and B would produce *n*-type semiconductors because the conduction is produced by electrons. Conduction in these non-stoichiometric oxides is not easily described by band theory, for the reasons given earlier for their stoichiometric counterparts — the interelectronic repulsions have localized the electrons on the atoms. Therefore, it is easiest to think of the conduction electrons (or holes) localized or trapped at atoms or defects in the crystal instead of delocalized in bands throughout the solid. Conduction then occurs by jumping or **hopping** from one site to another under the influence of an electric field. In a perfect ionic crystal where all the cations are in the same valence state, this would be an extremely energetic process. However, when two valence states, such as $Zn^{2+}$ and $Zn^+$, are available as in these transition metal non-stoichiometric compounds, the electron jump between them does not take much energy. Although we cannot develop this theory here, we can note that the conduction in these so-called **hopping semiconductors** can be described by the equations of diffusion theory in much the same way as we did earlier for ionic conduction. We find that the mobility of a charge carrier (either an electron or a positive hole), $\mu$, is an activated process and we can write:

$$\mu \propto \exp(-E_a/kT) \tag{5.16}$$

where $E_a$ is the activation energy of the hop, and is approximately 0.1 to 0.5 eV. The hopping conductivity is given by the expression

$$\sigma = ne\mu \tag{5.17}$$

where $n$ is the number of mobile charge carriers per unit volume and $e$ is the electronic charge. (Notice that these equations are analogous to Equation (5.8) and Equation (5.7) in Section 5.3, describing ionic mobility, $\mu$, and ionic conductivity, respectively.) The density of mobile carriers, $n$, depends only on the composition of the crystal, and does not vary with temperature. From Equation (5.16), we can see that, as for ionic conductivity, the hopping electronic conductivity increases with temperature.

In the type C and D monoxides, we have demonstrated that the lack of metal is compensated by oxidation of neighbouring cations to $M^{3+}$. The $M^{3+}$ ions can be regarded as $M^{2+}$ ions associated with a positive hole. Accordingly, if sufficient energy is available conduction can be thought to occur via the positive hole hopping to another $M^{2+}$ ion and the electronic conductivity in these compounds will be $p$-type. MnO, CoO, NiO, and FeO are materials that behave in this way. This behaviour of hopping semiconduction was described for NiO in Chapter 4 when it was described in terms of electron hopping. Regarding the charge carriers as positive holes is simply a matter of convenience and the description of a positive hole moving from $Ni^{3+}$ to $Ni^{2+}$ is the same as saying that an electron moves from $Ni^{2+}$ to $Ni^{3+}$.

Non-stoichiometric materials can be listed which cover the whole range of electrical activity from metal to insulator. Here we have considered some metallic examples which can be described by band theory, such as TiO and VO, and others, such as MnO, which are better described as hopping semiconductors. Other cases, such as $WO_3$ and $TiO_2$, fall in between these extremes and a different description again is needed. We also met non-stoichiometric compounds, such as calcia-stabilized zirconia and β-alumina, which are good ionic conductors. Indeed, stabilized zirconia exhibits *both* electronic and ionic conductivity though, fortunately for its industrial usefulness, electronic conduction only occurs at low oxygen pressures. It is thus difficult to make generalizations about this complex behaviour, and each case is best treated individually.

Semiconductor properties are extremely important to the modern electronics industry, which is constantly searching for new and improved materials. Much of their research is directed at extending the composition range and thus the properties of these materials. The composition range of a non-stoichiometric compound is often quite narrow (a so-called line phase), so to extend it (and thus extend the range of its properties also) the compound is doped with an impurity. Here is one example: If we add $Li_2O$ to NiO and then heat to high temperatures in the presence of oxygen, $Li^+$ ions become incorporated in the lattice and the resulting black material has the formula $Li_xNi_{1-x}O$ where $x$ lies in the range 0 to 0.1. The equation for the reaction (using stoichiometric NiO for simplicity) is given by:

$$\tfrac{1}{2}xLi_2O + (1-x)NiO + \tfrac{1}{4}xO_2 = Li_xNi_{1-x}O$$

To compensate for the presence of the $Li^+$ ions, $Ni^{2+}$ ions will be oxidised to $Ni^{3+}$ or the equivalent of a high concentration of positive holes located at Ni cations.

This process of creating electronic defects is called **valence induction**, and it increases the conductivity range of 'NiO' tremendously. Indeed, at high Li concentrations, the conductivity approaches that of a metal (although it still exhibits semiconductor behaviour in that its conductivity increases with temperature).

## 5.11 CONCLUSIONS

In this chapter, we have tried to give some idea of the size and complexity of this subject and also its fascination, without it becoming overwhelming. The main point that emerges from our explorations is that the concept of random, isolated point

**TABLE 5.9**
**Defect concentration data for CsI**

| T/K | $\dfrac{n_S}{N}$ |
|-----|------------------|
| 300 | $1.08 \times 10^{-16}$ |
| 400 | $1.06 \times 10^{-12}$ |
| 500 | $2.63 \times 10^{-10}$ |
| 600 | $1.04 \times 10^{-8}$ |
| 700 | $1.43 \times 10^{-7}$ |
| 900 | $4.76 \times 10^{-6}$ |

defects does *not* explain the complex structures of non-stoichiometric compounds, but that defects cluster together and become ordered, or even eliminated, in many different ways. It is these defects which can be exploited to give solids properties useful in many technologically useful devices.

## QUESTIONS

1. If $\Delta H_m^{\ominus}$ for the formation of Schottky defects in a certain MX crystal is 200 kJ mol$^{-1}$, calculate $n_S/N$ and $n_S$ per mole for the temperatures 300, 500, 700, and 900 K.
2. Table 5.9 gives the variation of defect concentration with temperature for CsI. Determine the enthalpy of formation for one Schottky defect in this crystal.
3. If we increase the quantity of impurity (say $CaCl_2$) in the NaCl crystal, how will this affect the plot given in Figure 5.8? How is the transition point in the graph affected by the purity of the crystal?
4. In NaCl, the cations are more mobile than the anions. What effect, if any, do you expect small amounts of the following impurities to have on the conductivity of NaCl crystals: (a) AgCl, (b) $MgCl_2$, (c) NaBr, (d) $Na_2O$?
5. Unlike fluorides, pure oxides with the fluorite structure ($MO_2$) show high anion conduction only at elevated temperatures, above about 2300 K. Suggest a reason for this.
6. Figure 5.42 shows the fluorite structure with the tetrahedral environment of one of the anions shaded. (The anion behind this tetrahedron has been omitted for clarity.) Suppose this anion jumps to the octahedral hole at the body centre. Describe, and sketch, the pathway it takes in terms of the changing coordination by cations.
7. Make a simple estimate of the energy of defect formation in the fluorite structure: (a) describe the coordination by nearest neighbours and next-nearest neighbours of an anion both for a normal lattice site and for an interstitial site at the centre of the unit cell shown in Figure 5.3(a).

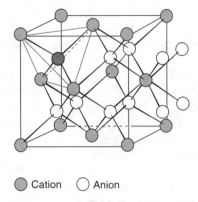

Cation ○ Anion

**FIGURE 5.42** The fluorite structure, showing the coordination tetrahedron around one of the anions (shown in colour).

(b) Use the potential energy of two ions, given by:

$$E = - \frac{e^2 Z}{4\pi\varepsilon_0 r}$$

$$= - (2.31 \times 10^{-28} \text{ J m}) Z/r$$

where $Z$ is the charge on the other ion and $r$ is the distance between them, to estimate the energy of defect formation in fluorite; $a = 537$ pm.

8. The zinc blende structure of γ-AgI, a low-temperature polymorph of AgI, is illustrated in Figure 5.43. Discuss the similarities and differences between this structure and that of α-AgI. Why do you think the conductivity of the Ag⁺ ions is lower in γ-AgI?

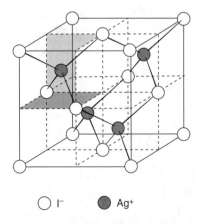

○ I⁻  ● Ag⁺

**FIGURE 5.43** The zinc blende structure of γ-AgI.

9. The compounds in Table 5.4 mostly contain either $I^-$ ions or ions from the heavier end of Group 16. Explain.

10. Undoped β-alumina demonstrates a maximum conductivity and minimum activation energy when the sodium excess is around 20 to 30 mole%. Thereafter, further increase in the sodium content causes the conductivity to decrease. By contrast, β-alumina crystals doped with $Mg^{2+}$ have a much higher conductivity than do undoped crystals. Explain these observations.

11. Confirm the presence of iron vacancies for a sample of wustite which has a unit cell dimension of 428.2 pm, an Fe:O ratio of 0.910 and an experimental density of $5.613 \times 10^3$ kg m$^{-3}$.

12. How does the change in lattice parameter of 'FeO' with iron content corroborate the iron vacancy model and refute an oxide interstitial model?

13. How would you expect the formation of colour centres to affect the density of the crystal?

14. Figure 5.44 depicts the central section of a possible defect cluster for FeO. (a) Determine the vacancy:interstitial ratio for this cluster. (b) Assuming that this section is surrounded by Fe ions and oxide ions in octahedral sites as in the Koch–Cohen cluster, determine the formula of a sample made totally of such clusters. (c) Determine the numbers of $Fe^{2+}$ and $Fe^{3+}$ ions in octahedral sites.

15. Use Figure 5.28 to confirm that TiO is a one-sixth defective NaCl structure, by counting up the atoms in the monoclinic cell.

16. Figure 5.29 illustrates the layers in the $TiO_{1.25}$ structure with both the Ti and O sites marked. Use this and Figure 5.30 to demonstrate that the unit cell shown has the correct stoichiometry for the crystal.

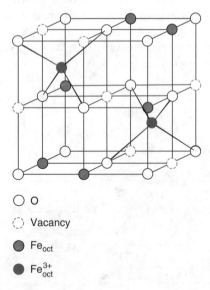

○ O

◌ Vacancy

● Fe$_{oct}$

● Fe$_{oct}^{3+}$

**FIGURE 5.44** A possible cluster in $Fe_{1-x}O$.

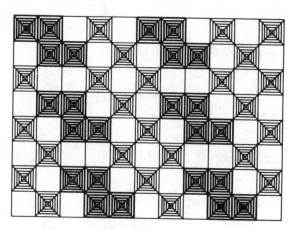

**FIGURE 5.45** A member of the $W_nO_{3n-1}$ homologous series.

17. How would you expect charge neutrality to be maintained in $TiO_{1.25}$?

18. Take a simple case where *two* metal oxide octahedra wish to eliminate oxygen by sharing. How does the formula change as they (i) share a corner, (ii) share an edge, (iii) share a face?

19. Figure 5.45 shows a member of the homologous series, $W_nO_{3n-1}$. To what formula does it correspond?

20. ZnO is a type B (excess metal) material. What do you expect to happen to its electronic properties if it is doped with $Ga_2O_3$ under reducing conditions?

# 6 Carbon-Based Electronics

## 6.1 INTRODUCTION

We tend to think of organic compounds as insulators, but research has demonstrated that organic materials can be semiconductors, metallic conductors, or even superconductors. The 2000 Nobel Prize for chemistry was awarded to Alan Heeger, Hideki Shirakawa, and Alan McDiarmid for the discovery of the electrically conducting polymer, polyethyne or polyacetylene $(-CH=)_n$ based on their work in the 1960s and 1970s. Following this work, other less air-sensitive conducting polymers were discovered. Polypyrroles, for example, are commercially available and used in gas sensors. Other possible applications of these materials include light-emitting diodes (LEDs) and due to their low weight and cost, it has been suggested they could be used for throwaway luggage tags.

Crystalline organic metals are generally composed of organic molecules with delocalised ring systems together with another organic molecule or inorganic species. These materials are electrical conductors and demonstrate increasing conductivity with decreasing temperature — the hallmark of a metallic conductor. Some of these organic metals are even superconductors at low temperatures.

By contrast, the organic polymers used in lithium batteries have negligible electronic conductivity but are ionic conductors.

Finally, we look at carbon itself in the form of graphite, which finds extensive use as an electrode material, and the relatively recently discovered allotropes, the fullerenes. We begin with polyacetylene and related polymers.

## 6.2 POLYACETYLENE AND RELATED POLYMERS

### 6.2.1 THE DISCOVERY OF POLYACETYLENE

Extension of ideas of delocalised electrons in conjugated organic molecules led to a suggestion that a conjugated polymer might be an electrical conductor or semiconductor. First attempts to make polyacetylene and similar solids, however, resulted in short chain molecules or amorphous, unmeltable powders. Then in 1961, Hatano and co-workers in Tokyo managed to produce thin films of polyacetylene, by polymerising ethyne (acetylene). Polyacetylene exists in two forms *cis* and *trans*, as depicted in Figure 6.1, of which the *trans* form is the more stable. Ten years after Hatano's work, Shirakawa, and Ikeda made films of *cis*-polyacetylene which could be converted into the *trans* form. They achieved this by directing a stream of ethyne gas on to the surface of a Ziegler–Natta catalyst (a mixture of triethyl aluminium and titanium tetrabutoxide). To make a large film, the catalyst solution can be spread in a thin layer over the walls of a reaction vessel (Figure 6.2), and then ethyne gas

$$n(H—C≡C—H)$$
acetylene

*trans*

*cis*

polyacetylene

**FIGURE 6.1** *cis-* and *trans*-polyacetylene.

allowed to enter. The polyacetylene produced in this way has a smooth shiny surface on one side and a sponge-like structure. It can be converted to the thermodynamically more stable *trans* form by heating. The conversion is quite rapid above 370 K. After conversion, the smooth side of the film is silvery in appearance, becoming blue when the film is very thin.

A way of improving the conductivity was found when Shirakawa visited McDiarmid and Heeger in Pennsylvania later in the 1970s. The Americans had been working on smaller conjugated molecules to which they added an electron acceptor in order to make them conducting. It was a natural step to try this approach with polyacetylene. If an electron acceptor such as bromine is added to polyacetylene forming $[(CH)^{\delta+}Br_\delta^-]_n$, its conductivity is greater than that of the undoped material.

**FIGURE 6.2** A film of polyacetylene forms on the inner surface of the reaction vessel, after ethyne gas passes over the catalyst solution of the walls. The paper-thin flexible sheet of polyacetylene is then stripped from the walls before doping. (Photograph by James Kilkelly. Originally published in *Scientific American*.)

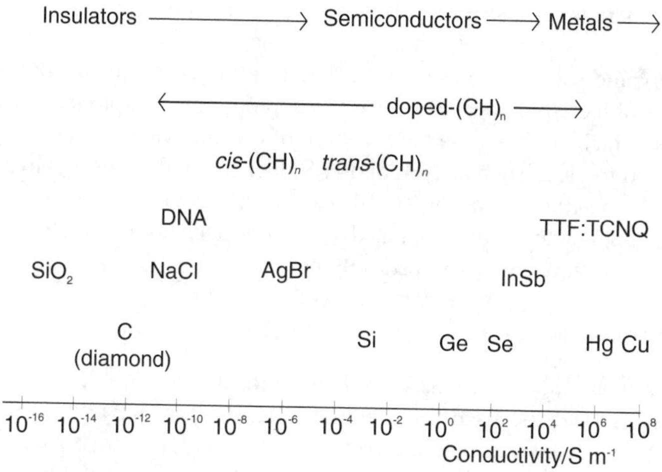

**FIGURE 6.3** Conductivities of doped polyacetylenes: conductivities of insulators, semiconductors and metals are given for comparison.

Other examples of dopants that can oxidise polyacetylene are $I_2$, $AsF_5$ and $HClO_4^-$. The effect of these dopants can be to raise the conductivity from $10^{-3}$ S m$^{-1}$ to as much as $10^5$ S m$^{-1}$ using only small quantities of dopant.

The conductivity of polyacetylene is also increased by dopants that are electron donors. For example, the polymer can be doped with alkali metals to give, for example, $[Li_\delta^+(CH)^{\delta-}]_n$. The wide range of conductivities produced by these two forms of doping is illustrated in Figure 6.3.

Polyacetylene is very susceptible to attack by oxygen. The polymer loses its metallic lustre and becomes brittle when exposed to air. However, other conjugated polymers were found. Polypyrrole, polythiophene, polyaniline, polyphenylenevinylene (Figure 6.4), and others are conjugated polymers whose bonding and con-

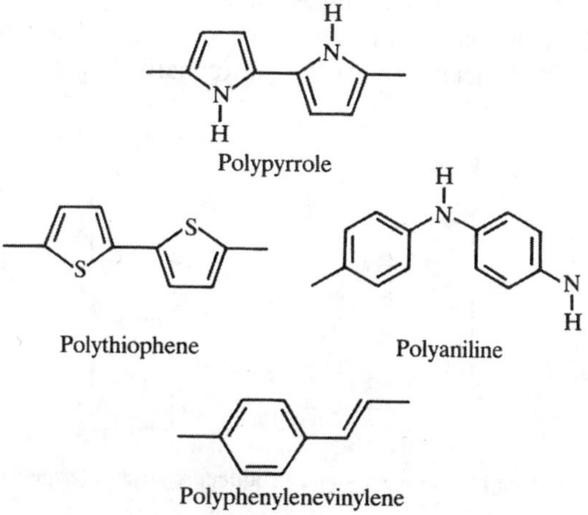

Polypyrrole

Polythiophene

Polyaniline

Polyphenylenevinylene

**FIGURE 6.4** Repeating units of some conducting polymers.

ductivity are similar to those of polyacetylene. These polymers are, however, less sensitive to oxygen and by attaching suitable side chains can be made soluble in nonpolar organic solvents and thereby easier to process. As for polyacetylene, the conductivity of these polymers is sensitive to doping. This is exploited in polypyrrole gas sensors, which are based on the variation of conductivity of a thin polymer film when exposed to gases such as $NH_3$ and $H_2S$. Doped conducting polymers can also be used as a metallic contact in organic electronic devices. The most promising commercial applications of conjugated polymers such as the LED described in Section 6.2.3, however, use undoped polymer.

### 6.2.2 BONDING IN POLYACETYLENE AND RELATED POLYMERS

In small conjugated alkenes such as butadiene with alternate double and single bonds, the $\pi$ electrons are delocalised over the molecule. If we take a very long conjugated olefin, we might expect to obtain a band of $\pi$ levels, and if this band were partly occupied, we would expect to have a one-dimensional conductor. Polyacetylene is just such a conjugated long chain polymer. Now if polyacetylene consisted of a regular evenly spaced chain of carbon atoms, the highest occupied energy band, the $\pi$ band, would be half full and polyacetylene would be an electrical conductor. In practice, polyacetylene demonstrates only modest electrical conductivity, comparable with semiconductors such as silicon: the *cis* form has a conductivity of the order $10^{-7}$ S m$^{-1}$ and the *trans* form, $10^{-3}$ S m$^{-1}$. The crystal structure is difficult to determine accurately, but diffraction measurements indicate that an alternation in bond lengths of about 6 pm occurs. This is much less than would be expected for truly alternating single and double bonds (C—C, 154 pm in ethane; C=C, 134 pm in ethene). Nonetheless, this does indicate that the electrons are tending to localise in double bonds instead of being equally distributed over the whole chain.

What in fact is happening is that two bands, bonding and anti-bonding, are forming with a band gap where nonbonding levels would be expected. Just enough electrons are available to fill the lower band. As illustrated in Figure 6.5, this leads to a lower energy than the half-full single band.

This splitting of the band is an example of **Peierls' theorem**, which asserts that a one-dimensional metal is always electronically unstable with respect to a

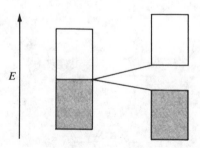

**FIGURE 6.5** The band gap in polyacetylene produced by the alternation of long and short bonds.

non-metallic state — there is always some way of opening an energy gap and creating a semiconductor. Thus although a very simple bonding picture of such solids would suggest a half-filled band and metallic conductivity, the best that can be expected is a semiconducting polymer.

If an electron acceptor is added, it takes electrons from the lower π bonding band. The doped polyacetylene now has holes in its valence band and, like *p*-type semiconductors, has a higher conductivity than the undoped material. Electron donor dopants add electrons to the upper π band, making this partly full, and so producing an *n*-type semiconductor.

### 6.2.3  ORGANIC LEDs

Polymer light-emitting diodes are being developed for flat panel displays in mobile phones, laptops, and televisions. A typical polymer LED is built up on a glass substrate. At the bottom is a layer of semitransparent metallic conductor, usually indium-doped tin oxide (ITO), which acts as one electrode (Figure 6.6). On top of this is a layer of undoped conjugated polymer such as polyphenylenevinylene and on top of that an easily ionised metal such as Ca, Mg/Ag, or Al. The metal forms the second electrode. To act as an LED, a voltage is applied across the two electrodes such that the ITO layer is positively charged. At the ITO/polymer interface, electrons from the polymer move into the ITO layer attracted by the positive charge. This leaves gaps in the lower energy band. At the same time, electrons from the negatively charged second electrode move into the polymer and travel toward the ITO layer. When such an electron reaches a region of the polymer where vacancies exist in the lower energy band, it can jump down to this band emitting light as it does so. The wavelength of the light, as in semiconductor LEDs (Chapter 8), depends on the band gap. Polyphenylenevinylene LEDs emit green/yellow light, polyphenylene-based LEDs emit blue light, and some polythiophenes emit red light.

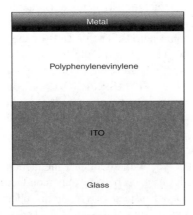

**FIGURE 6.6** Section through a polymer LED.

## 6.3  MOLECULAR METALS

### 6.3.1  ONE-DIMENSIONAL MOLECULAR METALS

Molecular metals are molecular solids with metallic electrical conductivity. Most molecular metals contain one-dimensional chains composed of stacks of two unsaturated cyclic molecules arranged alternately. The first one of this class discovered was TTF-TCNQ. The two molecules involved are tetrathiafulvalene (TTF) and tetracyanoquinonedimethane (TCNQ). These are depicted in Figure 6.7, which also shows how they are stacked in the crystal.

The electrical conductivity of TTF:TCNQ is of the order of $10^2$ S m$^{-1}$ at room temperature and increases with decreasing temperature until around 80 K when the conductivity drops as the temperature is lowered. TCNQ is a good electron acceptor and, for example, accepts electrons from alkali metal atoms to form ionic salts. In TTF-TCNQ, the columns of each type of molecule interact to form delocalised orbitals. Some electrons from the highest energy filled band of TTF move across to partly fill a band of TCNQ, so that both types of columns have partially occupied bands. The number of electrons transferred corresponds to about 0.69 electrons per molecule. This partial transfer only occurs with molecules such as tetrathiafulvalene whose electron donor ability is neither too small nor too large. With poor electron donors, no charge transfer occurs. With very good electron donors such as alkali metals, one electron per TCNQ is transferred and the acceptor band is full. Thus, K$^+$(TCNQ)$^-$ is an insulator.

At low temperatures, TTF-TCNQ suffers a periodic distortion and so its conductivity drops.

### 6.3.2  TWO-DIMENSIONAL MOLECULAR METALS

Polyacetylene and TTF-TCNQ are all pretty well described as one-dimensional electronic conductors, because little interaction occurs between chains in the crystal. Another class of molecular metals exists, however, which, while appearing to resemble these solids, are less unambiguously defined as one-dimensional.

For example, tetramethyl-tetraselenofulvalene (TMTSF), Figure 6.8, forms a series of salts with inorganic anions. The crystals of these salts contain stacks of TMTSF molecules and the TMTSF molecules carry a fractional charge (0.5+). As expected, these salts have high electronic conductivities at room temperatures. Unlike TTF-TCNQ, however, $(TMTSF)_2^+(ClO_4)^-$ and other similar salts remain highly conducting at low temperatures and indeed at very low temperatures become superconducting. The reason for this appears to be that significant overlap occurs between stacks. As a result, the one-dimensional model is not as valid as for TTF-TCNQ, and in particular, Peierls' theorem no longer holds.

Other materials of this type with significant interaction between chains include $(SN)_x$ and $Hg_{3-x}AsF_6$, both of which become superconducting at low temperatures.

Other similar compounds, instead of containing chain-like stacks of organic molecules, contain flat organic ring compounds stacked so that molecular orbitals on different molecules overlap to form bands. The organic molecules form planes

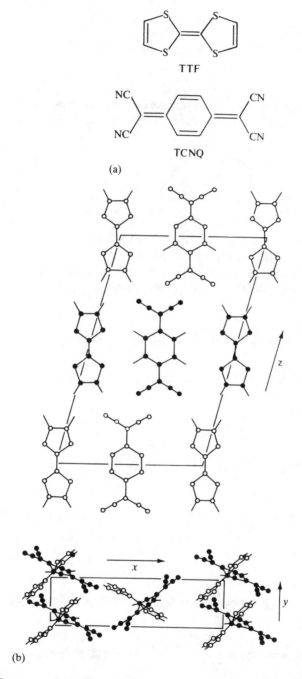

TTF

TCNQ

(a)

(b)

**FIGURE 6.7** Structures of (a) TTF and TCNQ and (b) solid TTF-TCNQ, showing alternate stacks of TTF and TCNQ molecules.

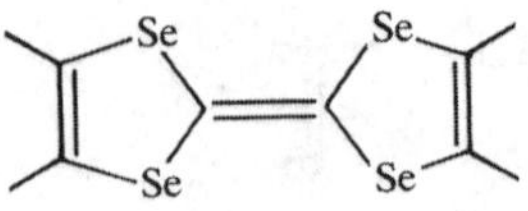

**FIGURE 6.8** Structure of tetramethyl-tetraselenofulvalene (TMTSF).

separated by layers of anions. Electrons are delocalised over the planes of organic molecules. A typical example is bis(ethylenedithio)tetrathiofulvalene (BEDT-TTF).

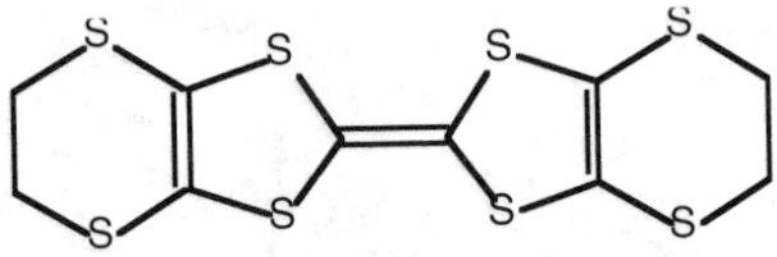

This forms molecular metals (BEDT-TTF)$_2$X with a variety of anions including [Cu(NCS)$_2$]$^-$, I$_3^-$, AuI$_2^-$, IBr$_2^-$, [Cu(N(CN)$_2$)X]$^-$ (X = Cl or Br), [Cu(CN)$_3$]$^-$, GaCl$_4^-$, SF$_5$CH$_2$CF$_2$SO$_3^-$. The packing arrangement varies with the anion. Figure 6.9 depicts the arrangement for β-(BEDT-TTF)$_2$I$_3$, which contains layers of I$_3^-$ ions.

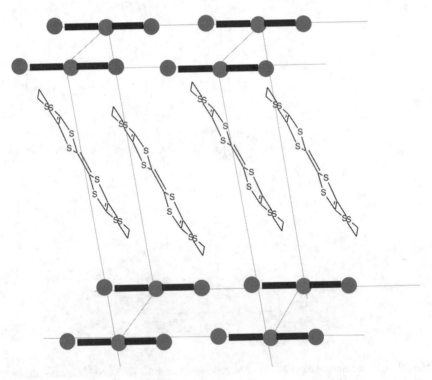

**FIGURE 6.9** Structure of β-(BEDT-TTF)$_2$I$_3$.

Because of the presence of the anions, the BEDT-TTF layers are positively charged, with a formal charge of 0.5 per molecule. Thus, the highest occupied band is only partially filled and the crystals will conduct electricity. Many of these crystals become superconducting at low temperatures (typically 2 to 12 K at normal pressures). Despite the low values of the critical temperatures, the superconductivity of these materials is of the same type as that of the high-temperature superconductors (see Chapter 10).

## 6.4 POLYMERS AND IONIC CONDUCTION — RECHARGEABLE LITHIUM BATTERIES

Whittingham in the seventies developed a battery that operated at room temperature based on the intercalation of Li in $TiS_2$. In the lithium/titanium disulfide battery, one electrode is lithium metal and the other is titanium disulfide bonded to a polymer such as teflon. The electrolyte is a lithium salt dissolved in an organic solvent. Typically, the solvent is a mixture of dimethoxyethane (DME) and tetrahydrofuran (THF). The setup is illustrated in Figure 6.10.

When the circuit is complete, lithium metal from the lithium electrode dissolves giving solvated ions, and solvated ions in the solution are deposited in the titanium disulfide. These ions intercalate into the disulfide, and electrons from the external circuit balance the charge. Thus, the two electrode reactions are:

$$Li(s) = Li^+(solv) + e^- \tag{6.1}$$

and

$$x Li^+(solv) + TiS_2(s) + xe^- = Li_x TiS_2(s) \tag{6.2}$$

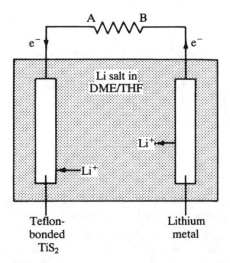

**FIGURE 6.10** The Li-TiS$_2$ battery during a discharge phase.

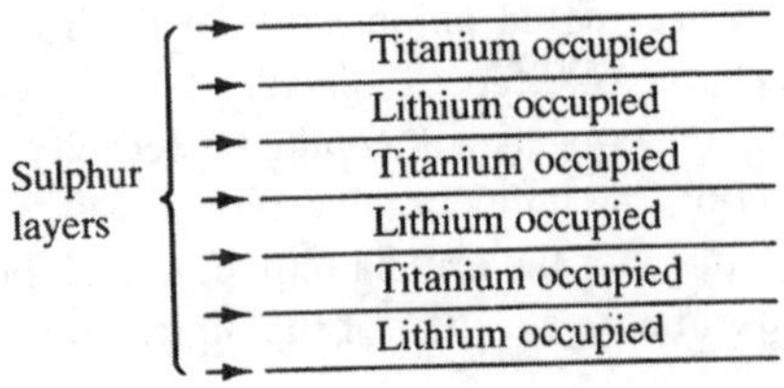

| Sulphur layers | Titanium occupied |
| | Lithium occupied |
| | Titanium occupied |
| | Lithium occupied |
| | Titanium occupied |
| | Lithium occupied |

**FIGURE 6.11** Occupation pattern of octahedral holes between close-packed layers of sulfur atoms in $LiTiS_2$.

giving an overall reaction

$$xLi(s) + TiS_2(s) = Li_xTiS_2(s) \qquad (6.3)$$

The intercalation compound formed has layers of a variable number of lithium ions between sulfide layers as given in Figure 6.11.

When the titanium disulfide has completely discharged, applying a voltage across AB can recharge the battery. A voltage high enough to overcome the free energy of the cell reaction will cause lithium ions to return to the solvent and lithium to be deposited on the metal electrode, thus restoring the battery to its original condition. The usefulness of the lithium/titanium disulphide electrode depends on the electrode material being a good conductor and on the electrode reaction being reversible. $Li_xTiS_2$ is a particularly good electrical conductor, because not only is the electronic conductivity increased by the donation of electrons to the conduction band, but also the lithium ions themselves act as current carriers and so the compound is both an electronic and an ionic conductor. (Ionic conductors were described in Chapter 5.). Compounds other than $TiS_2$ have proved more useful for commercial batteries. For example, in 1990, Sony marketed a camcorder battery based on $LiCoO_2$ as the electrode.

Lithium polymer batteries are similar in principle to the lithium batteries described above but the electrolyte is a polymer. The advantage of these batteries is the absence of liquid in the cell and so the batteries do not leak. The polymer electrolyte is a polymer–alkali metal salt complex. The best known such electrolytes are complexes of poly(ethylene)oxide (PEO).

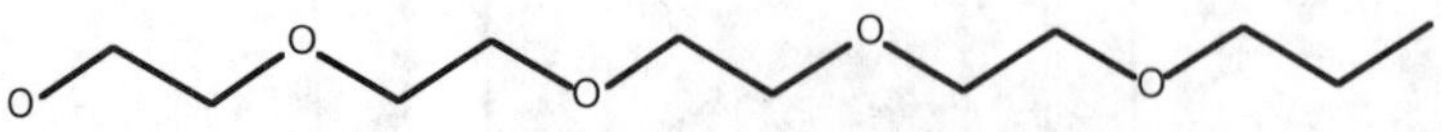

Poly(propylene)oxide, poly(ethylene)succinate, poly(epichlorohydrin), and poly(ethylene imine) have also been investigated as possible polymer bases for polymer electrolytes. These are all polymers containing electronegative elements, oxygen, or nitrogen, in the chain which will coordinate to metal cations. Electrical current is carried by the ions of the salts moving through the polymer. The conductivity depends on the polymer molecular weight and the polymer:salt ratio, and

**TABLE 6.1**
**Conductivities of polymer-metal salt complexes**

| Polymer | Molecular mass | Salt | Conductivity/Sm$^{-1}$ | Temperature range/K |
|---------|----------------|------|------------------------|---------------------|
| PEO | 400 | LiBF$_4$ | $3.1 \times 10^{-4} - 3.1 \times 10^{-5}$ | 298 – 450 |
| PEO | $5 \times 10^6$ | LiBF$_4$ | $10^{-5} - 10^{-7}$ | 298 – 450 |
| PEO | $4 \times 10^6$ | CH$_3$COOLi (O/Li = 4) | $3.16 \times 10^{-4} - 10^{-9}$ | 298 – 420 |
| PEO | $4 \times 10^6$ | CH$_3$COOLi (O/Li = 9) | $10^{-4} - 10^{-9}$ | 298 – 420 |
| PEO | $4 \times 10^6$ | CH$_3$COOLi (O/Li = 18) | $10^{-4} - 10^{-8}$ | 194 – 420 |
| PEO (cross-linked) | 3000 | LiClO$_4$ | $10^{-2} - 10^{-9}$ | 253 – 273 |
| PEO (linear) | 3000 | LiClO$_4$ | $10^{-8} - 10^{-11}$ | 253 – 273 |

experiments suggest that the amorphous part of the polymer produces high ionic conductivity. Some values for Li$^+$ conductivity in various polymer-lithium salt complexes are given in Table 6.1. Conductivity of approximately $10^{-5}$ Sm$^{-1}$ is suitable for batteries.

In lithium polymer batteries, one electrode is lithium foil, or in some cases another electrically conducting material such as graphite, and the other is a reversible intercalation compound as in liquid electrolyte lithium batteries. Compounds used as intercalation electrodes include LiCoO$_2$ and V$_6$O$_{13}$. The cell developed in the Anglo-Danish project, which ran from 1979 to 1995, was

$$\text{Li(s)|PEO-LiClO}_4\text{(s)|V}_6\text{O}_{13}\text{(composite)(s)}$$

Test cells operating between 80 and 140°C achieved nearly 100% of theoretical capacity during the first few cycles but performance dropped off on repeated cycling.

Currently, several companies market lithium polymer batteries for use in, for example, phones.

## 6.5 CARBON

### 6.5.1 GRAPHITE

Graphite is, of course, a very familiar substance, with many uses. The lead in lead pencils is graphite, and finely divided forms of graphite are used to absorb gases and solutes. Its absorption properties find a wide range of applications from gas masks to decolouring food. Graphite formed by pyrolysis of oriented organic polymer fibres is the basis of carbon fibres. Graphite is also used as a support for several industrially important catalysts. Its electronic conductivity is exploited in several

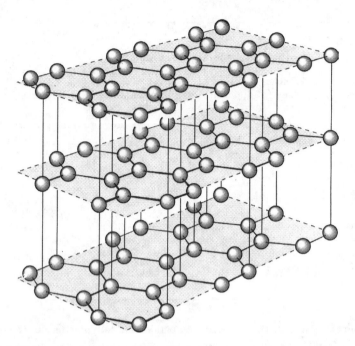

**FIGURE 6.12** Hexagonal structure of graphite layers.

industrial electrolysis processes where it is used as an electrode. Crystals of graphite are, however, only good conductors in two dimensions, and it is the two-dimensional aspect that we are concerned with here.

Crystals of graphite contain layers of interlocking hexagons, as discussed in Chapter 1 and reproduced in Figure 6.12.

This figure illustrates the most stable form in which the layers are stacked ABAB with the atoms in every other layer directly above each other. Another form also contains layers of interlocking hexagons but in which the layers are stacked ABCABC.

Each carbon can be considered as having three single bonds to neighbouring carbons. This leaves one valence electron per carbon in a $p$ orbital at right angles to the plane of the layer. These $p$ orbitals combine to form delocalised orbitals which extend over the whole layer. If the layer contains n carbon atoms, then n orbitals are formed, and $n$ electrons are available to fit in them. Thus, half of the delocalised orbitals are filled. If the orbitals were to form one band, this would explain the conductivity of graphite very nicely because there would be a half-filled band confined to the layers. The situation is close to this, but not quite as simple. The delocalised orbitals in fact form two bands, one bonding and one anti-bonding. (This is reminiscent of diamond, where as discussed in Chapter 4, the $s–p$ band split with a gap where non-bonding orbitals would lie.) The lower band is full and the upper band empty. Graphite is a conductor because the band gap is zero, and electrons are thus readily promoted to the upper band. The band structure for graphite is illustrated in Figure 6.13.

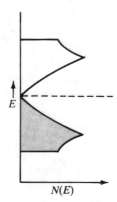

FIGURE 6.13 Band structure of graphite.

Because the density of states is low at the Fermi level, the conductivity is not as high as that for a typical metal. It can, however, be increased as you will now see.

### 6.5.2 Intercalation Compounds of Graphite

Because the bonding between layers in graphite is weak, it is easy to insert molecules or ions into the spaces between layers. The solids produced by reversible insertion of such guest molecules into lattices are known as intercalation compounds, and although originally applied to layered solids the term is now taken to include other solids with similar host–guest interactions. Since the 1960s, attention has been paid to intercalation compounds as of possible importance as catalysts and as electrodes for high energy–density batteries (see previous section).

Many layered solids form intercalation compounds, but graphite is particularly interesting because it forms compounds with both electron donors and electron acceptors. Amongst electron donors, the most extensively studied are the alkali metals. The alkali metals enter graphite between the layers and produce strongly coloured solids in which the layers of carbon atoms have moved farther apart. For example, potassium forms a golden compound $KC_8$ in which the interlayer spacing is increased by 200 pm. The potassium donates an electron to the graphite (forming $K^+$) and the conductivity of the graphite now increases because it has a partially full anti-bonding band.

The first intercalation compound was made in 1841. This contained sulfate, an electron acceptor. Since then, many other electron acceptor intercalation compounds have been made with, for example, $NO_3^-$, $CrO_3$, $Br_2$, $FeCl_3$, and $AsF_5$. In these compounds, the graphite layers donate electrons to the inserted molecules or ions, thus producing a partially filled bonding band. This increases the conductivity and some of these compounds have electrical conductivity approaching that of aluminium.

In graphite, the current is carried through the layers by delocalised $p$ electrons. Layered structures in which the current is carried by $d$ electrons are common in transition metal compounds, such as the disulfides of Ti, Zr, Hf, V, Nb, Ta, Mo, and W, and mixed lithium metal oxides. At one time, there was considerable interest in the disulfides because some of their intercalates were found to be superconductors at very low temperatures. It was hoped that by altering the interlayer spacing, a compound would be found that was superconducting at higher temperatures. Unfortunately, it soon transpired that altering the spacing by inserting different molecules had very little effect on the temperature at which superconductivity appeared. It was concluded that the superconductivity was confined to the layers. When high temperature superconductors were discovered, these were also layer structures. As discussed in Chapter 10, however, the crucial ingredient was a CuO layer.

### 6.5.3 BUCKMINSTER FULLERENE

This polymorph of carbon was only discovered in 1985 by Sir Harry Kroto at the University of Sussex while looking for carbon chains. It is made by passing an electric arc between two carbon rods in a partial atmosphere of helium. Kroto was awarded the Nobel Prize in chemistry in 1996, along with two American researchers (Robert F. Curl Jr. and Richard E. Smalley). The molecule has the formula $C_{60}$ and has the same shape as a soccer ball — a truncated icosahedron; it takes its name from the engineer and philosopher Buckminster Fuller who discovered the architectural principle of the hollow geodesic dome that this molecule resembles (a geodesic dome was built for EXPO 67 in Montreal). The structure is depicted in Figure 6.14.

The molecule is extremely symmetrical with every carbon atom in an identical environment; it consists of 12 pentagons of carbon atoms joined to 20 hexagons. The carbon–carbon distances between adjacent hexagons is 139 pm, and where a hexagon joins a pentagon, the carbon–carbon distance is 143 pm — similar to that found in the graphite layers. Indeed, we can think of this structure as a carbon sheet with graphite-type delocalized bonding, which bends back on itself to form a polyhedron.

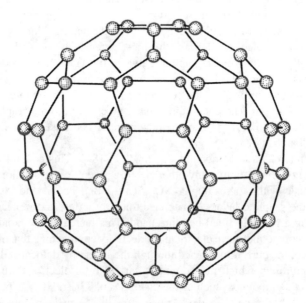

**FIGURE 6.14** The structure of buckminster fullerene, $C_{60}$.

(Other fullerenes are known, such as $C_{70}$; they all have 12 pentagons of carbon atoms linked to different numbers of hexagons.) The molecules, commonly called buckyballs, pack together in a cubic close-packed array in the crystals. Fullerene chemistry is currently a very active research area. The use of fullerenes has been suggested for electronic devices and even in targeted cancer therapy. One of the interesting features is the formation of salts with the alkali metals, known as buckides. For instance, potassium buckide ($K_3C_{60}$) has a *ccp* array of buckyballs, with all the octahedral holes, and tetrahedral holes filled by potassiums: this is a metallic substance which becomes superconducting below 18 K. Other alkali metal buckides become superconducting at even higher temperatures.

In 1991, Sumio Iijima produced fullerene-type molecules in the form of long tubes capped at each end. These were named nanotubes. Nanotubes are discussed in Chapter 11.

## QUESTIONS

1. Which orbitals would you expect to combine to form a delocalised band in polyphenylenevinylene?

2. Would doping of polyacetylene with (a) Rb, (b) $H_2SO_4$ give an *n*-type or a *p*-type conductor?

3. When polyacetylene is doped with chloric (VII) acid, $HClO_4$, part of the acid is used to oxidise the polyacetylene and part to provide a counter-anion. The oxidation reaction for the acid is given in Equation (6.6). Write a balanced equation for the overall reaction.

$$ClO_4^- + 8H^+ + 8e^- = Cl^- + 4H_2O \qquad (6.4)$$

4. $(SN)_x$ remains a metallic conductor down to very low temperatures. Why does this suggest that interaction occurs between the chains in this solid?

5. HMTTF-TCNQ is metallic but $HMTTF-TCNQF_4$ is not. Suggest a possible explanation for the difference. HMTTF = hexamethylenetetrathiafulvalene, TCNQ = tetracyanoquinodimethane, $TCNQF_4$ = tetracyanotetrafluoroquinodimethane.

6. Undoped polythiophene is red. What does this suggest about the size of the conduction band/valence band energy gap in this polymer? Would you expect a polythiophene LED to emit red light?

7. Sketch the band structure of graphite intercalated with (a) an electron donor and (b) an electron acceptor.

8. Fullerenes consist of 12 pentagons and a number of hexagons; 20 in the case of $C_{60}$ and 25 in the case of $C_{70}$. What would be the formula of the smallest possible fullerene?

# 7 Zeolites and Related Structures

## 7.1 INTRODUCTION

Zeolites occur naturally in large deposits, much of it in China. Because of their many uses, they are also manufactured synthetically and represent a large proportion of the chemical industry. Zeolites are used as cation exchangers for water softening, and as molecular sieves for separating molecules of different sizes and shapes (e.g., as drying agents). Recent research has focused on their ability to act as catalysts in a wide variety of reactions, many of them highly specific, and they are now used extensively in industry for this purpose. Approximately 56 naturally occurring zeolites have now been characterized, but in the quest for new catalysts, more than 150 synthetic structures have been developed. Natural zeolites are mainly used as lightweight building materials, but also find use as adsorbers for cleaning wastewaters, and for pet litter. About 60% of the synthetically produced zeolites are used in detergents for softening water by ion exchange (removing calcium ions mainly); specialized catalysts and adsorbers mainly make up the rest of the market.

Zeolites were first described as a mineral group by the Swedish mineralogist Baron Axel Cronstedt in 1756. They are a class of crystalline **aluminosilicates** based on rigid anionic frameworks with well-defined **pores (channels)** running through them, which intersect at **cavities (cages)**. These cavities contain exchangeable metal cations ($Na^+$, $K^+$, etc.), and can hold removable and replaceable guest molecules (water in naturally occurring zeolites). It is their ability to lose water on heating that earned them their name; Cronstedt observed that on heating with a blowtorch they hissed and bubbled as though they were boiling and named them zeolites from the Greek words *zeo*, to boil, and *lithos*, stone. With cavity sizes falling between 200 and 2000 pm, zeolites are classified as **microporous** substances.

## 7.2 COMPOSITION AND STRUCTURE

The general formula for the composition of a zeolite is:

$$M_{x/n}[(AlO_2)_x(SiO_2)_y] \, mH_2O$$

where cations M of valence $n$ neutralize the negative charges on the aluminosilicate framework.

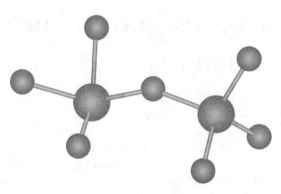

**FIGURE 7.1** The zeolite building units. Two $SiO_4/AlO_4$ tetrahedra linked by corner sharing.

## 7.2.1 FRAMEWORKS

The primary building units of zeolites are $[SiO_4]^{4-}$ and $[AlO_4]^{5-}$ tetrahedra (Chapter 1) linked together by **corner sharing**, forming oxygen bridges (Figure 7.1). The oxygen bridge is not usually linear — the Si/Al—O—Si/Al linkage is very flexible and the angle can vary between $120°$ and $180°$. Silicon-oxygen tetrahedra are electrically neutral when connected together in a three-dimensional network as in quartz, $SiO_2$ (Figure 7.2). The substitution of Si(IV) by Al(III) in such a structure, however, creates an electrical imbalance, and to preserve overall electrical neutrality, each $[AlO_4]$ tetrahedron needs a balancing positive charge. This is provided by exchangeable cations such as $Na^+$, held electrostatically within the zeolite.

It is possible for the tetrahedra to link by sharing two, three, or all four corners, thus forming a variety of different structures. The linked tetrahedra are usually illustrated by drawing a straight line to represent the oxygen bridge connecting two tetrahedral units. In this way, the six linked tetrahedra in Figure 7.3(a) and Figure 7.3(b) are simply represented by a hexagon (Figure 7.3(c)). This is known as a **6-ring**, and a tetrahedrally coordinated atom occurs at each intersection between two straight lines. As we see later, many different ring sizes are found in the various zeolite structures.

Many zeolite structures are based on a secondary building unit that consists of 24 silica or alumina tetrahedra linked together; here we find 4- and 6-rings linked together to form a basket-like structure called the **sodalite unit** (also known as the **β-cage**) depicted in Figure 7.4, and which has the shape of a **truncated octahedron** (Figure 7.5). Several of the most important zeolite structures are based on the sodalite unit (Figure 7.6).

The mineral **sodalite** is composed of these units, with each 4-ring shared directly by two β-cages in a primitive array. Note that the **cavity** or **cage** enclosed by the eight sodalite units depicted in Figure 7.6(a) is actually a sodalite unit (i.e., sodalite units are space-filling). In this three-dimensional structure, a tetrahedral Si or Al atom is located at the intersection of four lines because oxygen bridges are made by corner-sharing from all four vertices of the tetrahedron. Sodalite is a highly symmetrical structure and the cavities link together to form **channels** or **pores**, which run parallel to all three cubic crystal axes, the entrance to these pores governed by the 4-ring window.

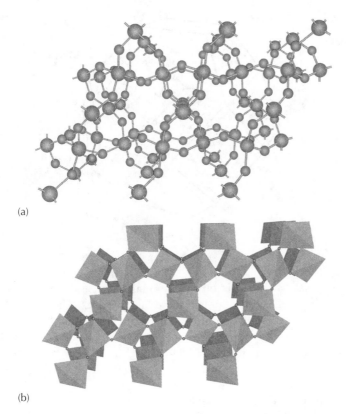

(a)

(b)

**FIGURE 7.2** The structure of quartz (a) as a ball and stick representation and (b) as linked [SiO$_4$] tetrahedra.

A synthetic zeolite, **zeolite A** (also called Linde A) is illustrated in Figure 7.6(b). Here the sodalite units are again stacked in a primitive array, but now they are linked by oxygen bridges between the 4-rings. A three-dimensional network of linked cavities each with a **truncated cuboctahedron** shape (Figure 7.5) through the structure is thus formed; the truncated cuboctahedra are also space-filling (each

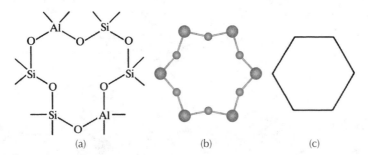

(a)                          (b)                    (c)

**FIGURE 7.3** (a) 6-ring containing two Al and four Si atoms, (b) computer model of the 6-ring, and (c) shorthand version of the same 6-ring.

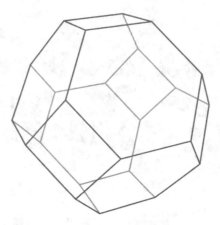

**FIGURE 7.4** The sodalite unit.

shares its octagonal face with six others), forming channels which run parallel to the three cubic axial directions through these large cavities. The computer-drawn models in Figure 7.7 are of the zeolite A framework, illustrating the cavity and its 8-ring window more clearly as well as how it links to a sodalite cage. The formula of zeolite A is given by: $Na_{12}[(SiO_2)_{12}(AlO_2)_{12}].27H_2O$. In this typical example, the Si/Al ratio is unity, and we find that in the crystal structure the Si and Al atoms strictly alternate.

The structure of **faujasite**, a naturally occurring mineral, is illustrated in Figure 7.6(c). The sodalite units are linked by oxygen bridges between four of the eight 6-rings in a tetrahedral array. The tetrahedral array encloses a large cavity (sometimes known as the $\alpha$-cage) entered through a 12-ring window. The synthetic zeolites

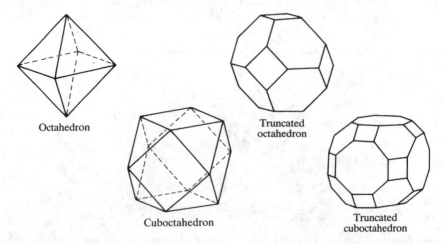

Octahedron

Truncated octahedron

Cuboctahedron

Truncated cuboctahedron

**FIGURE 7.5** The relationship between an octahedron, a truncated octahedron, a cuboctahedron, and a truncated cuboctahedron.

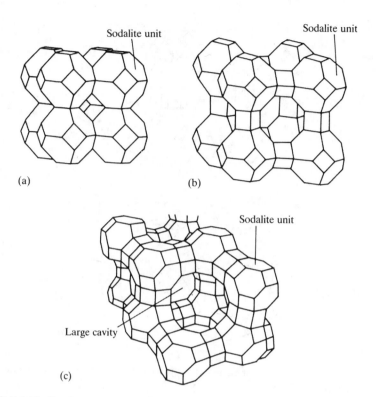

**FIGURE 7.6** Zeolite frameworks built up from sodalite units: (a) sodalite (SOD), (b) zeolite A (LTA), and (c) faujasite (zeolite X and zeolite Y) (FAU).

**zeolite X** and **zeolite Y**   (Linde X and Linde Y) also have this basic underlying structure. Zeolite X structures have a Si/Al ratio between 1 and 1.5, whereas zeolite Y structures have Si/Al ratios between 1.5 and 3.

## 7.2.2 NOMENCLATURE

The naming of zeolites and related structures has been somewhat unsystematic. Some structures were named after the parent minerals (e.g., sodalite, faujasite), while others were named by researchers, or after the projects which synthesized them (e.g., ZSM [Zeolite Socony Mobil]). Unfortunately, this led to the same zeolites synthesized by different routes and bearing different names — in some cases, up to 20 different trade names!

The International Union of Pure and Applied Chemistry (IUPAC) introduced a three-letter structure code to try and simplify matters; zeolite A and the more silicon-rich zeolite, ZK-4, have the same framework structure and are designated LTA. Similarly, ZSM-5 and its silicon-rich relation, silicalite, have the same framework and are both designated MFI.

Zeolites are also often written as M-[zeolite], where M refers to the particular cation in the structure (e.g., Ca-zeolite A).

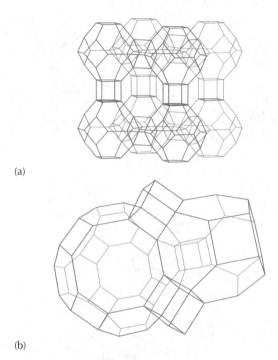

(a)

(b)

**FIGURE 7.7** (a) The zeolite A framework; (b) a sodalite unit in zeolite A, illustrating the linkage to the truncated cuboctahedral cavity.

### 7.2.3 Si/Al Ratios

We saw that zeolite A has a Si/Al ratio of 1. Some zeolites have quite high Si/Al ratios: zeolite ZK-4 (LTA), with the same framework structure as zeolite A, has a ratio of 2.5. Many of the new synthetic zeolites that have been developed for catalysis are highly siliceous: **ZSM-5** (MFI) can have a Si/Al ratio which lies between 20 and $\infty$ (the latter, called silicalite (see Section 7.2.2) being virtually pure $SiO_2$); this far outstrips the ratio of 5.5 found in mordenite, which is the most siliceous of the naturally occurring zeolite minerals.

Clearly, changing the Si/Al ratio of a zeolite also changes its cation content; the fewer aluminium atoms, the fewer cations will be present to balance the charges. The highly siliceous zeolites are inherently hydrophobic in character, and their affinity is for hydrocarbons.

### 7.2.4 Exchangeable Cations

The zeolite Si/Al—O framework is rigid, but the cations are not an integral part of this framework and are often called **exchangeable cations**: they are fairly mobile and readily replaced by other cations (hence their use as cation exchange materials).

The presence and position of the cations in zeolites is important for several reasons. The cross section of the rings and channels in the structures can be altered by changing the size or the charge (and thus the number) of the cations, and this

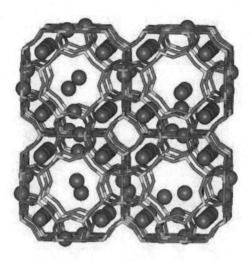

**FIGURE 7.8** Framework and cation sites in the Na⁺ form of zeolite A (LTA). See colour insert following page 356. (Courtesy of Dr. Robert Bell, Royal Institution of Great Britian, London.)

significantly affects the size of the molecules that can be adsorbed. A change in cationic occupation also changes the charge distribution within the cavities, and thus the adsorptive behaviour and the catalytic activity.

The balancing cations in a zeolite can have more than one possible location in the structure. Figure 7.8 depicts the available sites in the Na⁺ form of zeolite A. Some sites occupy most of the centres of the 6-rings, while the others are in the 8-ring entrances to the β-cages. The presence of cations in these positions effectively reduces the size of the rings and cages to any guest molecules that are trying to enter. To alter a zeolite to allow organic molecules, for instance, to diffuse into or through the zeolite, a divalent cation such as $Ca^{2+}$ can be exchanged for the univalent Na⁺ or K⁺ ions, not only halving the number of cations present, but also replacing them with a smaller ion. As the divalent cations tend to occupy the sites in the 6-rings, this opens up the 8-ring windows, thus leaving the channels free for diffusion.

The normal crystalline zeolites contain water molecules which are coordinated to the exchangeable cations. These structures can be dehydrated by heating under vacuum, and in these circumstances, the cations move position at the same time, frequently settling on sites with a lower coordination number. The dehydrated zeolites are extremely good drying agents, absorbing water to get back to the preferred hydrated condition.

### 7.2.5 CHANNELS AND CAVITIES

The important structural feature of zeolites, which can be exploited for various uses, is the network of linked cavities forming a system of channels throughout the structure. These cavities are of molecular dimensions and can adsorb species small enough to gain access to them. A controlling factor in whether molecules can be adsorbed in the cavities is the size of the **window** or **port** into the channel, thus the

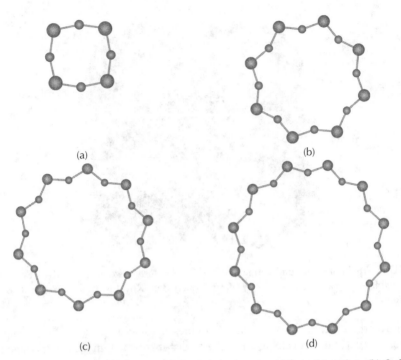

(a)

(b)

(c)

(d)

**FIGURE 7.9** Computer models of various window sizes in zeolites: (a) 4-ring, (b) 8-ring, (c) 10-ring, and (d) 12-ring.

importance of the number of tetrahedra forming the window (i.e., the ring size). Figure 7.9 illustrates how the window sizes can vary.

The windows to the channels thus form a three-dimensional sieve with mesh widths between about 300 and 1000 pm, thus the well-known name **molecular sieve** for these crystalline aluminosilicates. Zeolites thus have large internal surface areas and high sorption capacities for molecules small enough to pass through the window into the cavities. They can be used to separate mixtures such as straight-chain and branched-chain hydrocarbons.

The zeolites fall into three main categories. The channels may be parallel to: (i) a single direction, so that the crystals are fibrous; (ii) two directions arranged in planes, so that the crystals are lamellar; or (iii) three directions, such as cubic axes, in which strong bonding occurs in three directions. The most symmetrical structures have cubic symmetry. By no means do all zeolites fall neatly into this classification; some, ZSM-11 for instance, have a dominant two-dimensional structure interlinked by smaller channels. A typical fibrous zeolite is edingtonite (EDI), $Ba[(AlO_2)_2(SiO_2)_3]$. $4H_2O$, which has a characteristic chain formed by the regular repetition of five tetrahedra. Lamellar zeolites occur frequently in sedimentary rocks, e.g., phillipsite (PHI), $(K/Na)_5[(SiO_2)_{11}(AlO_2)_5]$.$10H_2O$ is a well-known example. In terms of their useful properties, zeolites are conveniently discussed by pore size.

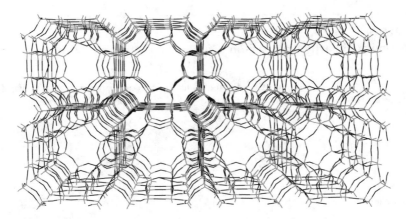

**FIGURE 7.10** Computer model of the zeolite A (LTA) structure, illustrating the channels and 8-ring windows.

## Small-Pore Zeolites

A cavity in sodalite (the β-cage) is bounded by a 4-ring with a diameter of 260 pm, and although this is a very small opening it can admit water molecules (Figure 7.6).

The channels in zeolite A run parallel to the three cubic axial directions, and are entered by a port of diameter 410 pm, determined by an 8-ring window; this is still considerably smaller than the diameter of the internal cavity, which measures 1140 pm across. The computer model of zeolite A in Figure 7.10 clearly illustrates the 8-ring windows, the channels running through the structure and the cavities created by their intersection.

Small-pore zeolites can accommodate linear chain molecules, such as straight-chain hydrocarbons and primary alcohols and amines, but not branched chain molecules. As discussed in the previous section, the port size can be enlarged to about 500 pm in diameter by replacing sodium ions with calcium ions.

Values for channel and cavity sizes for various zeolites and zeotypes are listed in Table 7.1.

## Medium-Pore Zeolites

In the mid-1970s, some completely novel zeolite structures were synthesized which led to significant new developments. This family of framework structures comprising the zeolites synthesized by the oil company Mobil, including ZSM-5 and ZSM-11 (MEL), silicalite (MFI), and some closely related natural zeolites, have been given the generic name **pentasil**.

ZSM-5 is a catalyst now widely used in the industrial world. Its structure is generated from the pentasil unit depicted in Figure 7.11 (as are the others of this group). These units link into chains, which join to make layers. Appropriate stacking of these layers gives the various pentasil structures. Both ZSM-5 and ZSM-11 are characterized by channels controlled by 10-ring windows with diameters of about 550 pm. The pore systems in these zeolites do not link big cavities, but they do contain intersections where larger amounts of free space are available for molecular

**TABLE 7.1**
**Window and cavity diameters in zeolites**

| Zeolite/zeotype | Structure type code | No. of tetrahedra in ring | Window diameter (pm) | Cavity diameter (pm) |
|---|---|---|---|---|
| Sodalite | SOD | 4 | 260 | 600 |
| Zeolite A | LTA | 8 | 410 | 1140 |
| Erionite-A | ERI | 8 | 360 × 520 | — |
| ZSM-5 | MFI | 10 | 510 × 550 | — |
|  |  |  | 540 × 560 |  |
| Faujasite | FAU | 12 | 740 | 1180 |
| Mordenite | MOR | 12 | 670 × 700 | — |
|  |  |  | 290 × 570 |  |
| Zeolite-L | LTL | 12 | 710 | — |
| ALPO-5 (see Section 7.6) | — | 12 | 800 | — |
| VPI-5 | — | 18 | 1200–1300 | — |

interactions to take place. Figure 7.12 depicts computer models of the structures of ZSM-5 and ZSM-11. Figure 7.13(a) shows the pore system of ZSM-5 with nearly circular zigzag channels intersecting with straight channels of elliptical cross section. This contrasts with the ZSM-11 structure which just has intersecting straight channels of almost circular cross section (Figure 7.13(b)). In both cases, the two-dimensional system of pores is linked by much smaller channels.

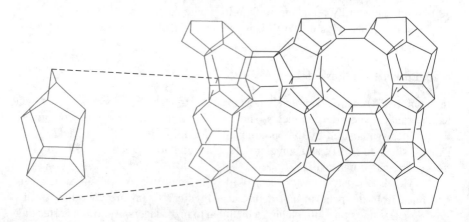

**FIGURE 7.11** A pentasil unit together with a slice of the structure of ZSM-5, illustrating a linked chain of pentasil units highlighted.

(a)

FIGURE 7.12 (a) Computer model of orthorhombic ZSM-5, illustrating the straight elliptical channels along the $y$ direction; (b) computer model of ZSM-11 showing the near-circular straight channels which run along both the $x$ and $y$ directions in the tetragonal crystals. Both structures are depicted as line drawings and with their van der Waals radii. (continued)

## Large-Pore Zeolites

Faujasite is a large-pore zeolite based on four sodalite cages in a tetrahedral configuration, linked via oxygen bridges through the 6-ring windows. This leads to a structure with large cavities of diameter 1180 pm entered by 12-ring windows of diameter 740 pm, which give a three-dimensional network of channels depicted in (Figure 7.14).

The channel system for **mordenite (MOR)** is illustrated in Figure 7.15. Mordenite has an orthorhombic structure and two types of channels are available, governed

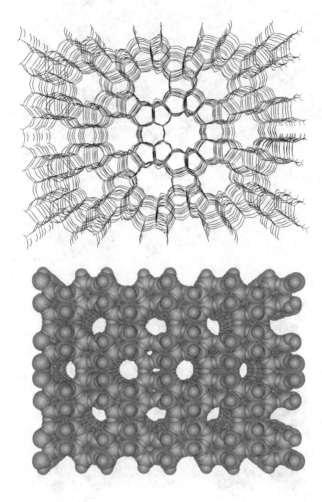

(b)

**FIGURE 7.12** (*continued*)

by 8- and 12-ring windows, respectively, all running parallel to each other and interconnected only by smaller 5- and 6-ring systems.

## 7.3 SYNTHESIS OF ZEOLITES

It was the pioneering work of Richard Barrer at the University of Aberdeen that led the way in the synthesis of zeolites. He prepared zeolites using reactive silica and alumina reagents such as sodium silicates and aluminates, $[Al(OH)_4]^-$, under hydrothermal conditions, at high pH obtained by using an alkali metal hydroxide and/or an organic base. A gel forms by a process of copolymerization of the silicate and aluminate ions. The gel is then heated gently (60–100°C) in an autoclave for several days, producing a condensed zeolite. The product obtained is determined by the synthesis conditions: temperature, time, pH, and mechanical movement are all possible

ZSM-5    ZSM-11

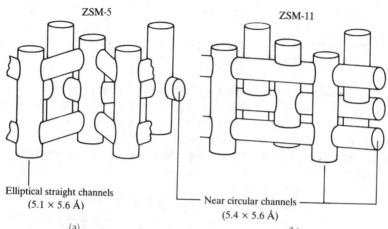

Elliptical straight channels
(5.1 × 5.6 Å)

Near circular channels
(5.4 × 5.6 Å)

(a)                              (b)

**FIGURE 7.13** The interconnecting channel systems in ZSM-5 and ZSM-11.

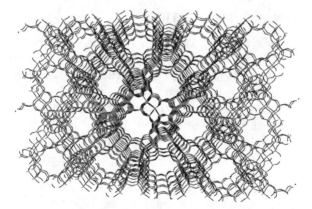

**FIGURE 7.14** The cubic faujasite structure, illustrating the channels which lie parallel to each of the face diagonals (oxygen bridges are included in this model).

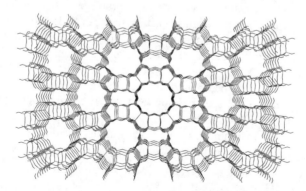

**FIGURE 7.15** The channel structure of mordenite, illustrating the larger channels running along the $z$ direction.

variables. The presence of organic bases is useful for synthesizing silicon-rich zeolites.

The formation of novel silicon-rich synthetic zeolites has been facilitated by the use of **templates**, such as large quaternary ammonium cations instead of $Na^+$. For instance, the tetramethylammonium cation, $[(CH_3)_4N]^+$, is used in the synthesis of ZK-4. The aluminosilicate framework condenses around this large cation, which can subsequently be removed by chemical or thermal decomposition. ZSM-5 is produced in a similar way using the tetra-*n*-propyl ammonium ion. Only a limited number of large cations can fit into the zeolite framework, and this severely reduces the number of $[AlO_4]$ tetrahedra that can be present, producing a silicon-rich structure.

The preparation of silicon-rich zeolites, such as zeolite Y, can be achieved by varying the composition of the starting materials but can also be done by subsequent removal of aluminium from a synthesized aluminosilicate framework using a chemical treatment. Several different methods are available, including extraction of the aluminium by mineral acid, and extraction using complexing agents.

## 7.4  STRUCTURE DETERMINATION

The structures of the zeolite frameworks have been determined by X-ray and neutron diffraction techniques. Some of the naturally occurring minerals were characterized in the 1930s, and the synthetic zeolites have been investigated from 1956 onward. Unfortunately, it is extremely difficult for diffraction techniques to determine a structure unequivocally because Al and Si are next to each other in the Periodic Table and thus have very similar atomic scattering factors (Chapter 2). It is possible to determine the overall shape of the framework with accurate atomic positions but not to locate the Si and Al atoms precisely.

The positions of Si and Al atoms have always been assigned by applying **Loewenstein's rule**, which forbids the presence of an Al—O—Al linkage in the structure. A corollary of this rule is that when the Si/Al ratio is 1, the amount of aluminium in a zeolite is at its maximum, and the Si and Al atoms alternate throughout this structure.

When it comes to locating the cations, other problems arise. Not every cation site is completely occupied, so although the cation sites can be located, their occupancy is averaged. Furthermore, zeolites are usually microcrystalline and for successful diffraction studies larger single crystals are needed (although Rietveld powder techniques have been successfully applied, especially with neutrons, see Chapter 2). One of the techniques currently being used to elucidate zeolite structures successfully is magic angle spinning NMR spectroscopy (MAS-NMR) (see Chapter 2). Five peaks can be observed for the $^{29}Si$ spectra of various zeolites, which correspond to the five possible different Si environments. Four oxygen atoms coordinate each Si, but each oxygen can then be attached either to a Si or to an Al atom giving five possibilities:

1. $Si(OAl)_4$
2. $Si(OAl)_3(OSi)$
3. $Si(OAl)_2(OSi)_2$

4. $Si(OAl)(OSi)_3$

5. $Si(OSi)_4$

Characteristic ranges of these shifts are assigned to each coordination type. Determining the $^{27}Al$ MAS-NMR spectrum of a zeolite can distinguish three different types of aluminium:

1. Octahedrally coordinated $[Al(H_2O)_6]^{3+}$ trapped as a cation in the pores with a peak at about 0 ppm ($[Al(H_2O)_6]^{3+}$(aq) is used as the reference).
2. Tetrahedral Al, $Al(OSi)_4$, which gives a single resonance with characteristic Al chemical shift values for individual zeolites in the range of 50 to 65 ppm.
3. Tetrahedral $[AlCl_4]^-$, which gives a peak at about 100 ppm; such a peak can occur when a zeolite has been treated with $SiCl_4$ to increase the Si/Al ratio in the framework and should disappear with washing.

An important feature of $^{29}Si$ MAS-NMR spectra is that measurement of the intensity of the observed peaks allows the Si/Al ratio of the framework of the sample to be calculated. This can be extremely useful when developing new zeolites for catalysis, as much research has concentrated on making highly siliceous varieties by replacing the Al in the framework. Conventional chemical analysis only gives an overall Si/Al ratio, which includes trapped octahedral Al species and $[AlCl_4]^-$, which has not been washed away. At high Si/Al ratios, $^{27}Al$ MAS-NMR results are more sensitive and accurate and so are preferred.

Even with this information, it is still an extremely complicated procedure to decide where each linkage occurs in the structure. The cation positions can give useful information because they tend to be as close as possible to the negatively charged Al sites.

High resolution electron microscopy (HREM) is also used extensively for structural examination of zeolites, particularly for intergrowths and faults. EXAFS has been used to determine the local coordination geometry of the exchangeable cations and how this changes on reaction or dehydration.

# 7.5 USES OF ZEOLITES

## 7.5.1 Dehydrating Agents

Normal crystalline zeolites contain water molecules which are coordinated to the exchangeable cations. As noted previously, heating under vacuum can dehydrate these structures. In these circumstances, the cations move position, frequently settling on sites with a much lower coordination number. The dehydrated zeolites are very good drying agents, adsorbing water to get back to the preferred high coordination condition. Zeolite A is a commonly used drying agent, and can be regenerated by heating after use. Highly siliceous zeolites have far fewer cations and so tend to be much less hydrophilic, but can be used to absorb small organic molecules.

## 7.5.2 Zeolites As Ion Exchangers

The cations $M^{n+}$ in a zeolite will exchange with others in a surrounding solution. In this way, the $Na^+$ form of zeolite A can be used as a water softener: the $Na^+$ ions exchange with the $Ca^{2+}$ ions from the hard water. The water softener is reusable because it can be regenerated by running through a very pure saline solution; this is a familiar procedure to anyone who has used a dishwasher. Zeolite A was added to detergents as a water softener for many years, replacing the polyphosphates which have given concern about possible ecological damage; the less dense zeolite MAP, the first new synthetic zeolite prepared by Barrer, is now widely used for this purpose. It is possible to produce drinking water from seawater by desalinating it through a mixture of Ag and Ba zeolites. This is such an expensive process, however, that it is only useful in emergency.

Some zeolites have a strong affinity for particular cations. Clinoptilolite (HEU) is a naturally occurring zeolite which sequesters caesium, and is used by British Nuclear Fuels (BNFL) to remove [137]Cs from radioactive waste, exchanging its own $Na^+$ ions for the radioactive $Cs^+$ cations. Similarly, zeolite A can be used to recover radioactive strontium. Zeolites were heavily used in the clean up operations after the Chernobyl and Three Mile Island incidents.

## 7.5.3 Zeolites As Adsorbents

Because dehydrated zeolites have very open porous structures, they have large internal surface areas and are capable of adsorbing large amounts of substances other than water. The ring sizes of the windows leading into the cavities determine the size of the molecules that can be adsorbed. An individual zeolite has a highly specific sieving ability that can be exploited for purification or separation. This was first noted for chabazite (CHA) as long ago as 1932, when it was observed it would adsorb and retain small molecules such as formic acid and methanol, but would not adsorb benzene and larger molecules. Chabazite has been used commercially to adsorb polluting $SO_2$ emissions from chimneys. Similarly, the 410 pm pore opening in zeolite A (determined by an 8-ring and much smaller than the 1140 pm diameter of the cavity), can admit a methane molecule, but excludes a larger benzene molecule.

The zeolites that are useful as molecular sieves do not demonstrate an appreciable change in the basic framework structure on dehydration although the cations move to positions of lower coordination. After dehydration, zeolite A and others are remarkably stable to heating and do not decompose below about 700°C. The cavities in dehydrated zeolite A amount to about 50% of the volume.

The selective sorbent properties of zeolites have many uses, of which we will describe just a few. Table 7.2 gives a brief summary of the industrial uses that are made. The zeolites are regenerated after use by heating, evacuation, or flushing with pure gases.

It is possible to 'fine-tune' the pore opening of a zeolite to allow the adsorption of specific molecules. As discussed earlier, one method is to change the cation. This method can be used to modify zeolite A so as to separate branched and cyclic

## TABLE 7.2
## Applications of molecular sieves in industrial adsorption processes

| Fields of application | Uses | | |
|---|---|---|---|
| | Drying | Purification | Separations |
| Refineries and petrochemical industry | Paraffins, olefins, acetylenes, reformer gas, hydrocracking gas, solvents | Sweetening[a] of 'liquid petrol gas' and aromatics, removal of $CO_2$ from olefin-containing gases, purification of synthesis gas | Normal and branched-chain alkanes |
| Industrial gases | $H_2$, $N_2$, $O_2$, Ar, He, $CO_2$, natural gas | Sweetening and $CO_2$ removal from natural gas, removal of hydrocarbons from air, preparation of protective gases | Aromatic compounds |
| Industrial furnaces | Exogas, cracking gas, reformer gas | Removal of $CO_2$ and $NH_3$ from exogas and from ammonia fission gas | Nitrogen and oxygen |

[a]'Sweetening' is the removal of sulfur-containing compounds.

hydrocarbons from the straight-chain alkanes (paraffins). When Na[+] ions are replaced by Ca[2+] ions, the effective aperture increases. Once Ca[2+] has replaced approximately one-third of the Na[+] ions, many straight chain alkanes can be adsorbed (Figure 7.16). However, all branched-chain, cyclic, and aromatic hydrocarbons are excluded because their cross-sectional diameters are too large. This process can be useful industrially for separating the long straight-chain hydrocarbons required as the starting materials in the manufacture of biodegradable detergents. Petrol can also be upgraded by removing from it the straight-chain, low-octane-number alkane constituents which produce 'pinging' ('knocking'), small explosions which are harmful to engines.

At −196°C, oxygen is freely adsorbed by Ca-zeolite A, whereas nitrogen is essentially excluded. The two molecules are not very different in size: $O_2$ has a diameter of 346 pm whereas that of $N_2$ is 364 pm. As the temperature rises, the adsorption of $N_2$ increases to a maximum at around −100°C. The main reason is probably due to the thermal vibrations of the oxygen atoms in the window. Over a range of 80 to 300 K, a variation of vibrational amplitude of 10 to 20 pm could well be expected. Thus, a variation of 30 pm in the window diameter is not unreasonable. This would make the window just small enough at lower temperatures to exclude the $N_2$ molecules. Zeolite A can be used to separate or purify these gases.

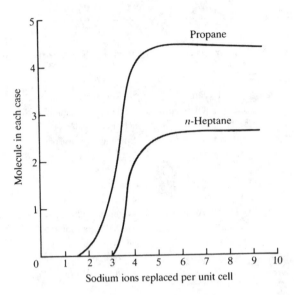

**FIGURE 7.16** Effect of calcium exchange for sodium in zeolite A on hydrocarbon adsorption. Replacement of four sodium ions by two calcium ions permits easy diffusion of *n*-alkanes into the zeolite channels.

The other method used for 'fine-tuning' the pore openings is to change the Si/Al ratio. An increase in the proportion of Si will:

- Slightly decrease the unit cell size and thus the size of the cavities.
- Decrease the number of cations, thus freeing the channels.
- Make the zeolite more hydrophobic (literally 'water-hating') in character. Hydrophobic zeolites can potentially be used to remove organic molecules from aqueous solution; possible uses range from the removal of toxic materials from blood, to the production of nonalcoholic beverages by the selective removal of alcohol, and to the decaffeination of coffee.

### 7.5.4  ZEOLITES AS CATALYSTS

Zeolites are very useful catalysts displaying several important properties that are not found in traditional amorphous catalysts. Amorphous catalysts have always been prepared in a highly divided state in order to give a high surface area and thus a large number of catalytic sites. The presence of the cavities in zeolites provides a very large internal surface area that can accommodate as many as 100 times more molecules than the equivalent amount of amorphous catalyst. Zeolites are also crystalline and so can be prepared with improved reproducibility: they tend not to demonstrate the varying catalytic activity of amorphous catalysts. Furthermore, their molecular sieve action can be exploited to control which molecules have access to (or which molecules can depart from) the active sites. This is generally known as **shape-selective catalysis**.

The catalytic activity of decationized zeolites is attributed to the presence of acidic sites arising from the [AlO$_4$] tetrahedral units in the framework. These acid sites may be Brønsted or Lewis in character. Zeolites as normally synthesized usually have Na$^+$ ions balancing the framework charges, but these can be readily exchanged for protons by direct reaction with an acid, giving surface hydroxyl groups — the **Brønsted sites**. Alternatively, if the zeolite is not stable in acid solution it is common to form the ammonium, NH$_4^+$, salt, and then heat it so that ammonia is driven off leaving a proton. Further heating removes water from the Brønsted site, exposing a tricoordinated aluminum ion, which has electron-pair acceptor properties; this is identified as a **Lewis acid site**. A scheme for the formation of these sites is shown in Figure 7.17. The surfaces of zeolites can thus display either Brønsted or Lewis acid sites, or both, depending on how the zeolite is prepared. Brønsted sites are converted into Lewis sites as the temperature is increased above 600°C and water is driven off.

Not all zeolite catalysts are used in the decationized or acid form; it is also quite common to replace the Na$^+$ ions with lanthanide ions such as La$^{3+}$ or Ce$^{3+}$. These ions now place themselves so that they can best neutralize three separated negative charges on tetrahedral Al in the framework. The separation of charges causes high electrostatic field gradients in the cavities which are sufficiently large to polarize

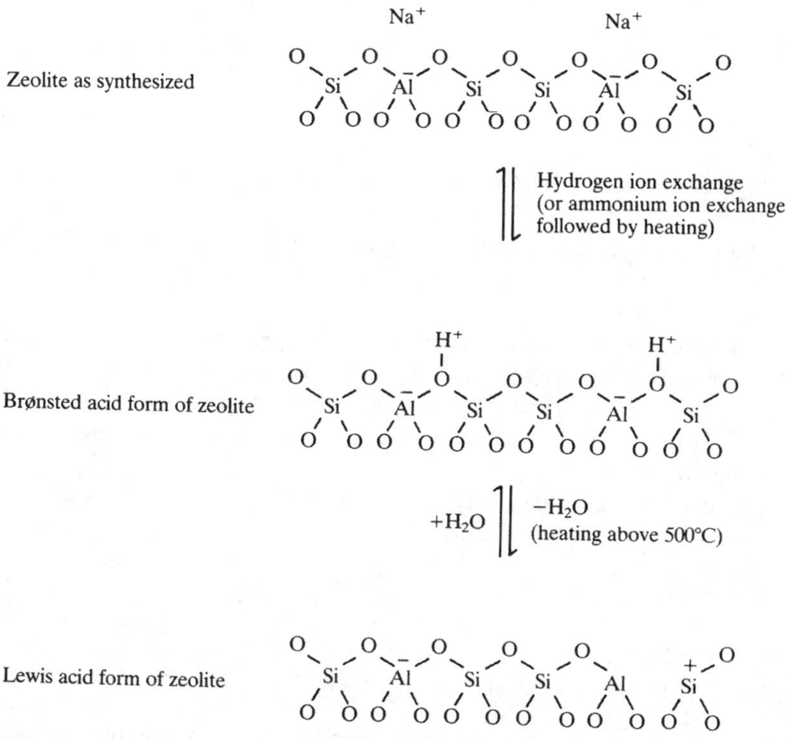

**FIGURE 7.17** Scheme for the generation of Brønsted and Lewis acid sites in zeolites.

C—H bonds or even to ionize them, enabling reaction to take place. This effect can be strengthened by a reduction in the aluminium content of the zeolite so that the $[AlO_4]$ tetrahedra are farther apart. If one thinks of a zeolite as a solid ionizing solvent, the difference in catalytic performance of various zeolites can be likened to the behaviour of different solvents in solution chemistry. It was a rare-earth substituted form of zeolite X that became the first commercial zeolite catalyst for the cracking of petroleum in the 1960s. Crude petroleum is initially separated by distillation into fractions, and the heavier gas-oil fraction is cracked over a catalyst to give petrol (gasoline). A form of zeolite Y, that has proved more stable at high temperatures, is now used. These catalysts yield ~20% more petrol than earlier catalysts, and do so at lower temperatures. The annual catalyst usage for catalytic cracking is worth billions of £s.

A third way of using zeolites as catalysts is to replace the $Na^+$ ions with other metal ions such as $Ni^{2+}$, $Pd^{2+}$, or $Pt^{2+}$ and then reduce them *in situ* so that metal atoms are deposited within the framework. The resultant material displays the properties associated with a supported metal catalyst and extremely high dispersions of the metal can be achieved. Another technique for preparing a zeolite-supported catalyst involves the physical adsorption of a volatile inorganic compound, followed by thermal decomposition: $Ni(CO)_4$ can be adsorbed on zeolite X and decomposed with gentle heating to leave a dispersed phase of nearly atomic nickel in the cavities; this has been demonstrated as a good catalyst for the conversion of carbon monoxide to methane:

$$CO + 3H_2 = CH_4 + H_2O$$

Several types of shape-selective catalysis are used:

- **Reactant shape–selective catalysis**: only molecules with dimensions less than a critical size can enter the pores and reach the catalytic sites, and so react there. This is illustrated diagrammatically in Figure 7.18(a) in which a straight-chain hydrocarbon is able to enter the pore and react but the branched-chain hydrocarbon is not.
- **Product shape–selective catalysis**: only products less than a certain dimension can leave the active sites and diffuse out through the channels, as illustrated in Figure 7.18(b) for the preparation of xylene. A mixture of all three isomers is formed in the cavities but only the *para* form is able to escape.
- **Transition-state shape–selective catalysis**: certain reactions are prevented because the transition state requires more space than is available in the cavities, as depicted in Figure 7.18(c) for the transalkylation of dialkylbenzenes.

## Reactant Shape–Selective Catalysis

Reactant shape-selective catalysis is demonstrated in the dehydration of butanols. If butan-1-ol (*n*-butanol) and butan-2-ol (*iso*-butanol) are dehydrated

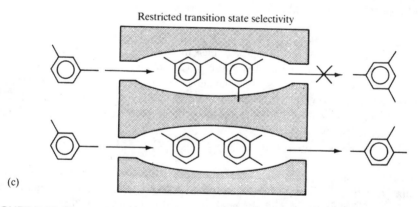

FIGURE 7.18 Shape-selective catalysis: (a) reactant, (b) product, and (c) transition state.

either over Ca-zeolite A or over Ca-zeolite X, we see a difference in the products formed.

$$CH_3CH_2CH_2CH_2OH = CH_3CH_2CH=CH_2 + H_2O$$

$$\underset{\underset{OH}{|}}{CH_3CH_2CHCH_3} = CH_3CH=CHCH_3 + H_2O$$

Zeolite X has windows large enough to admit both alcohols easily, and both undergo conversion to the corresponding alkene. Over zeolite A, however, the dehydration of

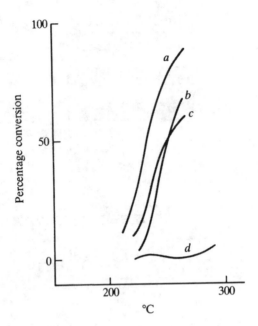

**FIGURE 7.19** Dehydration of (a) *iso*-butanol on Ca-X, (b) *n*-butanol on Ca-X, (c) *n*-butanol on Ca-A, and (d) *iso*-butanol on Ca-A.

the straight-chain alcohol is straightforward but virtually none of the branched-chain alcohol is converted, as it is too large to pass through the smaller windows of zeolite A. These results are summarized in Figure 7.19. Notice that, at higher temperatures, curve *d* begins to rise. This is because the lattice vibrations increase with temperature, making the pore opening slightly larger and thus beginning to admit butan-2-ol. The very slight conversion at lower temperatures is thought to take place on external sites.

## Product Shape–Selective Catalysis

One of the industrial processes using ZSM-5 provides us with an example of product shape–selective catalysis: the production of 1,4-(*para*-)xylene. *Para*-xylene is used in the manufacture of terephthalic acid, the starting material for the production of polyester fibres such as 'Terylene'.

Xylenes are produced in the alkylation of toluene by methanol:

$CH_3OH +$

toluene         =         (*ortho*-)1,2-xylene     +     (*meta*-)1,3-xylene     +     (*para*-)1,4-xylene     $+ H_2O$

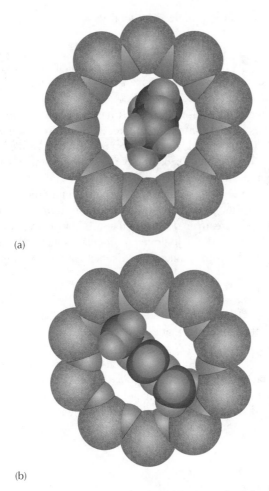

(a)

(b)

**FIGURE 7.20** Computer models illustrating how (a) *para*-xylene fits neatly in the pores of ZSM-5 whereas (b) *meta*-xylene is too big to diffuse through. See colour insert following page 356.

The selectivity of the reaction over ZSM-5 occurs because of the difference in the rates of diffusion of the different isomers through the channels. This is confirmed by the observation that selectivity increases with increasing temperature, indicating the increasing importance of diffusion limitation. The diffusion rate of *para*-xylene is approximately 1000 times faster than that of the other two isomers. The computer models in Figure 7.20 demonstrate why *meta*- and *ortho*-xylene cannot enter a 10-ring window easily and so cannot diffuse along the channels. The xylenes isomerize within the pores, and so *para*-xylene diffuses out while the *ortho*- and *meta*-isomers are trapped and have more time to convert to the *para*- form before escaping. Up to 97% of selective conversion to *para*-xylene has been achieved by suitable treatment of this catalyst. Using a zeolite catalyst makes this a much greener process than using the conventional alkylation reaction using $CH_3I$ over an $AlCl_3$ or $FeCl_3$

catalyst, as a pure product is produced in good yield, and the zeolite catalyst can be regenerated.

ZSM-5 is also used as the catalyst in the production of monoethyl benzene from benzene and ethene; it is highly selective for the mono-substituted product, which is the precursor of styrene, and again does not use the harmful aluminium chloride, which the old Friedel-Crafts process employed.

ZSM-5 is also the catalyst used to convert methanol into hydrocarbons — the *methanol to gasoline* or *MTG* process. This research received a great impetus at the end of the 1970s when oil was in short supply and prices rose sharply. Subsequently, more oil was released, prices dropped and much of the research was put into abeyance. However, the political instability of the Middle East, and concern over long-term oil reserves make this an important process. New Zealand, with no oil reserves of its own, but with a supply of natural gas ($CH_4$) brought the first MTG plant on stream in 1986. Methane is first converted to methanol, $CH_3OH$, which is then partially dehydrated to dimethyl ether and water. This then dehydrates further to give ethene and other alkenes which react rapidly to give a range of hydrocarbons.

$$2CH_3OH = CH_3OCH_3 + H_2O \xrightarrow{-H_2O} C_2 \text{ to } C_5 \text{ alkenes} \longrightarrow$$
$$\text{alkanes, cycloalkanes, aromatics}$$

The conversion over ZSM-5 into hydrocarbons produces mainly branched-chain and aromatic hydrocarbons in the $C_9$-$C_{10}$ range, which is ideal for high octane unleaded fuel. It is thought that the methanol molecules enter ZSM-5 through the more restricted zigzag channels, react at the cavities where the pores intersect, and then the larger hydrocarbons exit through the straight channels — a so-called **molecular traffic control**.

### Transition-State Shape–Selective Catalysis

In the acid-catalysed transalkylation of dialkylbenzenes, one of the alkyl groups is transferred from one molecule to another. This bimolecular reaction involves a diphenylbenzene transition state. When the transition state collapses, it can split to give either the 1,2,4-isomer or the 1,3,5-isomer, together with the monoalkylbenzene. When the catalyst used for this reaction is mordenite (Figure 7.18(c)), the transition state for the formation of the symmetrical 1,3,5-isomer is too large for the pores, and the 1,2,4-isomer is formed in almost 100% yield. (This compares with the equilibrium mixtures, in which the symmetrically substituted isomers tend to dominate.)

The synthesis of the open-framework zeolites improved the number of accessible active sites for catalysis dramatically. It is estimated that ZSM-5 has a turnover of more than 300 molecules per active site per minute during the cracking process, and that other processes such as xylene isomerization are even faster, with turnovers of up to $10^7$ molecules per active site per minute.

## 7.6 MESOPOROUS ALUMINOSILICATE STRUCTURES

In 1992, scientists at Mobil Research and Development Corporation developed a family of silicate and aluminosilicate materials, M41S, which had much larger pores than conventional zeolites. The pores in these materials lie in the range of 1.5 to 10 nm, and give rise to the name **mesoporous** solids. There are considerable advantages to having larger pores: the surface areas of the materials are very large, diffusion through the material could be easier and faster, the pores do not block up so easily, and they can be used for liquid-phase reactions, unlike zeolites which are usually used for gases and vapours. The pores are nevertheless small enough to provide shape-selective effects, and the walls of the pores can be modified by having functional groups added. The possibilities for such solids are very exciting because, as well as the more obvious roles in shape-selective catalysis, sorption and molecular sieving, incorporation of metals into the framework may lead to new and better catalysts. Applications as hosts for nanoclusters with useful magnetic, optical and electronic properties, and the formation of molecular wires in the channels are also being researched.

The mesoporous material that has received most attention so far is MCM-41. It has a highly ordered hexagonal array of uniformly sized mesopores, and it can have a huge surface area of 1200 $m^2$ per gram. It is made by a templating technique, where the silicate or aluminosilicate walls of the mesopores instead of forming around a single molecule or ion, form around an assembly of molecules known as a **micelle**. In a solution of silicate or silicate and aluminate anions, cationic long-chain alkyl trimethylammonium surfactants, $[CH_3(CH_2)_n(CH_3)_3N^+]X^-$ form rod-like micelles (Figure 7.21), with the hydrophobic tails clustering together inside the rods and the cationic heads forming the outside; the silicate/aluminate ions forms a cladding around the micelles. These silicate-coated micelles pack together along the axes of the rods, earning the synthesis technique the name **liquid crystal templating**, and under hydrothermal conditions, this mesoporous structure precipitates out of solution. Calcination of the filtered solid in air at temperatures up to 700°C removes the template and produces the mesoporous solid. As you would expect, the alkyl chain length determines the size of the pores; where $n = 11$, 13, and 15, the pore diameters are 300, 340, and 380 nm, respectively, but when $n = 5$, micelles are not formed and zeolites such as ZSM-5 are formed around a single ion template instead. Adding trimethylbenzene to the synthesis mixture increases the pore diameter by 2 nm. The hexagonal arrangement of the pores in MCM-41 is found to be very ordered, although considerable disorder occurs within the actual walls. Other mesoporous structures, such as the cubic MCM-48 and MCM-50, a lamellar structure, have also been produced.

The mesoporous solids have great potential for use as heterogeneous catalysts, which have the advantage of being simpler to use than homogeneous catalysts as they can be separated by filtration at the end of a reaction. A great deal of research has gone into various modifications of the structure to make them suitable for particular reactions. The aluminosilicate walls of mesoporous materials can be modified in various ways:

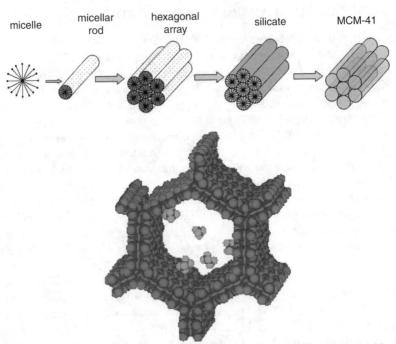

**FIGURE 7.21** (a) Possible preparative route for the formation of MCM-41 by liquid crystal templating; (b) computer graphic of ethane and methane molecules inside one of the hexagonal pores of MCM-41. See colour insert following page 356. (Courtesy of Vladimir Gusev, Romania.)

- by direct synthesis;
- by grafting on to the surface using alcohols, silanols, or organosilicon compounds;
- further modification of a functionalized surface by chemical reaction, or by heating.

Organically modified MCM-41 can be prepared directly by using alkoxysilanes or organosiloxanes in the synthesis mixture thus coating the internal wall of the pores with functional groups. An example of a condensation reaction of an alcohol with the surface silanol groups to modify the pore wall is shown in Figure 7.22.

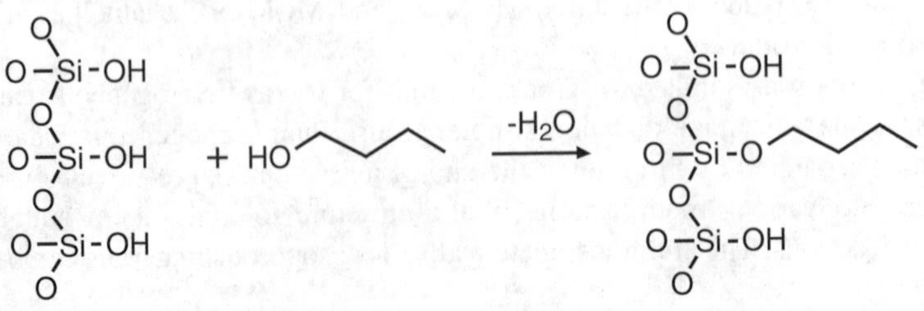

**FIGURE 7.22** Formation of a Si—O—C bond at the silicate surface.

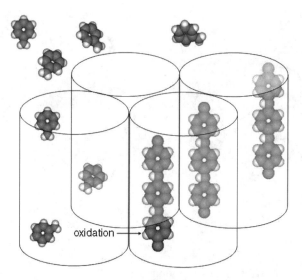

**FIGURE 7.23** The encapsulation of aniline, and formation of polyaniline in MCM-41.

Transition metals and their complexes can be immobilized in the mesopores or incorporated in the structure to make silica-supported metal catalysts. For instance, titanium catalysts for selective oxidation can be formed by modifying the mesoporous structure with either Ti grafted on the surface (Ti↑MCM-41) or Ti substituted into the framework (Ti→MCM-41). The grafted version makes the better catalyst for the epoxidation of alkenes using peroxides, and has good resistance to leaching of the metal.

Small metal clusters can be incorporated into the pores of MCM-41 by encapsulating an organometallic compound by absorption and then decomposing it at low temperatures (2–300°C). Nanometre size Sn-Mo clusters have been made in this way.

Polyaniline has been formed in the pores of Cu- or Fe-exchanged MCM-41 by adsorption of aniline vapour and subsequent oxidative polymerization (Figure 7.23), and these **molecular wires** demonstrate significant electronic conduction, although less than that of bulk polyaniline. Pyrolysis of polyacrylonitrile in the pores produces a graphite-like carbon chain, which exhibits microwave conductivity ten times that of bulk carbonized polyacrylonitrile. Such materials have potential for use in information processing as storage capacitors.

## 7.7   OTHER FRAMEWORK STRUCTURES

Other framework structures based on zeolites have also been synthesized which contain atoms other than aluminium and silicon, such as boron, gallium, germanium, and phosphorus, which are tetrahedrally coordinated by oxygen. Such compounds are known as **zeotypes**. Pure aluminium phosphate, commonly called ALPO, and its derivatives, can take the same structural forms as some of the zeolites such as sodalite (SOD), faujasite (FAU), and chabazite (CHA) (e.g., ALPO-20 is isostructural

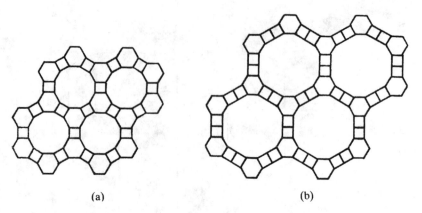

(a)                                                                      (b)

**FIGURE 7.24** Schematic structural projection drawings of (a) ALPO-5 and (b) VPI-5.

with sodalite and ALPO-17 with erionite), as well as novel structures, with many new compounds now synthesized. The metal-aluminium phosphates can be formed with metals, such as Li, Be, Mg, Mn, Fe, Co, Ni, and Zn, replacing some of the aluminium; these are usually designated MeALPOs, where Me stands for metal. If the compound contains silicon, or silicon and a metal, partially replacing aluminium or phosphorus, then they become known as SAPOs. In the same way that the replacement of Si(IV) by Al(III) in the zeolite structures leads to the formation of Brønsted acid sites, so does the replacement of Al(III) by Me(II) in ALPOs; readily detachable protons are loosely bound to a framework oxygen atom near the divalent metal. Consequently, there has been much interest in recent years in these compounds as possible heterogeneous catalysts which might be at least as good, if not better, than the zeolites. Indeed, SAPOs have been found to be very good catalysts for the conversion of methanol to olefins.

The aluminophosphate compounds have been formed using templates in a similar fashion to many of the zeolites, but under more acidic conditions. Typical templates for forming large pores are the tetra-*n*-propyl ammonium ion and tri-*n*-propylamine. One of the new channel structures formed is ALPO-5, which is depicted in projection in Figure 7.24(a); ALPO-5 has 12-ring windows with a diameter of about 800 pm, similar to that found in zeolite Y. More recently, extra-large-pore materials have been prepared. The first of these was an ALPO designated VPI-5 (Figure 7.24(b)), which has no zeolite analogue, and is based on β-cages separated by double four-membered rings to give 18-ring windows and a pore diameter of between 1200 and 1300 pm. Others such as cloverite, a gallophosphate with a 3000 pm supercage accessed through six cloverleaf-shaped channels (Figure 7.25), have followed. As with the mesoporous aluminosilicates, some of the interest in these large channelled structures lies in their potential for allowing catalytic conversions to take place using molecules which are too large to enter the conventional zeolitic–size channels; one example of this, the hydrogenation of cyclooctene using rhodium-substituted VPI-5, has been reported. Other possibilities lie in the formation of new materials.

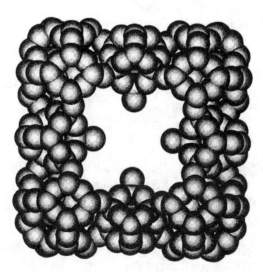

**FIGURE 7.25** The framework of the synthetic gallophosphate, cloverite, represented as shaded spheres of van der Waals radii. (From C.R.A. Catlow (ed.) (1992) *Modelling of Structure and Reactivity in Zeolites*, Academic Press, London. Reproduced with permission of Academic Press Ltd.)

## 7.8 NEW MATERIALS

Research in zeolites has also branched out to try to prepare new materials by incorporating various molecules and ions in the cages of these microporous and mesoporous structures. An early example of this was the preparation of the pigment ultramarine used in many paints and colourants. It is based on the zeolite sodalite (SOD) structure and contains $S_3^-$ ions trapped in the cages; this is the same anion found in the mineral lapis lazuli, to which it imparts the beautiful deep blue colour. Treatment of zeolites such as Na-zeolite Y with sodium vapour traps $Na_4^{3+}$ ions in the cavities, which impart a deep red colour.

One area of this research has focussed on depositing semiconductor materials in zeolite cages. The resulting nanometre-size particles (sometimes called **quantum dots**) exhibit quantum-size effects in their physical properties (i.e., they have interesting electronic, magnetic, and optical properties which are a consequence of their size rather than their chemical composition). One example of this is the narrow band gap semiconductor CdS, which in low concentrations forms discrete cubic $(CdS)_4$ clusters in the sodalite cages of zeolites A, X, and Y. These clusters start to absorb in the UV at 350 nm, compared with 540 nm in bulk CdS, and do not exhibit the intense emission of the bulk compound. As the pore filling of the zeolite increases, the nanoparticles become interconnected forming superclusters, and materials are produced with properties intermediate between those of discrete particles and bulk semiconductors; the absorption edge shifts to 420 nm, and strong emission also appears. Other semiconductors such as Se, Te, and GaP have also been encapsulated in this way, with the hope that this research will lead to possibilities for optical transistors and optical data storage devices. Other guests that can be incorporated into zeolite cages and into the mesoporous structures, include the alkali metals, silver and silver salts, selenium, and various conducting polymers.

Suffice it to say that these new materials are being investigated with respect to interesting physical properties which then have the possibility of commercial exploitation — properties such as semi-, photo-, and fast-ion conductivity, luminescence, colour, and quantum-size effects.

## 7.9 CLAY MINERALS

Similar to the zeolites, the smectite clays are a class of naturally occurring aluminosilicate minerals. They are often found as components of soils and sediments, and some large deposits, such as bentonite (consisting mainly of montmorillonite and beidellite), are found in Cornwall, United Kingdom, and in the state of Wyoming, USA. The smectite clays all have a basic layer structure, which is depicted in Figure 7.26; it consists of parallel layers of tetrahedral silicate $[SiO_4]$ sheets and octahedral aluminate $[Al(O,OH)_6]$ sheets. The silicate layers contain $[SiO_4]$ tetrahedra each linked through three corners to form an infinite layer. The aluminate layers contain a plane of octahedrally coordinated edge-linked aluminium ions, sandwiched between two inward pointing sheets of corner-linked $[SiO_4]$ tetrahedra. If there is no replacement of Si or Al, then the layers are electrically neutral, producing the mineral pyrophyllite. In the smectite clays, different structures are formed because substitution of silicon and aluminium by metal ions can take place in both the tetrahedral and octahedral layers, the resulting negative charge is distributed on the oxygens of the layer surface, and any charge balance is restored by interlayer cations (usually $Na^+$ or $Ca^{2+}$). For instance, in montmorillonite, approximately one-sixth of the $Al^{3+}$ ions have been replaced by $Mg^{2+}$, whereas in beidellite about one-twelfth of the $Si^{4+}$ have been replaced by $Al^{3+}$. The different members of the smectite group of clays are distinguished by the type and position of the cations in the framework.

The peculiar layer structure of these clays gives them cation exchange and intercalation properties that can be very useful. Molecules, such as water, and polar organic molecules, such as glycol, can easily intercalate between the layers and cause the clay to swell. Water enters the interlayer region as integral numbers of complete layers. Calcium montmorillonite usually has two layers of water molecules but the sodium form can have one, two, or three water layers; this causes the interlayer spacing to increase stepwise from about 960 pm in the dehydrated clay to 1250, 1550, and 1900 pm as each successive layer of water forms.

The $Na^+$ and $Ca^{2+}$ cations which make up the charge balance due to the substitution of Si and Al in the layers by other metals, are usually hydrated and are located in the interlayer regions, loosely bound to the layer surfaces. They are known as **exchangeable cations**, and can be replaced easily by other cations using ion exchange methods or by protons in the form of $H_3O^+$, to form an acidic clay. Such acidic clays form very useful catalysts when the reactant molecules enter the interlayer regions. For many years, modified clays were used as the catalyst for petroleum cracking to produce petrol, although they have since been replaced by more thermally stable and selective zeolites. However, smectite clays can be used for the dimerization of oleic acid, the conversion of hexene to dihexyl ethers, and the formation of ethyl

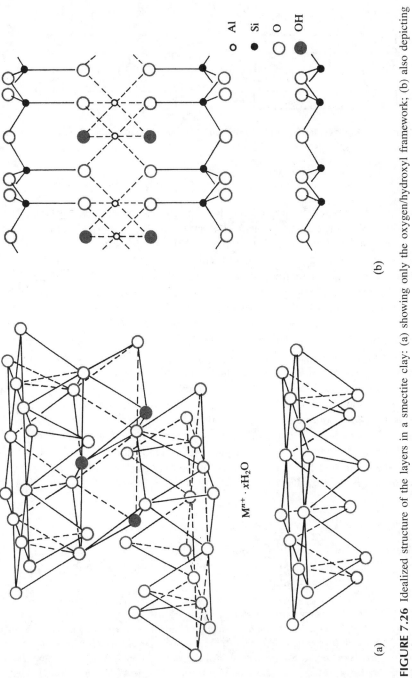

**FIGURE 7.26** Idealized structure of the layers in a smectite clay: (a) showing only the oxygen/hydroxyl framework; (b) also depicting aluminium and silicon positions.

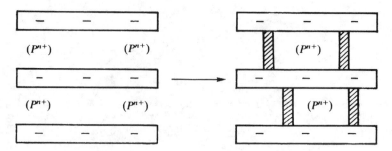

**FIGURE 7.27** Representation of the formation of pillars in a smectite clay where $P^{n+}$ is the pillaring cation.

acetate from ethanoic acid and ethene. Recent research on clay catalysts has concentrated on the introduction of metal complexes into the interlayer region to act as the catalytic centre and on the formation of robust molecular props to hold the interlayer regions apart — so-called **pillaring**.

One of the problems with using clays as catalysts is that at high temperatures (> 200°C), they start to dehydrate and the interlayer region collapses. Much effort has concentrated on forming temperature-stable pillars that will hold. the layers apart even at high temperature and in the absence of a swelling solvent, to form a large interlayer region. The pillaring leads to a porous network between the layers analogous to the zeolite channels (Figure 7.27) and the hope is that by judicious adjustment of the size and spacings of the pillars, the pore sizes can be made bigger than those of the faujasitic zeolites. Because the layers are negatively charged, the molecular props that have been used for pillaring are large cations such as tetraalkyl ammonium ions and the 'Al$_{13}$' Keggin ion, [Al$_{13}$O$_4$(OH)$_{28}$$^{3+}$]; once the ion is incorporated, the clay is heated to dehydrate it, resulting in columns of small clusters of alumina (Al$_2$O$_3$) which are thermally stable up to 500°C. Other pillars can be formed from silica, and from the oxides of various metals, such as iron, zirconium, and tin.

## 7.10 POSTSCRIPT

We end this chapter with a quotation from a lecture given by Professor J.M. Thomas. It should please those of you who are beginning to love the patterns and symmetry of crystalline solids.

" . . . [Figure 7.28] shows, on the right, a projected structure of the zeolite we have been studying, a synthetic catalyst discovered in New Jersey in 1975. Its structure was elucidated some 6 years ago. On the left, I reproduce a pattern made on the wall of a mosque in Baku in the Soviet state of Azerbaijan in 1086 AD, the year that the Domesday Book appeared by order of William the Conqueror. These two structures have exactly the same pattern. 'There is nothing new under the sun.' "

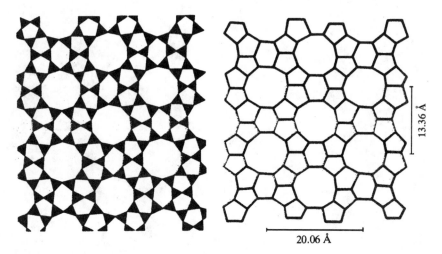

**FIGURE 7.28** Around a central 10-membered ring in the synthetic zeolite catalyst, presented schematically on the right, eight 5-membered and two 6-membered rings are present. In exactly the same sequence, such 5- and 6-membered rings circumscribe a central 10-membered ring in the pattern on the left.

## QUESTIONS

1. Zeolite A has a single peak in the $^{29}$Si MAS NMR spectrum at 89 ppm and has a Si/Al ratio of 1. Comment on these observations.
2. A series of zeolites was synthesized with an increasing Si:Al ratio. It was found that the catalytic activity also increased until the ratio was about 15:1 after which it declined. Suggest a reason for this behaviour
3. Zeolite A (Ca form), when loaded with platinum, has been found to be a good catalyst for the oxidation of hydrocarbon mixtures. However, if the mixture contains branched-chain hydrocarbons, these do not react. Suggest a possible reason.
4. Both ethene and propene can diffuse into the channels of a particular mordenite catalyst used for hydrogenation. Explain why only ethane is produced.
5. Explain why, when 3-methyl pentane and *n*-hexane are cracked over zeolite A (Ca form) to produce smaller hydrocarbons, the percentage conversion for 3-methyl pentane is less than 1% whereas that for *n*-hexane is 9.2%.
6. When toluene is alkylated by methanol with a ZSM-5 catalyst, increase in the crystallite size from 0.5 to 3 μm approximately doubles the amount of *para*-xylene produced. Suggest a possible explanation.
7. The infrared stretching frequency of the hydroxyl associated with the Brønsted sites in decationized zeolites, falls in the range 3600 to 3660 cm$^{-1}$. As the Si/Al ratio in the framework increases, this frequency tends to decrease. What does this suggest about the acidity of the highly siliceous zeolites?

# 8 Optical Properties of Solids

## 8.1 INTRODUCTION

Perhaps the most well-known example of solid state optical devices is the **laser** and, with the widespread use of CDs, DVDs, and laser printers, lasers have become commonplace. Of interest to the solid state chemist are two types of laser, typified by the **ruby laser** and the **gallium arsenide laser**. Because laser light is more easily modulated than light from other sources, it is increasingly used for sending information: light travelling along optical fibres and replacing electrons travelling along wires in, for example, telecommunications. To transmit light over long distances, the optical fibres must have particular absorption and refraction properties, and the development of suitable substances has become an important area of research.

Applications of solid state optical properties were known and commercially important before lasers were developed, and examples of these are also considered. **Light-emitting diodes (LEDs)** are used for displays including those on digital watches and scientific instruments. The mechanism by which light is produced in LEDs is similar to that in the gallium–arsenide laser. Another important group of solids are the light-emitting solids known as **phosphors**, which are used on television screens and fluorescent light tubes. In the past two decades, a new type of material, photonic crystals, with the potential to form the basis of optical integrated circuits has been the subject of much interest. These are discussed briefly in this chapter.

Very broadly speaking, two situations have to be considered in explaining devices such as those we have mentioned. In the first, which is relevant to the ruby laser and to phosphors for fluorescent lights, the light is emitted by an impurity ion in a host lattice. We are concerned here with what is essentially an atomic spectrum modified by the lattice. In the second case, which applies to LEDs and the gallium arsenide laser, the optical properties of the delocalised electrons in the bulk solid are important.

We shall begin with the first case and take atomic spectra as our starting point.

## 8.2 THE INTERACTION OF LIGHT WITH ATOMS

When an atom absorbs a photon of light of the correct wavelength, it undergoes a transition to a higher energy level. To a first approximation in many cases, we can think of one electron in the atom absorbing the photon and being excited. The electron will only absorb the photon if the photon's energy matches that of the energy difference between the initial and final electronic energy level, and if certain rules, known as selection rules, are obeyed. In light atoms, the electron cannot change its spin and its orbital angular momentum must change by one unit; in terms of quantum numbers $\Delta s = 0$, $\Delta l = \pm 1$. (One way of thinking about this is that the photon has zero spin and one unit of angular momentum. Conservation of spin and angular momentum then produces these rules.) For a sodium atom, for example, the $3s$ electron can absorb one photon and go to the $3p$ level. (There is no restriction on changes of the principal quantum number.) The $3s$ electron will not, however, go to the $3d$ or $4s$ level. Figure 8.1 illustrates allowed and forbidden transitions.

The spin and orbital angular momenta are not entirely independent and coupling between them allows forbidden transitions to occur, but the probability of an electron absorbing a photon and being excited to a forbidden level is much smaller than the probability of it being excited to an allowed level. Consequently, spectral lines corresponding to forbidden transitions are less intense than those corresponding to allowed transitions.

An electron that has been excited to a higher energy level will sooner or later return to the ground state. It can do this in several ways. The electron may simply emit a photon of the correct wavelength some time after it has been excited. This is known as **spontaneous emission**. Alternatively, a second photon may come along and instead of being absorbed may induce the electron to emit. This is known as **induced** or **stimulated emission** and plays an important role in the action of lasers. The emitted photon in this case is in phase with, and travelling in the same direction as, the photon inducing the emission; the resulting beam of light is said to be coherent. Finally, the atom may collide with another atom, losing energy in the

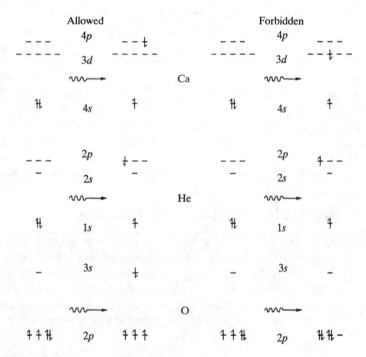

**FIGURE 8.1** Allowed and forbidden atomic transitions.

process, or give energy to its surroundings in the form of vibrational energy. These are examples of **non-radiative transitions**. Spontaneous and stimulated emissions obey the same selection rules as absorption. Non-radiative transitions have different rules. In a crystal (or, of course, a molecule), the atomic energy levels and selection rules are modified. Let us take as an example an ion with one d electron outside a closed shell ($Ti^{3+}$ for example). This will help us understand the ruby laser.

In the free ion, the five $3d$ orbitals all have the same energy. In a crystal, these levels are split; for example, if the ion occupied an octahedral hole, the $3d$ levels would be split into a lower, triply degenerate ($t_{2g}$) level and a higher, doubly degenerate ($e_g$) level. This is depicted in Figure 8.2.

An electronic transition between these levels is now possible. In the free ion, a transition from one $d$ level to another involves zero energy change, and so would not be observed even if it were allowed. In the crystal, the transition involves a change in energy, but is still forbidden by the selection rules. Lines corresponding to such transitions can, however, be observed, albeit with low intensity, because the crystal vibrations mix different electronic energy levels. Thus, the $3d$ levels may be mixed with the $4p$ giving a small fraction of 'allowedness' to the transition. Figure 8.3 depicts an absorption band due to a transition from $t_{2g}$ to $e_g$ for the ion $Ti^{3+}$, which has one $d$ electron. (This band is in fact two closely spaced bands as the excited state is distorted from a true octahedron and the $e_g$ level further split into two.)

We shall see now the role played by a similar forbidden transition in the operation of the ruby laser.

### 8.2.1 THE RUBY LASER

Ruby is corundum (one form of $Al_2O_3$) with 0.04 to 0.5% $Cr^{3+}$ ions as an impurity replacing aluminum ions. The aluminum ions, and hence the chromium ions, occupy

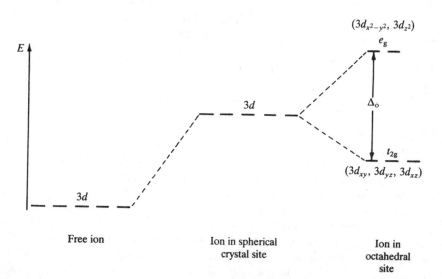

**FIGURE 8.2** Splitting of $d$ levels in an octahedral site in a crystal.

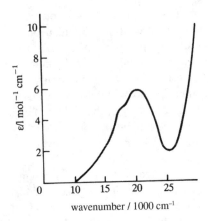

wavenumber / 1000 cm$^{-1}$

**FIGURE 8.3** The $t_{2g}$ to $e_g$ transition of Ti$^{3+}$. The band is, in fact, two overlapping bands. This is due to a further splitting of the $e_g$ levels.

distorted octahedral sites. As discussed previously, therefore, the 3$d$ levels of the chromium ion will be split. Cr$^{3+}$ has three 3$d$ electrons, and, in the ground state, these occupy separate orbitals with parallel spins. When light is absorbed, one of these electrons can undergo a transition to a higher energy 3$d$ level. This is similar to the transition discussed in the previous section, but with three electrons it is necessary to consider changes in the electron repulsion as well as changes in orbital energies. Including electron repulsion changes gives two transitions corresponding to the jump from the lower to the higher 3$d$ level.

Having absorbed light and undergone one of these transitions, the chromium ion could now simply emit radiation of the same wavelength and return to the ground state. However, in ruby, a fast, radiationless transition occurs, in which the excited electron loses some of its energy and the crystal gains vibrational energy. The chromium ion is left in a state in which it can only return to the ground state by a transition in which an electron changes its spin. Such a transition is doubly forbidden because it still breaks the rule that forbids 3$d$⇔3$d$ and is even less likely to occur than the original absorption process. The states involved are depicted schematically in Figure 8.4.

The ions absorb light and go to states 3 and 4. They then undergo a radiationless transition to state 2. Because the probability of spontaneous emission for state 2 is low, and no convenient non-radiative route is available to the ground state, a considerable population of state 2 can build up. When eventually (about 5 milliseconds later) some ions in state 2 return to the ground state, the first few spontaneously emitted photons interact with other ions in state 2 and induce these to emit. The resulting photons will be in phase and travelling in the same direction as the spontaneously emitted photons and will induce further emission as they travel through the ruby. The ruby is in a reflecting cavity so that the photons are reflected back into the crystal when they reach the edge. The reflected photons induce further emission and by this means, an appreciable beam of coherent light is built up. The mirror on one end can then be removed and a pulse of light emitted. The name laser is a reflection of this build up of intensity. It is an acronym that stands for 'Light

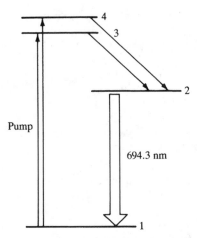

**FIGURE 8.4** The states of the $Cr^{3+}$ ion involved in the ruby laser transition.

Amplification by Stimulated Emission of Radiation'. (Similar devices producing coherent beams of microwave radiation are known as *masers*.) A typical arrangement for a pulsed ruby laser is depicted in Figure 8.5.

A high intensity flash lamp excites the $Cr^{3+}$ ions from level 1 to levels 3 and 4. The lamp can lie alongside the crystalline rod of ruby or be wrapped around it. At one end of the reflective cavity surrounding the ruby crystal is a Q switch, which switches from being reflective to transmitting the laser light and can be as simple as a rotating segmented mirror, but is usually a more complex device.

Ruby was the first material for lasers but several other crystals are now employed. The crystals used need to contain an impurity with an energy level such that return to the ground state is only possible by a forbidden transition in the infrared, visible, or near ultraviolet. It must also be possible to populate this level via an allowed (or at least less forbidden) transition. Research has tended to concentrate on transition metal ions and lanthanide ions in various hosts since these ions have suitable transitions of the

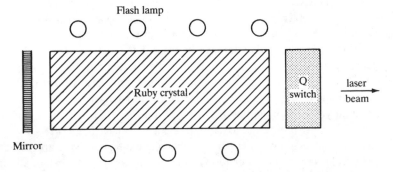

**FIGURE 8.5** Sketch of a ruby laser.

**TABLE 8.1**
**Impurity ions used in lasers**

| Ion | Host | Wavelength emitted (nm) |
|-----|------|-------------------------|
| $Nd^{3+}$ | Fluorite ($CaF_2$) | 1046 |
| $Sm^{3+}$ | Fluorite | 708.5 |
| $Ho^{3+}$ | Fluorite | 2090 |
| $Nd^{3+}$ | Calcium tungstate ($CaWO_4$) | 1060 |

right wavelength. Some examples are given in Table 8.1 with the wavelength of the laser emission.

## 8.2.2 PHOSPHORS IN FLUORESCENT LIGHTS

Phosphors are solids which absorb energy and re-emit it as light. As in the lasers we have just described, the emitter is usually an impurity ion in a host lattice. However for the uses to which phosphors are put it is not necessary to produce intense, coherent beams of light, and the emitting process is spontaneous instead of induced. Phosphors have many applications, for example, the colours of your television picture are produced by phosphors that are bombarded with electrons from a beam (cathode rays) or from a transistor (flat screen LCD displays). In terms of tonnage produced one of the most important applications is the fluorescent light tube.

Fluorescent lights produce radiation in the ultraviolet (254 nm) by passing an electric discharge through a low pressure of mercury vapour. The tube is coated inside with a white powder which absorbs the ultraviolet light and emits visible radiation. For a good fluorescent light, the efficiency of the conversion should be high and the emitted light should be such that the appearance of everyday objects viewed by it should resemble as closely as possible their appearance in daylight. Most phosphors for fluorescent lights have been based on alkaline earth halophosphates such as $3Ca_3(PO_4)_2.CaF_2$. As in lasers, the usual dopants are transition metal or lanthanide ions, but more than one impurity ion is needed to approximate the whole visible spectrum. Not all the impurity ions need to be capable of absorbing the exciting radiation, however, as the host lattice can act to transmit the energy from one site to another. For example, in a phosphor doped with $Mn^{2+}$ and $Sb^{3+}$ ions, the ultraviolet radiation from the mercury lamp is only absorbed by the antimony ($Sb^{3+}$) ions. The excited antimony ion drops down to a lower excited state via a non-radiative transition. Emission from this lower state produces a broad band in the blue region of the visible spectrum. Some of the energy emitted by the antimony travels through the host crystal and is absorbed by the manganese ions. The excited $Mn^{2+}$ ions emit yellow light and return to the ground state. The two emission bands together produce something close to daylight. Phosphors have been introduced which are more efficient and give a closer approximation to daylight. A good approximation is, for example, given by a combination of blue from barium magnesium aluminate doped with divalent europium ($Eu^{2+}$), green from an aluminate doped with cerium

($Ce^{3+}$) and terbium ($Tb^{3+}$) ions, and red from yttrium oxide doped with trivalent europium ($Eu^{3+}$).

Fluorescent lights also emit broadband radiation in the near infrared. Finding luminescent materials that will convert this to visible light has generated much interest. In this case, the incident light is of lower energy than the emitted light and the process is known as **upconversion**. Such processes are also exploited in upconversion lasers where a phosphor is used to produce shorter wavelength light from a red laser. Obviously, a ground state atom or ion cannot absorb a photon of one frequency radiation and then emit a photon of higher frequency radiation from the excited state reached by the absorption process. So how does upconversion work?

In upconversion systems absorption takes place in two stages. An ion absorbs a photon of the incident radiation and goes to an excited state. It then transfers most of the energy either to another state of that ion or to the excited state of another ion. If this second excited state is metastable, it has time to absorb another photon before it spontaneously emits radiation and returns to a lower state. Either the absorbing ion or one to which the energy has been transferred is now in a higher level, whose energy above the ground state is greater than the energy of the absorbed photon. Emission from this higher level produces shorter wavelength radiation than that absorbed. The ions doped into upconvertors are frequently lanthanide ions, or a transition metal ion and a lanthanide ion. For example, low concentrations of $Ho^{3+}$ in $Y_2O_3$ absorb red laser light and emit yellow-green light. $Ho^{3+}$ in the ground state, $^5I_8$, absorbs a photon and is excited to an excited state, $^5F_5$, from where it drops to a succession of lower energy states ending in the $^5I_7$ state, Figure 8.6. While in the $^5I_7$ state the ion absorbs another photon of red light and goes to the $^5F_3$ state from where it emits yellow-green light and returns to the ground state (Figure 8.6).

In phosphors and in the ruby laser, light was absorbed and emitted by electrons localised on an impurity site, but in other optical devices, delocalised electrons emit the radiation. In the next section, therefore, we shall consider the absorption and emission of radiation in solids with delocalised electrons, particularly in semiconductors.

## 8.3 ABSORPTION AND EMISSION OF RADIATION IN SEMICONDUCTORS

Radiation falling on a semiconductor will be absorbed by electrons in delocalised bands, particularly those near the top of the valence band, causing these electrons to be promoted to the conduction band. Because many closely packed levels are present in an energy band, the absorption spectrum is not a series of lines as in atomic spectra, but a broad peak with a sharp threshold close to the band gap energy. The absorption spectrum of GaAs, for example, is depicted in Figure 8.7.

Transitions to some levels in the conduction band are more likely than transitions to others. This is because transitions between valence band and conduction band levels, like those between atomic energy levels are governed by selection rules. The spin selection rule still holds; when promoted to the conduction band the electron

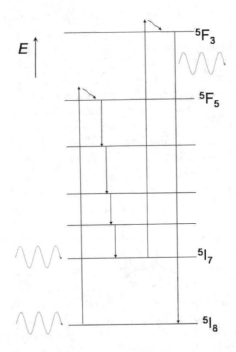

**FIGURE 8.6** Atomic states of $Ho^{3+}$.

does not change its spin. Orbital angular momentum rules are not appropriate for energy bands, however, and so the rule governing change in the quantum number, $l$, is replaced by a restriction on the wave vector, $kh$. As discussed in Chapter 4, the energy levels in a band are characterised by the wave vector, the momentum of the electron wave being given by $k\hbar$. The momentum of a photon with wavelength in the infrared, visible, or ultraviolet is very small compared to that of the electron in the band and so conservation of momentum produces the selection rule for transitions

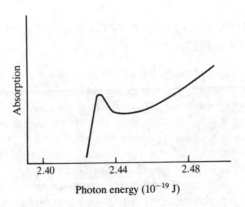

**FIGURE 8.7** Absorption spectrum of GaAs.

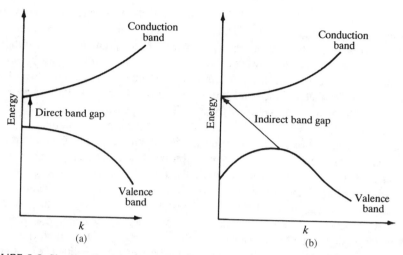

**FIGURE 8.8** Sketch of energy bands for (a) a solid with a direct band gap and (b) a solid with an indirect band gap. Note that in this diagram the horizontal axis is $k$, not the density of states. In this representation, a band is depicted as a line going from 0 to the maximum value of $k$ occurring for that band.

between bands. An electron cannot change its wave vector when it absorbs or emits radiation. Thus an electron in the valence band with wave vector, $k_j$, can only undergo allowed transitions to levels in the conduction band which also have wave vector, $k_j$. In some solids, for example GaAs, the level at the top of the valence band and that at the bottom of the conduction band have the same wave vector. Subsequently, an allowed transition occurs at the band gap energy. Such solids are said to have a **direct band gap**.

For other semiconductors, for example silicon, the direct transition from the top of the valence band to the bottom of the conduction band is forbidden. These solids are said to have an **indirect band gap**. Illustrations of band structures for solids with direct and indirect band gaps are given in Figure 8.8. Note that in these diagrams, the energy for a band in one direction is plotted against the wave number, $k$. Similar diagrams were given in Chapter 4 but with the energy plotted against the density of states.

The simple free electron model might suggest that the lowest energy orbital in any band is that with $k = 0$. Figure 8.9, however, illustrates two combinations of

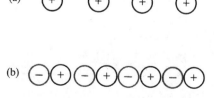

**FIGURE 8.9** A row of (a) $s$ orbitals and (b) $p$ orbitals, both with $k = 0$.

orbitals that will have $k = 0$ for a chain because all the atomic orbitals are combined in phase. The combination of $p$ orbitals is obviously antibonding and so would be expected to have the highest energy in its band; the combination of $s$ orbitals is bonding and would have the lowest energy in its band. If the $p$ band lies below the $s$ band, a transition between these levels would be allowed and would correspond to a direct transition across the band gap. In real solids, the highest and lowest levels in bands will contain contributions from different types of atomic orbital and it becomes difficult to predict whether a band gap will be direct or indirect.

One consequence of an indirect band gap is that an electron in the bottom level of the conduction band has only a small probability of emitting a photon and returning to the top of the valence band. This is of importance when selecting materials for some of the applications we are going to consider.

Transitions across the band gap are also responsible for the appearance of many solids. Because a solid is very concentrated, the probability that a photon with energy that corresponds to an allowed transition will be absorbed, is very high. Many such photons will therefore be absorbed at or near the surface of the solid. These photons will then be re-emitted in random directions so that some will be reflected back toward the source of radiation and some will travel farther into the solid. Those travelling into the solid stand a very good chance of being re-absorbed and then re-emitted, again in random directions. The net effect is that the radiation does not penetrate the solid but is reflected by its surface. If the surface is sufficiently regular, then solids which reflect visible radiation appear shiny. Thus, silicon, with a band gap that is at the lower end of the visible region and has allowed transitions covering most of the visible wavelengths, appears shiny and metallic. Many metals have strong transitions between the conduction band and a higher energy band, which lead to their characteristic metallic sheen. Some metals, such as tungsten and zinc, have a band gap in the infrared, and transitions in the visible are not so strong. These metals appear relatively dull. Gold and copper have strong absorption bands due to excitation of d band electrons to the $s/p$ conduction band. In these elements, the $d$ band is full and lies some distance below the Fermi level (Figure 8.10). The reflectivity peaks in the yellow part of the spectrum and blue and green light are less strongly absorbed, hence the metals appear golden. Very thin films of gold appear

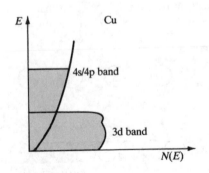

**FIGURE 8.10** Band structure of copper.

green because the yellow and red light is absorbed, and only the blue and green transmitted.

Insulators typically have band gaps in the ultraviolet, and unless a localised transition occurs in the visible region of the electromagnetic spectrum, appear colourless.

The devices which we consider involving band gap transitions are, however, concerned with emission of light instead of absorption or reflection; the electrons being initially excited by electrical energy.

### 8.3.1 LIGHT-EMITTING DIODES

Light-emitting diodes (LEDs) are widely used for displays. Similar to transistors, they are based on the $p$–$n$ junction but the voltage applied across the $p$–$n$ junction in this case leads to the emission of light.

Figure 8.11 is a $p$–$n$ junction in a semiconductor such as GaAs.

The band structure depicted is for the junction in the dark and with no electric field applied. Now suppose that an electrical field is applied so that the $n$-type is made negative relative to the $p$-type (i.e., in the reverse direction to the applied voltage in transistors; see Chapter 4). Electrons will then flow from the $n$-type to the $p$-type. An electron in the conduction band moving to the $p$-type side can drop down into one of the vacancies in the valence band on the $p$-type side, emitting a photon in the process. This is more likely to happen if the transition is allowed, so that semiconductors with direct band gaps are usually used in such devices. To use the LED as a display, for example, it is then wired so that an electric field is applied across the parts making up the required letters or numbers. Using semiconductors of differing band gap can produce different colours. GaP produces red light, but by mixing in various proportions of aluminum to form $Ga_{1-x}Al_xP$, green or orange light can be produced.

It should be noted that semiconductors with indirect band gaps are used for LEDs, but in these cases, impurity levels play an important role. Thus, GaP is used

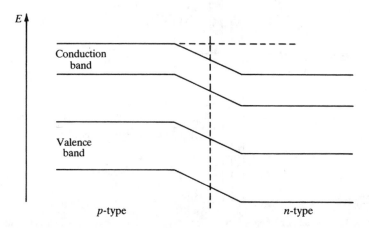

FIGURE 8.11 Energy bands near the junction in a $p$–$n$ junction.

although it has an indirect band gap. Silicon, however, is not suitable because a non-radiative transition is available to electrons at the bottom of the conduction band and these electrons donate thermal energy to the crystal lattice instead of emitting light when they return to the valence band.

When the electric field causes conduction electrons to move across the *p–n* junction, the resulting situation is one in which the population in the conduction band is greater than the thermal equilibrium population. An excess of electrons in an excited state is an essential feature of lasers, and several semiconductor lasers are based on the *p–n* junction. The best known of these is the gallium arsenide laser.

### 8.3.2 THE GALLIUM ARSENIDE LASER

The gallium arsenide laser actually contains a layer of GaAs sandwiched between layers of *p*- and *n*-type gallium aluminum arsenide ($Ga_{1-x}Al_xAs$). As depicted in Figure 8.12, the band gap of gallium aluminum arsenide is larger than that of gallium arsenide.

An electric field applied across the *p–n* junction, as in LEDs, produces an excess of electrons in the conduction band of the gallium arsenide. These electrons do not drift across into the gallium aluminum arsenide layer, however, because the bottom of the conduction band in this layer is higher in energy and the electrons would thus need to gain energy in order to move across. The excess conduction band electrons are therefore constrained to remain within the GaAs layer. Eventually, one of these electrons drops down into the valence band, emitting a photon as it does so. This photon induces other conduction band electrons to return to the valence band and

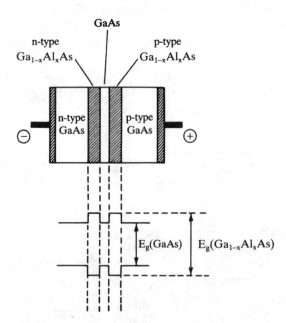

**FIGURE 8.12** The arrangement of the different semiconductor regions in a GaAs laser. The band gap profile is depicted below these regions.

thus a coherent beam of light begins to build up. As in the ruby laser, the initial burst of photons is reflected back by mirrors placed at the ends and induces more emission. Eventually a beam of infrared radiation is emitted.

Several such lasers have been developed, most of them based on III-V compounds (i.e., compounds of In, Ga, and/or Al with As, P, or Sb). It is possible to manufacture materials with band gaps over the range of 400 to 1300 nm by carefully controlling the ratios of the different elements.

Infrared semiconductor lasers are used, among many other things, for reading compact discs. A compact disc consists of a plastic disc coated with a highly reflective aluminum film and protected against mechanical damage by a layer of polymer. The original sound recording is split up into a number of frequency channels and the frequency of each channel given as a binary code (that is a series of 0s and 1s). This is converted to a series of pits in tracks on the disc spaced approximately 1.6 μm apart. A laser is focused on the disc and reflected on to a photodetector. The pits cause some of the light to be scattered thus reducing the intensity of the reflected beam. The signal read by the photodetector is read as 1 when a high intensity of light is present and as 0 when scattering reduces the intensity. The binary code is thus recovered and can be converted back to sound or pictures. DVDs are similar but use a laser of shorter wavelength which can read more closely spaced pits. This means more information can be packed into a given area.

### 8.3.3 Quantum Wells — Blue Lasers

Blue lasers allow higher resolution, and hence higher density of optical storage of information, on devices such as DVDs than the infrared GaAs lasers allow. The earliest blue lasers were based on ZnSe but their lifetime proved too short for commercial applications. Lasers based on gallium nitride (GaN), first demonstrated in 1995, have proved to have greater lifetimes. In these lasers, the photons are produced not in a bulk semiconductor but in quantum wells.

The active region of GaN lasers consists of of GaN containing several thin layers (30–40 Å thick) of indium doped-GaN, $In_xGa_{1-x}N$. The addition of indium reduces the band gap within the thin layers, so that the bottom of the conduction band is at lower energy than that in the bulk GaN. Electrons in this conduction band are effectively trapped because they need to gain energy from an external source to pass into the conduction band of the bulk GaN. Figure 8.13 illustrates schematically the bottom of the conduction band and the top of the valence band for a series of thin layers of $In_xGa_{1-x}N$ in GaN.

The trapped electrons behave like particles in a box (see Chapter 4) but with finite energy walls to the box. Such boxes are **quantum wells**. Within the well, the electron energy is quantised and the spacing of the energy levels depends on the (energy) depth and (spatial) width of the well. The depth of the wells is controlled by the extent of doping (i.e., the value of $x$). Although the electrons in the wells have insufficient energy to surmount the energy barrier to reach the next well, the probability exists that they will move to a similar energy level in the next well via quantum mechanical tunnelling. Electrons travelling from the bulk GaN enter a high level in the first well (Figure 8.13). From this level, the electron can emit a photon

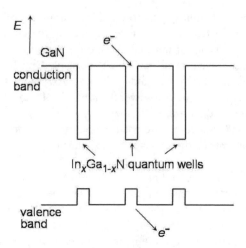

**FIGURE 8.13** An energy level diagram for the conduction band in a series of thin layers of $In_xGa_{1-x}N$ in GaN.

and go to a lower level or it can then tunnel through to the next well. Tunnelling is faster than photon emission so that the population of the higher level builds up. More electrons are now in the higher level than in the lowest levels, thus a population inversion exists — the requirement for laser action.

The active region containing the quantum wells is sandwiched between layers of *n*- and *p*-doped GaN and aluminum-doped GaN, $Al_yGa_{1-y}N$, which provide the electrons entering the quantum well and keep them confined to the active region. All these layers are built up on a substrate, for example, sapphire. The ends of the whole device are etched or cleaned to form a partial mirror that reflects the emitted photons allowing a coherent beam to build up.

One use of semiconductor lasers is as the light source in fibre optics.

## 8.4  OPTICAL FIBRES

Optical fibres are used to transmit light in the way that metal wires are used to transmit electricity. For example, a telephone call can be sent along an optical fibre in the form of a series of light pulses from a laser. The intensity, time between pulses, and the length of a pulse can be modified to convey the contents of the call in coded form. To transmit the information over useful distances (of the order of kilometres), the intensity of the light must be maintained so that a detectable signal still exists at the other end of the fibre. Thus, much of the art of making commercial optical fibres lies in finding ways of reducing energy loss.

The first requirement is that the laser beam keeps within the fibre. Laser beams diverge less than conventional light beams so that using laser light helps, but even so, some tendency exists for the beam to stray outside the fibre. Fibres are usually constructed therefore with a variable refractive index across the fibre. The beam is sent down a central core. The surrounding region has a lower refractive index than the core, so that light deviating from a straight path is totally internally reflected and hence remains in the core. This is illustrated in Figure 8.14.

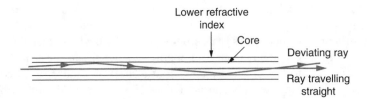

**FIGURE 8.14** Rays of light travelling along an optical fibre.

Adding selected impurities can vary the refractive index. In the case illustrated in Figure 8.14, the totally internally reflected rays travel a longer path than those that travel straight along the core. This will lead to a pulse being spread out in time. One way to keep the pulses together is to use very narrow cores so that essentially all the light travels the straight ray path. An alternative is the variable refractive index core. The refractive index is a measure of how fast light travels in a medium; the lower the refractive index, the faster the speed of light. If the outer parts of the core, therefore, have a lower refractive index, then the reflected light moves faster and this compensates for the longer path length.

To see why the refractive index is altered by composition, the atomic origin of the refractive index is considered briefly. Electromagnetic radiation has associated with it an oscillating electric field. Even when the radiation is not absorbed, this field has an effect on the electrons in the solid. If you think of an electrical field applied to an atom, you can imagine the electrons pulled by the field so that the atom is no longer spherical. The applied field produces a separation of the centres of positive and negative charge (i.e., it induces an electric dipole moment). (A molecule in a solid may also have a permanent electric dipole moment produced by an unequal distribution of bonding electrons between the nuclei but this is present in the absence of an applied field.)

The oscillating field of the radiation can be considered as pulling the electrons — alternately one way and then the other. The amount the electrons respond depends on how tightly bound the electrons are to the nucleus. This property is called the **polarisability** and is higher for large ions with low charge, for example $Cs^+$, than for small, highly charged ions such as $Al^{3+}$. If a high concentration of polarisable ions occurs in a solid, then the radiation will be slowed down, that is, the refractive index of the material will increase. Adjusting the refractive index by adding carefully selected impurities is also useful in other applications. For example, controlling the refractive index of glass is very important when making lenses for telescopes, binoculars, and cameras. Lead ions, $Pb^{2+}$, are highly polarisable and are used to produce glass of high refractive index.

Some imperfections in the fibre are bound to occur, and these are another source of energy loss. The imperfections cause scattering of the light of a type known as **Rayleigh scattering**. Rayleigh scattering does not cause any change in the wavelength of the light, only in its direction. The amount of scattering depends on $(1/\lambda^4)$, where $\lambda$ is the wavelength; therefore, much less scattering occurs for longer wavelengths. Even the reduction in going from blue to red light is significant and is responsible for the colour of the sky. To reduce Rayleigh scattering, the lasers employed for optical fibre systems usually emit infrared radiation.

A third source of energy loss is absorption of light by the fibre. In a fibre several kilometres long, a very small amount of impurity can give rise to substantial absorption. You can get an idea of this by looking at a sheet of window glass edge-on. Instead of being clear, the glass appears green. This is due to absorption by $Fe^{2+}$ ions in the glass. A windowpane is only about half a metre across, so that you can see that in a fibre of a few km in length, there would be considerable loss due to such absorption. In a glass, the spectrum of the impurity ions is similar to that in a crystal, but because the ions occupy several different types of site in a glass, the absorption bands are wider; each site giving rise to a band at a slightly different wavelength. In an optical fibre 3 km long operating at 1300 nm in the near infrared, the intensity of the $Fe^{2+}$ absorption is still such that a concentration of 2 parts in $10^{10}$ would reduce the amount of radiation by one half. Materials for optical fibres must therefore be very pure. One reason why silica has been widely used is that high purity silicon tetrachloride, developed for the semiconductor industry, is commercially available as a starting material.

Metal ions are not the only source of absorption, however. Using infrared radiation means there is likely to be loss due to absorption by molecular vibrations. In silica glass, the structure may contain dangling SiO bonds which easily react with water to form OH bonds. The vibrational frequencies of OH bonds are high and close to the frequencies used for transmission. It is important, therefore, to exclude water when manufacturing silica optical fibres. Even when water is excluded and no OH bonds are present, absorption by vibrational modes cannot be neglected. SiO bonds vibrate at lower frequencies than OH bonds, and so the maximum in the absorption does not interfere. However, the SiO absorption is very strong and the peak tails into the region of the transmission frequency. There has been some research into substances with lower vibrational frequencies than silica, particularly fluorides, but as yet such substances are not economically viable, being difficult to manufacture and more expensive than silica.

Losses still occur in the fibres developed for commercial use. Nonetheless, these have been reduced to a point where transmission over kilometres is possible. Along with transmission of information in telephone systems and similar applications, it has been suggested that optical devices may replace conventional electronics in more advanced applications such as computers. For such applications, it will be necessary to develop optical switches, amplifiers, and so on. In the last two decades, new materials have been developed that may form the basis of integrated optical circuits. These materials are photonic crystals.

## 8.5  PHOTONIC CRYSTALS

Photonic crystals have been hailed as the optical equivalent of semiconductors. Eli Yablonovitch at Bell Communications Research first developed the idea of such crystals in the 1980s.

A photonic crystal consists of a periodic arrangement of two materials of different refractive index. At each boundary between the two materials, light, or other electromagnetic radiation, will refract and partly reflect. The beams from the different interfaces will reinforce or cancel each other out depending on their relative phases. The wavelength of the radiation, its direction of travel, the refractive index of the photonic crystal materials, and the particular periodic arrangement determine whether two beams will be in phase. For certain wavelengths of radiation, refractive indices, and spacings of the materials, complete cancellation in all directions can occur so that the crystal does not transmit such wavelengths. The range of such forbidden wavelengths is known by analogy with semiconductors as the **photonic band gap**. Figure 8.15 illustrates how a forbidden wavelength can occur for a one-dimensional arrangement — a row of slabs of dielectric material.

As the light reaches each slab, some is reflected due to the change in refractive index. For the right spacing of the slabs, the reflected rays from each slab are in phase with each other but out of phase with the incident light. For such wavelengths, the incident and reflected rays cancel each other.

The original photonic crystal was produced by accurately drilling holes a millimetre in diameter in a block of material with refractive index of 3.6. This crystal had a photonic band gap in the microwave region. Similar structures with band gaps

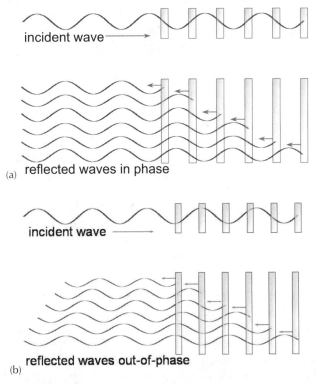

(a) reflected waves in phase

(b) reflected waves out-of-phase

**FIGURE 8.15** A row of dielectric slabs. Light travelling through can interfere destructively with reflected light giving rise to forbidden wavelengths. In (a) the reflected waves are all in phase with one another and destructive interference occurs. In (b) the spacing of the slabs is such that the waves are reflected with slightly different phases and the incident light travels through the array.

in the microwave and radio regions are being used to make antennae that direct radiation away from the heads of mobile phone users. Producing photonic crystals with band gaps at shorter wavelengths — infrared and visible — is less straigh forward. The hole size or lattice spacing needed for photonic crystals is roughly equal to the wavelength of the radiation divided by the refractive index. A GaAs laser produces radiation of wavelength 904 nm. To use this radiation with a photonic crystal composed of silica with a refractive index of 1.45 would require a spacing of 623 nm or 0.623 micrometres. Machining holes of this size is impractical, and the production of regular periodic structures with spacings of this order is a technical problem that needs to be overcome before integrated optical circuits and other devices can be manufactured. One approach to this problem is to suspend spheres (usually of silica) with a diameter of less than a micrometre in a colloidal suspension or hydrogel. The spheres arrange themselves into a close-packed structure.

Two-dimensional photonic materials (i.e., materials in which light is blocked within a plane but transmitted perpendicular to the plane) are useful as optical fibres. An ingenious method of constructing such fibres is to pack a series of hollow capillary tubes around a central glass core. The structure is then heated and drawn until it is only a few μm thick. The central core is now surrounded by a periodic array of tubes of the right diameter to have a photonic band gap in the near infrared. It is also possible to replace the central glass core by air and this enables very high power laser signals to be transmitted along the fibre without damage to the fibre material.

Other potential applications of photonic crystals include crystals with rows of holes to guide radiation around sharp bends (something that cannot be attained with conventional optical fibres), nanoscopic lasers formed from thin films, ultrawhite pigment formed from a regular array of submicron titanium dioxide particles, radiofrequency reflectors for magnetic resonance imaging (MRI) and LEDs.

Photonic crystals have only been studied in the laboratory for two decades, but naturally occurring examples exist, with the best known being the gemstone opal. Opals consist of tiny spheres of silica arranged in a face centred cubic structure. These are thought to have formed from colloidal silica solutions, and the colour depends on the size of the spheres.

## QUESTIONS

1. In the oxide MnO, $Mn^{2+}$ ions occupy octahedral holes in an oxide lattice. The degeneracy of the $3d$ levels of manganese are split into two, as for $Ti^{3+}$. The five $d$ electrons of the $Mn^{2+}$ ions occupy separate $d$ orbitals and have parallel spins. Explain why the absorption lines due to transitions between the two $3d$ levels are very weak for $Mn^{2+}$.

2. Figure 8.16 illustrates the energy levels of $Nd^{3+}$ in yttrium aluminium garnet ($Y_3Al_5O_{12}$), which are involved in the laser action of this crystal (known as the neodymium YAG laser). Describe the processes that occur when the laser is working.

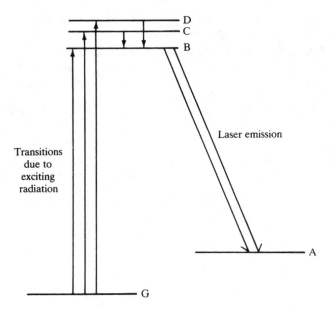

**FIGURE 8.16** Energy levels of Nd$^{3+}$ in yttrium aluminium garnet.

3. A phosphor commonly used on television screens is ZnS doped with Cu$^+$. This is much more efficient at transferring energy to the impurity sites for emission than are the phosphors based on phosphates as host. ZnS is a semiconductor. Suggest a reason for the efficiency of transfer in this solid.

4. Figure 8.17 shows two bands for a semiconductor. Is the band gap of this solid direct or indirect?

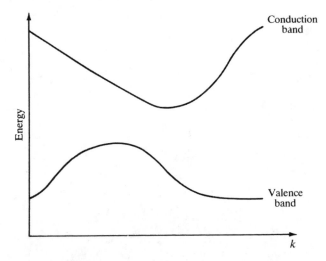

**FIGURE 8.17** A plot of energy against $k$ for two bands of a solid.

5. Explain why silicon is used for solar cells but not for LEDs.

6. The core of silica optical fibres contains some $B_2O_3$, $GeO_2$, and $P_2O_5$. How do these impurities increase the refractive index?

7. Red-orange opals contain larger spheres than blue-green opals. How do the wavelengths of the photonic band gap vary with the colour?

# 9 Magnetic and Dielectric Properties

## 9.1 INTRODUCTION

One consequence of the closeness of atoms in a solid is that properties of the individual atoms or molecules can interact cooperatively to produce effects not found in fluids. A well-known example of this is **ferromagnetism**. In a piece of iron used as a magnet, for example, the magnetism of the iron atoms aligns to produce a strong magnetic effect. Other cooperative magnetic effects lead to a cancelling (**antiferromagnetism**) or partial cancelling (**ferrimagnetism**) of the magnetism of different atoms. Ferro- and ferrimagnets have many commercial applications from compass needles and watch magnets to audio- and videotapes and computer memory devices.

Cooperative effects are not confined to magnetism; similar effects can occur for the response of a crystal to mechanical stress and to electric fields. The electrical analogue of ferromagnetism is the **ferroelectric effect**, in which the material develops an overall electrical polarisation, a separation of charge. Ferroelectric materials are important in the electronics industry as capacitors (for storing charge) and transducers (for converting, for example, ultrasound to electrical energy). Ferroelectric crystals are a subclass of piezoelectric crystals and piezoelectric crystals have commercial uses of their own. For example, quartz watches use piezoelectric quartz crystals as oscillators.

This chapter discusses the types of material that display cooperative magnetic and dielectric properties. To begin, however, we consider the weaker magnetic effects that can be found in all types of matter. Then we discuss the origin of the cooperative magnetic effects and their applications. Following magnetic effects we look at dielectric effects, starting with the **piezoelectric effect** and its applications and then considering ferroelectric materials and in particular the oxide barium titanate ($BaTiO_3$), which is widely used as a capacitor.

## 9.2 MAGNETIC SUSCEPTIBILITY

A magnetic field produces lines of force that penetrate the medium to which the field is applied. These lines of force appear, for example, when you scatter iron filings on a piece of paper covering a bar magnet. The density of these lines of force is known as the **magnetic flux density**. In a vacuum, the magnetic field and the magnetic flux density are related by the permeability of free space, $\mu_0$.

$$B = \mu_0 H \tag{9.1}$$

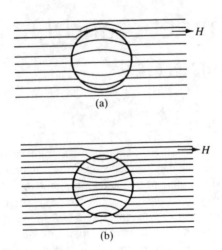

**FIGURE 9.1** Flux density in (a) a diamagnetic and (b) a paramagnetic sample.

If a magnetic material is placed in the field, however, it can increase or decrease the flux density. Diamagnetic materials reduce the density of the lines of force as depicted in Figure 9.1. Paramagnetic materials increase the flux density.

The field of the sample in the applied field is known as its **magnetisation**, $M$. The magnetic flux density is now given by Equation (9.2).

$$B = \mu_0(H + M) \qquad (9.2)$$

The magnetisation is usually discussed in terms of the **magnetic susceptibility**, $\chi$, where $\chi = M/H$.

Diamagnetism is present in all substances but is very weak so that it is not normally observed if other effects are present. It is produced by circulation of electrons in an atom or molecule. Atoms or molecules with closed shells of electrons are diamagnetic. Unpaired electrons, however, give rise to paramagnetism. Simple paramagnetic behaviour is found for substances such as liquid oxygen or transition metal complexes in which the unpaired electrons on different centres are isolated from each other. In a magnetic field, the magnetic moments on different centres tend to align with the field and hence with each other, but this is opposed by the randomising effect of thermal energy and in the absence of a field, the unpaired electrons on different centres are aligned randomly. The interplay of applied field and thermal randomization leads to the temperature dependence described by the **Curie law**

$$\chi = \frac{C}{T} \qquad (9.3)$$

where $\chi$ is the magnetic susceptibility, $C$ is a constant known as the Curie constant, and $T$ is the temperature in Kelvins. Different temperature dependence is observed when there is cooperative behaviour. The changeover from independent to cooperative behaviour is associated with a characteristic temperature. For ferromagnetism, the Curie law becomes

$$\chi = \frac{C}{T - T_C} \tag{9.4}$$

where $T_C$ is the **Curie temperature**. For antiferromagnetism, the temperature dependence is of the form

$$\chi = \frac{C}{T + T_N} \tag{9.5}$$

where $T_N$ is the **Néel temperature**. These two behaviours are illustrated in Figure 9.2.

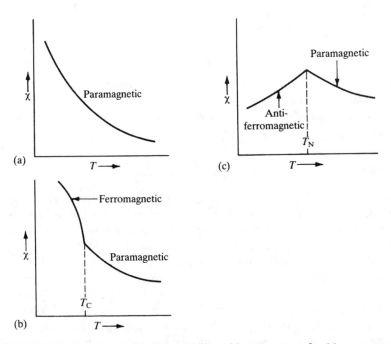

**FIGURE 9.2** Variation of magnetic susceptibility with temperature for (a) a paramagnetic substance, (b) a ferromagnetic substance, and (c) an antiferromagnetic substance.

**TABLE 9.1**
**Characteristics of the types of magnetism**

| Type | Sign of $\chi$ | Typical $\chi$ value (calculated using SI units) | Dependence of $\chi$ on H | Change of $\chi$ with increasing temperature | Origin |
|---|---|---|---|---|---|
| Diamagnetism | − | $-(1–600) \times 10^{-5}$ | Independent | None | Electron charge |
| Paramagnetism | + | 0–0.1 | Independent | Decreases | Spin and orbital motion of electrons on individual atoms |
| Ferromagnetism | + | $0.1–10^{-7}$ | Dependent | Decreases | Cooperative interaction between magnetic moments of individual atoms |
| Antiferromagnetism | + | 0–0.1 | May be dependent | Increases | |
| Pauli paramagnetism | + | $10^{-5}$ | Independent | None | Spin and orbital motion of delocalised electrons |

Ferrimagnetism has a more complicated form of temperature dependence with ions on different sites having different characteristic temperatures.

The characteristics of the various types of magnetism that we shall be concerned with are given in Table 9.1.

## 9.3 PARAMAGNETISM IN METAL COMPLEXES

In solids containing metal complexes such that the unpaired electrons on the different metal atoms are effectively isolated, the susceptibility can be discussed in terms of magnetic moments. The isolated metal complex can be thought of as a small magnet. Each complex in a solid will produce its own magnetic field due to the unpaired electrons. If the solid consists of one type of complex, then each complex produces the same magnitude magnetic field. However, thermal motion causes the orientation of these fields to be random. In the Curie Law (Equation (9.3), the temperature dependence is the result of this thermal motion but the constant, $C$, gives us information on the value of the magnetic field, known as the **magnetic moment**, $\mu$, of the complex. The magnetic moment, unlike the susceptibility does not normally vary with temperature. The dimensionless quantity $\chi$ in Equation (9.3) is the susceptibility per unit volume. To obtain the size of the magnetic field due to an individual complex, $\chi$ is divided by the specific gravity to give the susceptibility per unit mass of the sample and then multiplied by the relative molecular mass to obtain the molar susceptibility $\chi_m$. Assuming each complex has a fixed magnetic moment, $\mu$, and that the orientation is randomised by thermal motion, it can be demonstrated that $\chi_m$ is proportional to $\mu^2$ as in Equation (9.6)

$$\chi_m = \frac{N_A \mu_0}{3kT} \mu^2 \qquad (9.6)$$

where $N_A$ is Avogadro's number, $k$ is Boltzman's constant, $\mu_0$ is the permeability of free space, and $T$ is the temperature in Kelvin. In SI units, $\chi_m$ is in $m^3 \, mol^{-1}$ and the magnetic moment in joules per tesla ($J \, T^{-1}$). It is usual to quote $\mu$ in Bohr magnetons (B.M. or $\mu_B$) where one Bohr magneton has a value of $9.274 \times 10^{-24} \, J \, T^{-1}$.

The magnetic moment, $\mu$, is a consequence of the angular momentum of the unpaired electrons. The electrons possess both spin and orbital angular momentum. For the first row transition elements, the contribution from the orbital angular momentum is greatly reduced or 'quenched' because of the lifting of the fivefold degeneracy of the $3d$ orbitals. In complexes of these atoms, the magnetic moment is often close to that predicted for spin angular momentum only (Equation (9.7))

$$\mu_s = g \sqrt{S(S+1)} \qquad (9.7)$$

where $g$ is a constant which, for a free electron, has the value 2.00023, and $\mu_s$ is in Bohr magnetons.

The value of $S$ depends on the number of unpaired electrons, and Table 9.2 gives the values of $S$ and $\mu_s$ for the possible numbers of unpaired $3d$ electrons.

Contributions from orbital angular momentum cause deviations from these values. For complexes containing heavier metal ions, the interaction of spin and orbital angular momenta is greater. For the lanthanides, the magnetic moment depends on

**TABLE 9.2**
**Values of $S$ and $\mu_s$ for unpaired $3d$ electrons**

| Number unpaired electrons | Spin quantum number, $S$ | Magnetic moment $\mu_s$ in Bohr magnetons |
|:---:|:---:|:---:|
| 1 | $\frac{1}{2}$ | 1.73 |
| 2 | 1 | 2.83 |
| 3 | $\frac{3}{2}$ | 3.87 |
| 4 | 2 | 4.90 |
| 5 | $\frac{5}{2}$ | 5.92 |

the total angular momentum of the electrons, $\mathbf{J}$, not just or mainly on the spin angular momentum. The total angular momentum, $\mathbf{J}$, is the vector sum of the orbital, $\mathbf{L}$, and spin, $\mathbf{S}$, angular momenta.

$$\mathbf{J} = \mathbf{L} + \mathbf{S} \qquad (9.8)$$

If $\mathbf{J}$ is the quantum number for the total electronic angular momentum, then the magnetic moment is given by

$$\mu = g \sqrt{J(J+1)} \qquad (9.9)$$

where

$$g = 1 + \frac{J(J+1) + S(S+1) - L(L+1)}{2J(J+1)}$$

This can give rise to large magnetic moments, especially for shells that are more than half full. For example, $Tb^{3+}$ with an $f^8$ configuration has a magnetic moment from Equation (9.9) of 9.72 Bohr magnetons.

## 9.4 FERROMAGNETIC METALS

When discussing the electrical conductivity of metals, we described them in terms of ionic cores and delocalised valence electrons. The core electrons contribute a diamagnetic term to the magnetic susceptibility, but the valence electrons can give rise to paramagnetism or one of the cooperative effects we have described.

In filling the conduction band, we have implicitly put electrons into energy levels with paired spins. Even in the ground state of simple molecules, such as $O_2$, however, it can be more favourable to have electrons in different orbitals with parallel spins than in the same orbital with paired spins. This occurs when degenerate or nearly degenerate levels exist. In an energy band, many degenerate levels exist as well as many levels very close in energy to the highest occupied level. Therefore, it might well be favourable to reduce electron repulsion by having electrons with parallel spin singly occupying levels near the Fermi level. To obtain a measurable effect, however, the number of parallel spins would have to be comparable with the number of atoms; $10^3$ unpaired spins would not be noticed in a sample of $10^{23}$ atoms. Unless the density of states is very high near the Fermi level, a large number of electrons would have to be promoted to high energy levels in the band in order to achieve a measurable number of unpaired spins. The resulting promotion energy would be too great to be compensated for by the loss in electron repulsion. In the wide bands of the simple metals, the density of states is comparatively low, so that in the absence of a magnetic field, few electrons are promoted.

When a magnetic field is applied, the electrons will acquire an extra energy term due to interaction of their spins with the field. If the spin is parallel to the field, then its magnetic energy is negative (i.e., the electrons are at lower energy than they were in the absence of a field). For an electron with spin antiparallel to the field, it is now worthwhile to go to a higher energy state and change spin so long as the promotion energy is not more than the gain in magnetic energy. This will produce a measurable imbalance of electron spins aligned with and against the field and hence the solid will exhibit paramagnetism. This type of paramagnetism is known as **Pauli paramagnetism**. It is a very weak effect that gives a magnetic susceptibility much lower than the paramagnetism discussed in Section 9.3 (see Table 9.1).

For a very few metals, however, the unpaired electrons in the conduction band can lead to ferromagnetism. In the whole of the Periodic Table, only iron, cobalt, nickel, and a few of the lanthanides (Gd, Tb) possess this property. So, what is so special about these elements that confers this uniqueness on them? It is not their crystal structure; they each have different structures and the structures are similar to those of other non-ferromagnetic metals. Iron, cobalt, and nickel, however, all have a nearly full, narrow $3d$ band.

The $3d$ orbitals are less diffuse than the $4s$ and $4p$ (i.e., they are concentrated nearer the atomic nuclei). This leads to less overlap so that the $3d$ band is a lot narrower than the $4s/4p$ band. Furthermore, five $3d$ orbitals are present, so that for a crystal of $N$ atoms, $5N$ levels must be accommodated. With more electrons and a narrower band, the average density of states must be much higher than in $ns/np$ bands, particularly the density of states near the Fermi level is high. In this case, it is energetically favourable to have substantial numbers of unpaired electrons at the cost of populating higher energy levels. Thus, these elements have large numbers of unpaired electrons even in the absence of a magnetic field. For iron, for example, in a crystal of $N$ atoms up to $2.2N$ unpaired electrons are present, with their spins aligned parallel. Note the contrast with a paramagnetic solid containing transition metal complex ions where each ion may have as many as five unpaired electrons but, in the absence of a magnetic field, electrons on different ions are aligned randomly.

Ferromagnetism thus arises from the alignment of electron spins throughout the solid, and this occurs for partially filled bands with a high density of states near the Fermi level. The $4d$ and $5d$ orbitals are more diffuse than $3d$ and produce wider bands so that ferromagnetism is not observed in the second and third row transition elements. The $3d$ orbitals themselves become less diffuse across the transition series and lower in energy. In titanium the valence electrons are in the $4s/4p$ band with low density of states and, at the other end of the row in copper, the $3d$ band has dropped in energy so that the Fermi level is in the $4s/4p$ band. Thus, it is only at the middle of the series that the Fermi level is in a region of high density of states. Schematic band diagrams for Ti, Ni, and Cu are given in Figure 9.3. Shading indicates the occupied levels.

The high density of states found in the $3d$ bands of Fe, Co, and Ni leads to a reduction of the mean free path of the electrons in this band. This causes a decrease in their mobility and hence in the electrical conductivity of these elements compared to simple metals and copper where the conduction electrons are in $s/p$ bands.

The pure elements are not always suitable for applications requiring a metallic ferromagnet and many ferromagnetic alloys have been produced. Some of these

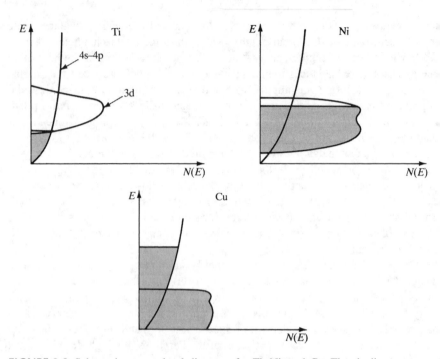

**FIGURE 9.3** Schematic energy level diagrams for Ti, Ni, and Cu. The shading represents occupied energy levels.

contain one or more ferromagnetic element and among these alloys of iron, cobalt, and nickel with the lanthanides, for example, $SmCo_5$ and $Nd_2Fe_{14}B$ have produced some of the most powerful permanent magnets known. In the lanthanide alloys, $f$ electrons contribute to the magnetism. Potentially, this could lead to a very high magnetisation because seven $f$ orbitals are present and, therefore, a maximum possible magnetisation corresponding to seven electrons per atom. The theoretical maximum magnetisation for the transition metals is five electrons per atom, as only five $d$ orbitals are present. In practice, this maximum is never reached. In the pure lanthanide metals, the overlap of $f$ orbitals is so small that they can be regarded as localised. In the ferromagnetic lanthanides, the magnetism is produced by delocalised $d$ electrons. The interaction between these $d$ electrons and the localised $f$ electrons causes alignment of the $d$ and $f$ electrons to reduce electron repulsion. Thus, $f$ electrons on different atoms are aligned through the intermediary of the $d$ electrons. In alloys, the $f$ electrons can align via the transition metal $d$ electrons, and although not all the $d$ and $f$ electrons are aligned, it can be observed that high values of the magnetisation could be achieved. It is not surprising, then, that it is these transition metal/lanthanide alloys that are the most powerful magnets. Other alloys can be made from nonmagnetic elements, such as manganese, and in these, the overlap of $d$ orbitals is brought into the range necessary for ferromagnetism by altering the interatomic distance from that in the element.

The usefulness of a particular ferromagnetic substance depends on factors such as the size of the magnetisation produced, how easily the solid can be magnetised and demagnetised and how readily it responds to an applied field. The number of unpaired electrons will determine the maximum field, but the other factors depend on the structure of the solid and the impurities it contains, as discussed next.

### 9.4.1 FERROMAGNETIC DOMAINS

A drawback to the previous explanation may have occurred to you. If $2.2N$ electrons are all aligned in any sample of iron, why are not all pieces of iron magnetic? The reason for this is that our picture only holds for small volumes (typically $10^{-24}$–$10^{-18}$ m$^3$) of metal within a crystal, called **domains**. Within each domain, the spins are all aligned, but the different domains are aligned randomly with respect to each other. It is actually possible to see these domains through a microscope by spreading finely divided iron powder on the polished surface of a crystal (Figure 9.4). What then causes these domains to form?

The spins tend to align parallel because of short-range **exchange interactions** stemming from electron–electron repulsion, but a longer-range **magnetic dipole interaction** also occurs, which tends to align the spins antiparallel. If you consider building up a domain starting with just a few spins, initially the exchange interactions dominate and so the spins all lie parallel. As more spins are added, an individual spin will be subjected to a greater and greater magnetic dipole interaction. Eventually, the magnetic dipole interaction overcomes the exchange interaction and the adjacent piece of crystal has the spins aligned antiparallel to the original domain. Thus, within domains exchange forces keep the spins parallel, whereas the magnetic dipole interaction keeps the spins of different domains aligned in different directions.

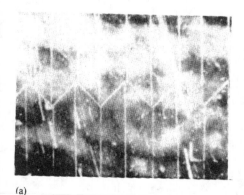
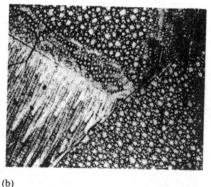

(a)                                     (b)

**FIGURE 9.4** (a) Domain patterns for a single crystal of iron containing 3.8% of silicon. The white lines illustrate the boundaries between the domains. (From R. Eisberg and R. Resnick (1985) *Quantum Physics of Atoms, Molecules, Solids, Nuclei and Particles*, John Wiley, New York. Courtesy of H.J. Williams, Bell Telephone Laboratories.); (b) magnetic domain patterns on the surface of an individual crystal of iron. (From W.J. Moore (1967) *Seven Solid States*, W.A. Benjamin Inc., New York.)

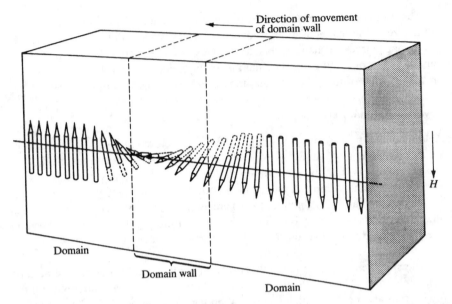

**FIGURE 9.5** Movement of a domain wall. The broken lines illustrate the domain wall, which separates two domains with magnetic moments lined up in opposite directions. The moments twist to align with the applied field, $H$, and the wall moves in the direction of the arrow.

When a magnetic field is applied to a ferromagnetic sample, the domains all tend to line up with the field. The alignment can be accomplished in two ways. First, a domain of correct alignment can grow at the expense of a neighbouring domain. Between the two domains is an area of finite thickness known as the domain wall. Change over from the alignment of one domain to that of the next is gradual within the wall. When the magnetic field is applied, the spins in the wall nearest the aligned domain alter their spins to line up with the bulk of the domain. This causes the next spins to alter their alignment. The net effect is to move the wall of the domain further out, as depicted in Figure 9.5. This process is reversible; the spins return to their former state after the magnetic field is removed,

If impurities or defects are present, it becomes harder for a domain to grow; there is an activation energy aligning the spins through the defect, and so a larger magnetic field is required. Once the domain has grown past the defect, however, it cannot shrink back once the magnetic field is removed because this will also need an energy input. In this case the solid retains magnetisation. The amount retained depends on the number and type of defects. Thus, steel (which is iron with a high impurity content) remains magnetic after the field is removed, whereas soft iron (which is much purer) retains hardly any magnetisation.

The second mechanism of alignment, which only occurs in strong magnetic fields, is that the interaction of the spins with the applied field becomes large enough to overcome the dipole interaction and entire domains of spins change their alignment simultaneously. The two mechanisms are compared in Figure 9.6.

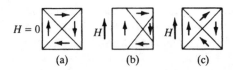

**FIGURE 9.6** Magnetization processes according to the domain model: (a) un-magnetized, (b) magnetized by domain growth, and (c) magnetized by domain rotation (spin alignment).

The magnetic behaviour of different ferromagnetic substances is demonstrated by their **hysteresis curves**. This is a plot of magnetic flux density, $B$, against applied magnetic field, $H$. If we start with a nonmagnetic sample in which all the domains are randomly aligned, then in the absence of a magnetic field, $B$ and $H$ are zero. As the field is increased, the flux density also increases. The plot of $B$ against $H$ is depicted in Figure 9.7. Initially, the curve is like 'oa', which is not simply a straight line because the magnetisation is increasing with the field. At point 'a', the magnetisation has reached its maximum value; all the spins in the sample are aligned. When the applied field is reduced, the flux density does not follow the initial curve. This is due to the difficulty of reversing processes where domains have grown through crystal imperfections. A sufficiently large field in the reverse direction to provide the activation energy for realignment through the imperfection must be applied before the magnetisation process can be reversed. At the point 'b', therefore, where $H$ is zero, $B$ is not zero because there is still a contribution from the magnetisation, $M$. The magnetisation at this point is known as the **remanent magnetisation**. The field that needs to be applied in the reverse direction to reduce the magnetisation to zero is the **coercive force** and is equal to the distance 'oc'.

### 9.4.2 PERMANENT MAGNETS

Substances used as permanent magnets need a large coercive force, so that they are not easily demagnetised and preferably should also have a large remanent magnetisation. These substances have fat hysteresis curves. They are often made from

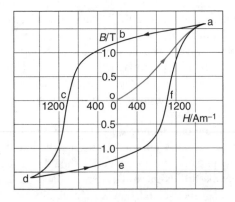

**FIGURE 9.7** $B–H$ curve for a typical hard steel.

alloys of iron, cobalt, or nickel which form with small crystals and include non-magnetic areas so that domain growth and shrinkage are difficult. Magnets for electronic watches, for example, are made from samarium/cobalt alloys. The best known of these alloys is $SmCo_5$, which has a coercive force of $6 \times 10^5$ A m$^-$ compared to 50 A m$^{-1}$ for pure iron.

## 9.5 FERROMAGNETIC COMPOUNDS — CHROMIUM DIOXIDE

Chromium dioxide $(CrO_2)$ crystallises with a rutile structure and is ferromagnetic with a Curie temperature of 392 K. $CrO_2$ has metal $3d$ orbitals which can overlap to form a band. In chromium dioxide, however, this band is very narrow and so, similar to Fe, Co, and Ni, chromium dioxide displays ferromagnetism. The dioxides later in the row have localised $3d$ electrons (e.g., $MnO_2$) and are insulators or semiconductors $TiO_2$ has no $3d$ electrons and is an insulator. $VO_2$ has a different structure at room temperature and it is a semiconductor. It does, however, undergo a phase transition to a metal at 340 K, when it becomes Pauli paramagnetic. Therefore, chromium dioxide occupies a unique position among the dioxides, similar to that of iron, cobalt, and nickel among the first row transition metals, in which dioxides of elements to the left have wide bands of delocalised electrons and elements to the right have dioxides with localised $3d$ electrons. Because the metal atoms are farther apart in the dioxides than in the elemental metals, the narrow bands that give rise to ferromagnetism occur earlier in the row than for the metallic elements.

### 9.5.1 AUDIOTAPES

Commercial interest in $CrO_2$ lies in its use as a magnetic powder on audiotapes. Recording tape usually consists of a polyester tape impregnated with needle-like crystals of a magnetic material such as chromium dioxide or $\gamma$-$Fe_2O_3$. The recording head has an iron core with a coil wrapped round it and a gap where the tape passes across it. Sound waves from the voice or music to be recorded hit a diaphragm in the microphone. The vibrations of the diaphragm cause a linked coil of wire to move in and out of a magnetic field. This causes a fluctuating electric current in the coil; the current depends on the frequency of the motion of the coil and hence on the frequency of the sound waves. The varying current is passed to the coil on the recording head, producing a varying magnetic field in the (soft) iron core. This in turn magnetises the particles on the tape and the strength and direction of this magnetisation is a record of the original sound. To play the tape, the whole process is reversed.

Materials for recording tape therefore need to retain their magnetisation so that the recording is not accidentally erased. Having needle-like crystals aligned with the recording field helps here, but the material should also have a high coercive force. Chromium dioxide fulfills these requirements and has a high magnetisation giving a large range of response and thus a high quality of reproduction. It does have some drawbacks, however; it has a relatively low Curie temperature, so that heating can erase recordings, and it is toxic.

## 9.6 ANTIFERROMAGNETISM — TRANSITION METAL MONOXIDES

These oxides have been discussed in an earlier chapter and you may remember that they all had the sodium chloride structure but had varying electrical properties. In this section we shall see that their magnetic properties are equally varied. In TiO and VO, the $3d$ orbitals are diffuse and form delocalised bands. These oxides were metallic conductors. The delocalised nature of the $3d$ electrons also determines the magnetic nature of these salts and, similar to the simple metals, they are Pauli paramagnetic. MnO, FeO, CoO, and NiO have localised $3d$ electrons and are paramagnetic at high temperatures. On cooling, however, the oxides become antiferromagnetic. The Néel temperatures for this transition are 122 K, 198 K, 293 K, and 523 K, respectively.

In antiferromagnetism, the spins on different nuclei interact cooperatively but in such a way as to cancel out the magnetic moments. Antiferromagnetic materials, therefore, show a drop in magnetic susceptibility at the onset of cooperative behaviour. The temperature which characterises this process is known as the Néel temperature (Figure 9.2(c)).

The appearance of cooperative behaviour suggests that $d$ electrons on different ions interact, but the electronic properties were explained by assuming that the $d$ electrons were localised. So how do we reconcile these two sets of properties?

The magnetic interaction in these compounds is thought to arise indirectly through the oxide ions, a mechanism known as **superexchange**. In a crystal of, for instance, NiO, a linear Ni—O—Ni arrangement exists. The $d_{z^2}$ orbital on the nickel can overlap with the $2p_z$ on oxygen, leading to partial covalency. The incipient NiO bond will have the $d_{z^2}$ electron and a $2p_z$ electron paired. The oxide ion has a closed shell and so there is another $2p_z$ electron, which must have opposite spin. This electron forms a partial bond with the next nickel and so the $d_{z^2}$ on this nickel pairs with the $2p$ electron of opposite spin. As shown in Figure 9.8, the net result is that adjacent nickel ions have opposed spins.

The alternating spin magnetic moments in antiferromagnets such as NiO can be observed experimentally using neutron diffraction. Because neutrons have a magnetic moment, a neutron beam used for diffraction responds not only to the nuclear positions but also to the magnetic moments of the atoms. X-rays on the other hand

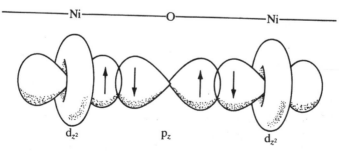

**FIGURE 9.8** Overlap between Ni $d_{z^2}$ orbitals and O $p_z$ orbitals in NiO.

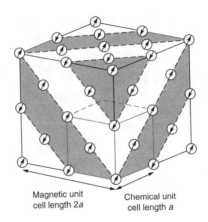

Magnetic unit
cell length 2a

Chemical unit
cell length a

**FIGURE 9.9** Magnetic unit cell of NiO with the crystallographic or chemical unit cell indicated.

have no magnetic moment and respond to electron density and hence to atomic positions. The structure of NiO as found by X-ray diffraction is a simple NaCl structure. When the structure is determined by neutron diffraction, however, extra peaks appear which can be interpreted as giving a magnetic unit cell twice the length of the X-ray determined unit cell. The positions of the nickel ions in this cell are shown in Figure 9.9. The normal crystallographic unit cell is bounded by identical atoms. The magnetic unit cell is bounded by identical atoms with identical spin alignment. The shading indicates layers of nickel ions parallel to the body diagonal of the cube. The spins of all nickel ions in a given layer are aligned parallel but antiparallel to the next layer (see Section 2.5.1).

## 9.7 FERRIMAGNETISM — FERRITES

The name ferrite was originally given to a class of mixed oxides having an **inverse spinel structure** and the formula $AFe_2O_4$ where A is a divalent metal ion. The term has been extended to include other oxides, not necessarily containing iron, which have similar magnetic properties.

The spinel structure is a common mixed oxide structure, typified by spinel itself $MgAl_2O_4$, in which the oxide ions are in a face-centred cubic close packed array (see Section 1.6.3 and Figure 1.43). For an array of $N$ oxide ions, there are $N$ octahedral holes, and the trivalent ions ($Al^{3+}$) occupy half of the octahedral sites (Figure 9.10).

In addition, there are $2N$ tetrahedral sites, and the divalent ions ($Mg^{2+}$) occupy one-eighth of these. In the inverse spinel structure, the oxide ions are also in a cubic close-packed arrangement, but the divalent metal ions occupy octahedral sites and the trivalent ions are equally divided amongst tetrahedral and octahedral sites.

Ions on octahedral sites interact directly with each other and their spins align parallel. The ions on octahedral sites also interact with those on tetrahedral sites but in this case, they interact through the oxide ions and the spins align antiparallel as in NiO.

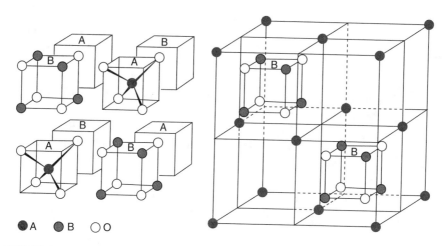

●A  ●B  ○O

**FIGURE 9.10** The spinel structure $AB_2O_4$.

In ferrites, $AFe_2O_4$, the $Fe^{3+}$ ions on tetrahedral sites are aligned antiparallel to those on octahedral sites, so that there is no net magnetisation from these ions. The divalent A ions, however, if they have unpaired electrons, tend to align their spins parallel with those of $Fe^{3+}$ on adjacent octahedral sites, and hence with those of other $A^{2+}$ ions. This produces a net ferromagnetic interaction for ferrites in which $A^{2+}$ has unpaired electrons. The magnetic structure of a ferrimagnetic ferrite is shown in Figure 9.11.

In magnetite $Fe_3O_4$, the divalent ions are iron and the interaction between ions on adjacent octahedral sites is particularly strong. One way of looking at the electronic structure of this oxide is to consider it as an array of $O^{2-}$ ions and $Fe^{3+}$ ions,

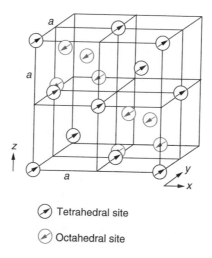

⊘ Tetrahedral site

⊘ Octahedral site

**FIGURE 9.11** Magnetic structure of ferrimagnetic inverse spinel.

with the electrons that would have made half the Fe ions divalent delocalised over all the ions on octahedral sites. $Fe^{3+}$ ions have five $3d$ electrons all with parallel spins. Since there can only be five $3d$ electrons of one spin on any atom, the delocalised spin must have the opposite spin. Being delocalised, it must also have the opposite spin to the $3d$ electrons on the next Fe ion. Hence, the two ions must have their spins aligned, and these spins must be aligned with those of all other Fe ions on octahedral sites. Delocalisation will be less for other ferrites.

Magnetite is the ancient lodestone used as an early compass. More recently, ferrites have found use as memory devices in computers, as magnetic particles on recording tapes and as transformer cores.

## 9.8 GIANT, TUNNELLING, AND COLOSSAL MAGNETORESISTANCE

### 9.8.1 GIANT MAGNETORESISTANCE

**Magnetoresistance** is the change in electric current flowing through a material as the result of applying a magnetic field to the sample. In 1988, it was found that for certain metallic multilayer materials, the application of a magnetic field of strength of 1 to 3 mT caused the resistivity to drop to half or less of its zero field value. This phenomenon was termed **giant magnetoresistance (GMR)**. Within a decade, commercial devices, such as computer hard disk read heads, based on GMR were available.

GMR is observed in metallic magnetic multilayers (i.e., stacks of nanometre-thick layers of different metals, particularly alternate layers of a ferromagnetic metal and a nonmagnetic metal [e.g., Fe/Cr, Co/Cu]). Within a single ferromagnetic layer, all the spins are aligned, but the coupling between adjacent ferromagnetic layers depends on the thickness of the intervening nonmagnetic layer. For certain thicknesses, adjacent ferromagnetic layers are coupled antiferromagnetically, and this increases the resistivity. When a magnetic field is applied, this aligns the layers in a ferromagnetic manner and causes a significant drop in electrical resistance.

GMR can be understood if we realise that in metals spin up and spin down electrons conduct electricity independently. In ferromagnetic materials, there is a much higher density of electrons of one spin state and these electrons will carry most of the current. One way of looking at this is to consider separate energy bands for spin up and spin down electrons. (This model can also account for the value 2.2 B.M. for the magnetic moment of metallic iron.) For one spin species (e.g., spin up), the $3d$ bands will be full. For the spin down electrons, the $3d$ band will only be partially full. Because the spin up $3d$ band is full, the spin up current carriers occupy a higher energy $s/p$ band, whereas the spin down current carriers are in the $3d$ band. The mean free path of electrons in the $s/p$ band is greater than in the $3d$ band and so the spin up electrons carry more current than the spin down electrons.

Now, suppose one ferromagnetic layer has a majority of spin up electrons. If the next layer also has a majority of spin up electrons (Figure 9.12(a)), then current can flow readily from one ferromagnetic layer to the next. However, if the next layer has a majority of spin down electrons (Figure 9.12(b)), then the current from the

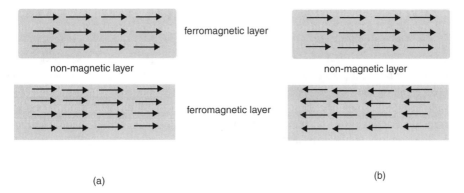

**FIGURE 9.12** Two magnetic layers separated by a nonmagnetic layer with (a) the spins in the two magnetic layers aligned ferromagnetically and (b) with the spins in the two magnetic layers aligned antiferromagnetically.

first layer has to be carried by the minority spin electrons in their $3d$ band, and thus the conductivity is reduced. Thus, the electrical resistance is much higher for two layers aligned antiferromagnetically than for two layers aligned ferromagnetically.

### 9.8.2 HARD DISK READ HEADS

In hard disk read heads, one ferromagnetic layer has its spin orientation fixed by coupling to an antiferromagnetic layer (Figure 9.13). A second ferromagnetic layer separated from the first by a nonmagnetic metal, is free to change its spin orientation when a field is applied.

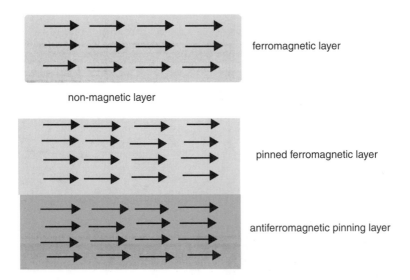

**FIGURE 9.13** Two ferromagnetic layers, as in a magnetic hard disk read head, illustrating pinning by an antiferromagnetic layer.

As the read head moves over the hard disk, magnetic fields on the disk cause the spins in the second layer to align either parallel or antiparallel to those in the first layer. The information on the hard disk is coded as a series of 0s and 1s corresponding to the different orientations of the magnetic field on the disk, and these give rise to a high or low current in the read head. GMR read heads can detect weaker fields than earlier read heads, enabling data to be packed more tightly on the hard disk. Their development has led to the high capacity hard disks found in current (2005) laptop computers. A similar principle is used for magnetic random access memory chips (MRAM).

### 9.8.3 TUNNELLING AND COLOSSAL MAGNETORESISTANCE

A related phenomenon, **tunnelling magnetoresistance (TMR)**, has aroused interest as the basis for magnetic sensor and storage devices. Here a thin insulating layer separates the two ferromagnetic layers. Electrons flow from one ferromagnetic layer to the next by quantum mechanical tunnelling.

In 1993, **colossal magnetoresistance (CMR)** was observed for certain compounds such as doped manganite perovskites (e.g., $La_{1-x}Ca_xMnO_3$). In these compounds, a change in electrical resistance of orders of magnitude is observed, but large magnetic fields of the order of several tenths of a Tesla (that is a hundred times stronger than those that produce giant magnetoresistance) or larger are needed.

Doped manganite perovskites exhibiting CMR have the general formula $RE_{1-x}M_xMnO_3$ where RE represents a rare earth element and M a divalent metal such as Cu, Cr, Ba, or Pb. The trivalent RE ions and divalent M ions occupy the A sites in the perovskite structure (Figure 9.14) and have 12-fold coordination to oxygen. The Mn ions occupy the octahedral B sites. $(1 - x)$ of the manganese ions

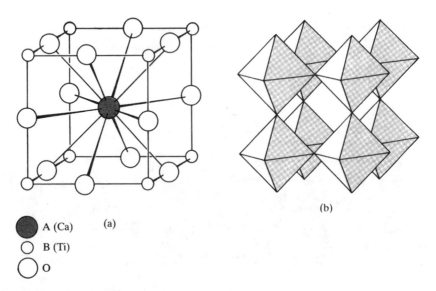

● A (Ca)    (a)

○ B (Ti)

◯ O

(b)

**FIGURE 9.14** Perovskite structure.

are $Mn^{3+}$ and $x$ are $Mn^{4+}$. The environment of the $Mn^{3+}(d^4)$ ions is distorted due to the Jahn–Teller effect and for small values of $x$, a cooperative distortion occurs.

As in other transition metal oxides such as NiO, the spins on the transition metal ions are coupled via the oxide ions. The $Mn^{4+}$ ions couple antiferromagnetically with each other via superexchange as in NiO. The coupling of the $Mn^{3+}$ ions varies and can be ferro- or antiferro-magnetic. $Mn^{3+}$ ions couple to $Mn^{4+}$ ions in a process known as double exchange. In this process, a simultaneous hop of an electron from $Mn^{3+}$ to an O $2p$ orbital and from the O $2p$ orbital to an $Mn^{4+}$ ion takes place.

$$Mn \uparrow\downarrow \ O \uparrow Mn \ \rightarrow \ Mn \uparrow O \downarrow\uparrow Mn$$

This produces ferromagnetic coupling.

Double exchange is strong in manganites with $x \approx {}^1/_3$, and these manganites form ferromagnetic phases at low temperatures. These phases can be described in terms of band theory. The $e_g$ orbitals on manganese and the O $2p$ orbitals combine to form two bands, one for spin-up electrons and one for spin-down electrons. Unusually there is a large band gap between the two so that at low temperatures, one band is empty and one partly full. (Such solids are referred to as half metals because only one type of spin is free to carry an electrical current.) The ferromagnetism disappears at the Curie temperature. Above this temperature, the $e_g$ electrons on manganese are better thought of as localised and the solid is paramagnetic and has a much higher electrical resistance. As the ferromagnetic phase approaches the Curie point, the electrical resistance rises as thermal energy starts to overcome the double exchange. It is in this region that the manganites exhibit CMR. A strong magnetic field applied to the manganite realigns the spins restoring the half metallic state and thus decreasing the resistivity.

## 9.9  ELECTRICAL POLARISATION

Although solids consist of charged particles (nuclei and electrons) a solid has no overall charge. For most solids, there is also no net separation of positive and negative charges; that is there is no net dipole moment. Even if a solid is composed of molecules with permanent dipole moments (e.g., ice), the molecules are generally arranged in such a way that the unit cell of the crystal has no net dipole moment and so the solid has none. If such a solid is placed in an electric field then a field is induced in the solid which opposes the applied field. This field arises from two sources, a distortion of the electron cloud of the atoms or molecules and slight movements of the atoms themselves. The average dipole moment per unit volume induced in the solid is the electrical polarisation, $P$, and is proportional to the field, $E$.

$$P = \varepsilon_0 \chi_e E \tag{9.10}$$

where $\varepsilon_0$ is the permittivity of free space (= 8.85 × $10^{-12}$ F m$^{-1}$) and $\chi_e$ is the (dimensionless) dielectric susceptibility. For most solids, the electric susceptibility lies between 0 and 10. The susceptibility is often determined experimentally by determining the capacitance of an electric circuit with and without the solid present. The ratio of these two capacitances is the relative permittivity or dielectric constant of the solid, $\varepsilon_r$.

$$\frac{C}{C_0} = \varepsilon_r \qquad (9.11)$$

where $C$ is the capacitance in farads in the presence of the solid and $C_0$ that in the absence of the solid. The dielectric constant $\varepsilon_r$ is related to the susceptibility by

$$\varepsilon_r = 1 + \chi_e \qquad (9.12)$$

If the experiment is performed using a high frequency alternating electric field, then the atoms cannot follow the changes in the field and only the effect due to electron displacement is measured. Electromagnetic radiation in the visible and ultraviolet regions provides such a field and the refractive index of a material is a measure of the electron contribution to the dielectric constant. Substances with high dielectric constants also tend to have high refractive indices.

Although most solids do not have a dipole moment in the absence of an electric field, the classes of solids that do are commercially important, and so form the subject matter of the rest of this chapter.

## 9.10 PIEZOELECTRIC CRYSTALS — α–QUARTZ

A **piezoelectric crystal** is one that develops an electrical voltage when subject to mechanical stress for example if pressure is applied to it, and conversely develops strain when an electric field is applied across it. Application of an electric field causes a slight movement of atoms in the crystal so that a dipole moment develops in the crystal. For it to be piezoelectric, a crystal must be made up from units that are non-centrosymmetric (i.e., they do not possess a centre of symmetry). Of the 32 crystal classes (see Chapter 1), 11 possess a centre of symmetry and one other cannot be piezoelectric because of other symmetry elements it possesses.

α-quartz is based on $SiO_4$ tetrahedra. Tetrahedra do not have centres of symmetry and in α-quartz, the tetrahedra are distorted so that each unit has a net dipole moment. However, these tetrahedra are arranged in such a way (Figure 9.15) that normally the crystal does not have an overall polarisation.

External stress changes the Si—O—Si bond angles between tetrahedra so that the dipole moments no longer cancel and the crystal has a net electrical polarisation. The effect in α-quartz is small; the output electrical energy is only 0.01 of the input

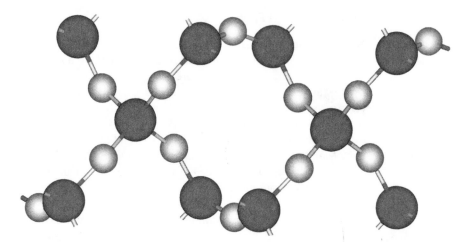

**FIGURE 9.15** Structure of α-quartz, viewed down a six-fold ring of silicons. Note that the oxygens around each silicon are arranged in a distorted tetrahedron and that the silicons themselves project a distorted hexagon.

strain energy, whereas for Rochelle salt (another commercially used piezoelectric crystal) the ratio of output to input energy is 0.81. α-quartz however is useful in applications where an oscillator of stable frequency is needed such as in quartz watches. An electric field causes distortion of quartz, and if an alternating electric field is applied, the crystal vibrates. When the frequency of the electric field matches a natural vibration frequency of the crystal, resonance occurs and a steady oscillation is set up with the vibrating crystal feeding energy back to the electrical circuit. The importance of α-quartz in devices such as watches is due to the fact that for some crystallographic cuts, a natural frequency of the crystal is independent of temperature and so the crystal will oscillate at the same frequency, and the watch will keep time, no matter how hot or cold the day is.

For other applications such as ultrasonic imaging, it is more important that the conversion of mechanical to electrical energy is high.

Some piezoelectric crystals are electrically polarised in the absence of mechanical stress; one example is gem-quality tourmaline crystals. Normally, this effect is unnoticed because the crystal does not act as the source of an electric field. Although there should be a surface charge, this is rapidly neutralised by charged particles from the environment and from the crystal itself. However, the polarisation decreases with increasing temperature and this can be used to reveal the polar nature of the crystal. If tourmaline is heated its polarisation decreases and it loses some of its surface charges. On rapid cooling it has a net polarisation and will attract small electrically charged particles such as ash. Such crystals are known as **pyroelectric**, and ferroelectric crystals are a special subclass of pyro-electric crystals.

## 9.11  THE FERROELECTRIC EFFECT

**Ferroelectric crystals** possess domains of different orientation of electrical polarisation that can be reorientated and brought into alignment by an electric field. Among the most numerous ferroelectrics are the perovskites (see Chapter 1) of which a classic example is barium titanate ($BaTiO_3$). This substance has a very large dielectric constant (around 1000) and is widely used in capacitors. Above 393 K, $BaTiO_3$ has a cubic structure as in Figure 1.44 (see Chapter 1) with $Ba^{2+}$ ions in the centre, $Ti^{4+}$ ions at the cube corners and an octahedron of $O^{2-}$ ions around each titanium ion. The $Ti^{4+}$ ion is small (75 pm radius) and there is room for it to move inside the $O_6$ cage. At 393 K, the structure changes to a tetragonal one in which the Ti atom moves off-centre along a Ti–O bond. At 278 K, a further change occurs in which the Ti moves off-centre along a diagonal between two Ti–O bonds and at 183 K, a rhombohedral phase is formed in which there is distortion along a cube diagonal. The three distortions are given in Figure 9.16. This figure magnifies the effect: the Ti atom is moved about 15 pm off-centre.

In these three phases, the $TiO_6$ octahedra have a net dipole moment. To illustrate how ferroelectricity arises, we use the tetragonal structure as an example. If all the Ti atoms were slightly off-centre in the same direction, then the crystal would have a net polarisation. Similar to ferromagnetics, however, ferroelectrics have domains within which there is a net polarisation but different domains have their polarisation in different directions thus giving a net zero polarisation. In the tetragonal phase of $BaTiO_3$, the Ti atom can be off-centre in six directions along any one of the Ti—O bonds. Neighbouring domains, as a result, have polarisations that are either at 90° or at 180° to each other (e.g., Figure 9.17).

A domain can be of the order of $10^{-5}$ m or even more. Several methods can be used to obtain pictures of the domains. Figure 9.18 is a photograph taken of a thin slice of barium titanate under a polarising microscope in which different domains can clearly be seen. Note the sharpness of the domain boundaries.

When an external electric field is applied, domains aligned favourably grow at the expense of others. As with ferromagnetics, the response to the field exhibits

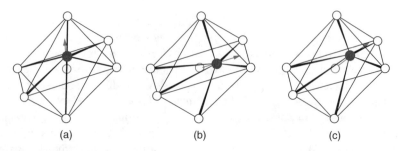

|      (a)       |       (b)       |       (c)       |

**FIGURE 9.16** Distortions of $TiO_6$ octahedra in (a) the tetragonal structure, (b) the orthorhombic, and (c) the rhombohedral structures of barium titanate.

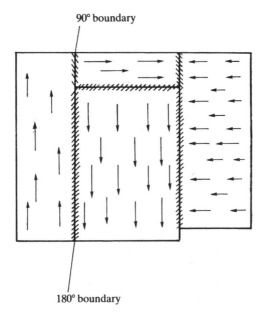

**FIGURE 9.17** Sketch of domains in barium titanate, illustrating 90° and 180° boundaries.

hysteresis; the polarisation grows until the whole crystal has its dipoles aligned, this polarisation remains while the field is reduced to zero and only declines as a field of opposite polarity is applied. Figure 9.19 is a hysteresis curve for barium titanate.

5 mm

**FIGURE 9.18** A photograph of a thin slice of barium titanate taken under the polarizing microscope, picturing domains of different polarisation. (From A. Guinier and R. Julien (1989) *The Solid State from Superconductors to Superalloys*, Oxford University Press/International Union of Crystallography, Oxford, Figure 2.9, p. 67. Reproduced by permission of Oxford University Press.)

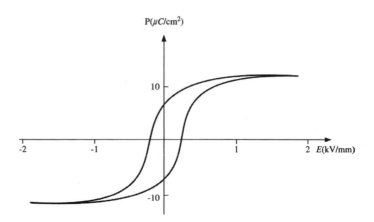

**FIGURE 9.19** A plot of polarization vs. applied electrical field for barium titanate of grain size $2 \times 10^{-5} - 1 \times 10^{-4}$ m.

The electric susceptibility and dielectric constant of ferroelectric substances obey a Curie law dependence on temperature (Equation (9.13):

$$\varepsilon_r = \varepsilon_\infty + C/(T - T_C) \tag{9.13}$$

where $\varepsilon_\infty$ is the permittivity at optical frequencies and $T_C$ is the Curie temperature. The origin of the Curie temperature in barium titanate is quite easy to see because the Curie temperature is the temperature, 393 K, at which barium titanate undergoes a phase transition to a cubic structure. Above the Curie temperature, the structure has a centre of symmetry and no net dipole moment. The dielectric constant is still high because the atoms can be moved off-centre by an applied electric field, but the polarisation is lost as soon as the field is removed.

In $PbZrO_3$, which also has a perovskite structure, the offset atoms are arranged alternately in opposite directions. This produces an antiferroelectric state. $PbZrO_3$ with some zirconium replaced by titanium gives the widely used ferroelectric materials, PZT ($PbZr_{1-x}Ti_xO_3$).

Remarkably some polymers are ferroelectric. Polyvinylene fluoride, $(-CH_2-CF_2-)_n$, as a thin film (approximately 25 μm thick) is used in shockwave experiments to measure stress. The pressure range over which this polymer operates is an order of magnitude larger than that of quartz or lithium niobate.

In the β-phase of polyvinylene fluoride (PVDF), the $CF_2$ groups in a polymer chain are all pointing in the same direction (Figure 9.20), so that a dipole moment

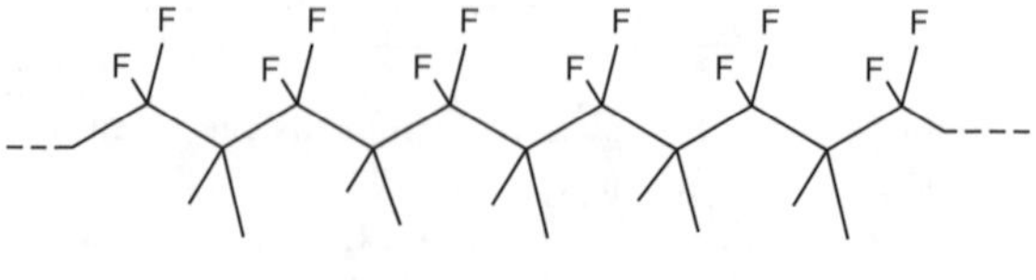

**FIGURE 9.20** Alignment of $-CF_2$ groups in β-polyvinylene fluoride.

is produced. Within a domain, the $CF_2$ groups on different chains are aligned. When an electric field is applied, the chains rotate through 60° increments to align with an adjacent domain.

### 9.11.1 MULTILAYER CERAMIC CAPACITORS

Capacitors are used to store charge. An electric field is applied to induce charge in the capacitor. The capacitor then remains charged until a current is required. To be useful in modern electronic circuits, including computers, space shuttles, TV sets, and many other applications, a capacitor must be small. In order to retain a high capacitance (i.e., to store a large amount of electrical energy while remaining small), a material needs a high permittivity. Thus, barium titanate with its very high permittivity has proved invaluable for this purpose. Pure barium titanate has a high permittivity (about 7000) close to the Curie temperature, but this rapidly drops with temperature to the room temperature value of 1 to 2000. Although this is still high, for electronic circuits it would be useful to retain the higher value so that the size of the capacitor could be reduced. It is also necessary for some applications, for the permittivity to be constant with temperature over a range of 180 K, from 218 K to 398 K. Barium titanate with some of the titanium substituted by zirconium or tin has a Curie temperature nearer room temperature and a flatter permittivity vs. temperature curve. Further improvements can be made by partially substituting the barium ions. Materials made in this way consist of several phases mixed together each with a different Curie temperature, and it is this that gives rise to the flatter permittivity versus temperature curve. Another factor that affects the dielectric properties of barium titanate is the grain size. On the surface of the tetragonal crystals, the structure is cubic, so that for small particles with a large surface to volume ratio there is a high proportion of cubic material. This leads to a higher room temperature permittivity but a smaller Curie temperature permittivity. For very small particles, no ferroelectric effect exists. It is therefore important to produce grains of a suitable size to give the properties needed. To manufacture multilayer capacitors, barium titanate of suitable grain size and appropriately doped is interleaved with conducting plates (Figure 9.21). This enables one device to be used in place of several single disc capacitors in parallel.

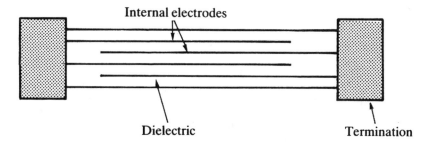

**FIGURE 9.21** A section through a multilayer capacitor.

Barium titanate is one example of a ferroelectric material. Other oxides with the perovskite structure are also ferroelectric (e.g., lead titanate and lithium niobate). One important set of such compounds, used in many transducer applications, is the mixed oxides PZT ($PbZr_{1-x}Ti_xO_3$). These, like barium titanate, have small ions in $O_6$ cages which are easily displaced. Other ferroelectric solids include hydrogen-bonded solids, such as $KH_2PO_4$ and Rochelle salt ($NaKC_4H_4O_6.4H_2O$), salts with anions which possess dipole moments, such as $NaNO_2$, and copolymers of polyvinylidene fluoride. It has even been proposed that ferroelectric mechanisms are involved in some biological processes such as brain memory and voltage-dependent ion channels concerned with impulse conduction in nerve and muscle cells.

Chapter 10 covers the exciting field of superconductors, including high-temperature superconductors, many of which have structures related to the perovskite structure.

## QUESTIONS

1. Although manganese is not ferromagnetic, certain alloys such as $Cu_2MnAl$ are ferromagnetic. The Mn–Mn distance in these alloys is greater than in manganese metal. What effect would this have on the $3d$ band of manganese? Why would this cause the alloy to be ferromagnetic?

2. The compound EuO has the NaCl structure and is paramagnetic above 70 K but magnetically ordered below it. Its neutron diffraction patterns at high and low temperatures are identical. What is the nature of the magnetic ordering?

3. $ZnFe_2O_4$ has the inverse spinel structure at low temperatures. What type of magnetism would you expect it to exhibit?

4. In transition metal pyrite disulfides, $MS_2$, the $M^{2+}$ ions occupy octahedral sites. If a $d$ band is formed, it will split into two as in the monoxides. Consider the information on some sulfides given next, and decide whether the $3d$ electrons are localised or delocalised, which band the electrons are in if delocalised, and, in the case of semiconductors, between which two bands the band gap of interest lies.
   $MnS_2$ antiferromagnetic ($T_N$ = 78 K), insulator: above $T_N$ paramagnetism fits five unpaired electrons per manganese.
   $FeS_2$ diamagnetic, semiconductor.
   $CoS_2$ ferromagnetic ($T_C$ = 115 K), metal.

5. In hydrogen-bonded ferroelectrics, the Curie temperature and permittivity alter when deuterium is substituted for hydrogen. What does this suggest about the origin of the ferroelectric transition in these compounds?

6. Pure $KTaO_3$ has a perovskite structure but is not ferroelectric or antiferroelectric. Replacing some K ions by Li does, however, produce a ferroelectric material. Explain why the substitution of Li might have this effect.

# 10 Superconductivity

## 10.1 INTRODUCTION

In the late 1980s, the amount of research effort in the field of superconductors increased dramatically due to the discovery of so-called 'high temperature' super-conductors by Bednorz and Muller in 1986. Their findings were thought to be so important that it was only a year later that they were awarded the Nobel Prize in Physics by the Royal Swedish Academy of Sciences.

Superconductors have two unique properties that have prompted commercial interest in them. First, they have zero electrical resistance and, therefore, carry current with no energy loss: this could revolutionize power transmission in cities, for instance, and is already exploited in the windings of superconducting magnets used in NMR spectrometers. Second, they expel all magnetic flux from their interior and so are forced out of a magnetic field. The superconductors can float or 'levitate' above a magnetic field: the Japanese have an experimental frictionless train that floats above magnetic rails and has achieved speeds of over 500 km h$^{-1}$ (300 m.p.h.). Until 1986, the highest temperature a superconductor operated at was 23 K, and so they all had to be cooled by liquid helium (boiling temperature (b.t.) ~4 K). This, of course, made any use of superconductors extremely expensive.

The discovery of a barium-doped lanthanum copper oxide which became super-conducting at 35 K led to a flood of new high temperature superconductors some of which were superconducting above the boiling temperature of nitrogen, 77 K. Over 50 high temperature superconductors, almost all containing copper oxide layers, are now known.

The past two decades have also seen the discovery of other unexpected types of superconductors. Low temperature superconductivity has been observed in organic polymers, and in doped $C_{60}$. In 2001, a record temperature of 40 K for the onset of conventional superconductivity was observed in $MgB_2$. Although this temperature is not high enough for liquid nitrogen cooling, it is technologically important as mains-operated cryocoolers are now available that can reduce tem-peratures to 30 K so that devices made from this material would not require liquid helium cooling.

Another unexpected discovery was of superconductors that are also ferromagnets — a combination of properties that is incompatible according to the accepted theory of conventional superconductors.

All these exciting developments are discussed, but we begin by looking at the properties of the low temperature superconductors that were discovered nearly a century ago.

## 10.2 CONVENTIONAL SUPERCONDUCTORS

### 10.2.1 THE DISCOVERY OF SUPERCONDUCTORS

In 1908, Kamerlingh Onnes succeeded in liquefying helium, and this paved the way
for many new experiments to be performed on the behaviour of materials at low
temperatures. For a long time, it had been known from conductivity experiments
that the electrical resistance of a metal decreased with temperature. In 1911, Onnes
was measuring the variation of the electrical resistance of mercury with temperature
when he was amazed to find that at 4.2 K, the resistance suddenly dropped to zero.
He called this effect **superconductivity** and the temperature at which it occurs is
known as the **(superconducting) critical temperature**, $T_c$. This effect is illustrated
for tin in Figure 10.1. One effect of the zero resistance is that no power loss occurs
in an electrical circuit made from a superconductor. Once an electrical current is
established, it demonstrates no discernible decay for as long as experimenters have
been able to watch!

For more than 20 years, little progress was made in the understanding of superconductors and only more substances exhibiting the effect were found. More than
20 metallic elements can be made superconducting under suitable conditions (Figure
10.2), as can thousands of alloys. It was not until 1933 that Meissner observed a
new effect.

### 10.2.2 THE MAGNETIC PROPERTIES OF SUPERCONDUCTORS

Meissner and Ochsenfeld found that when a superconducting material is cooled
below its critical temperature, $T_c$, it expels all magnetic flux from within its interior
(Figure 10.3(a)): the magnetic flux, $B$, is thus zero inside a superconductor. Because
$B = \mu_0 H(1 + \chi)$, when $B = 0$, $\chi$ must equal $-1$ (i.e., superconductors are perfect
diamagnets). If a magnetic field is applied to a superconductor, the magnetic flux is
excluded (Figure 10.3(b)) and the superconductor repels a magnet. This is pictured
in the photograph in Figure 10.4, where a magnet is seen floating in mid-air above
a superconductor.

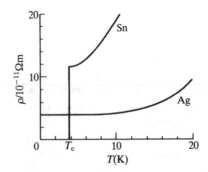

**FIGURE 10.1** A plot of resistivity, $\rho$, vs. temperature, $T$, illustrating the drop to zero at the
critical temperature, $T_c$, for a superconductor, and the finite resistance of a normal metal at
absolute zero.

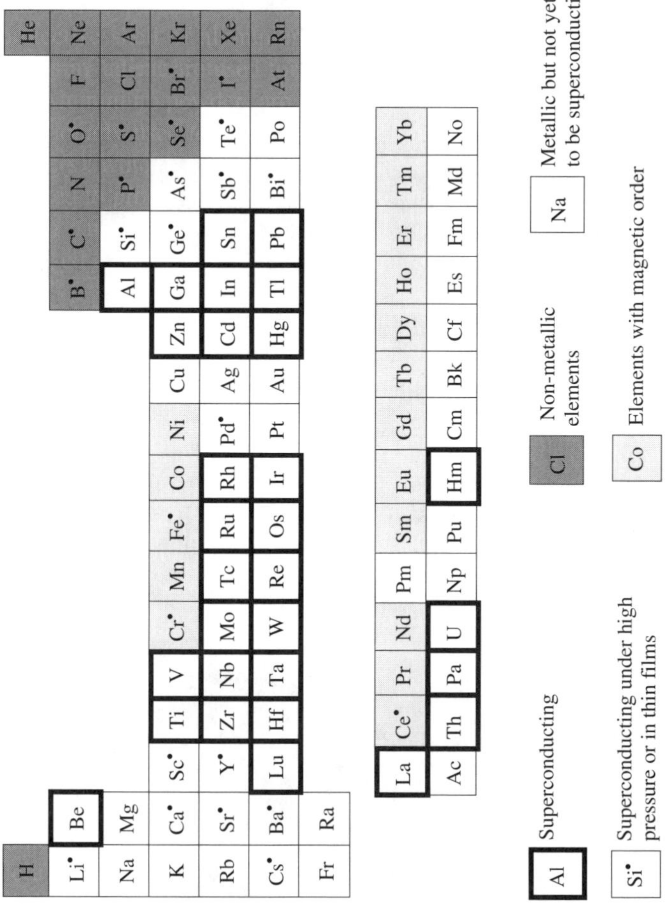


**FIGURE 10.2** The superconducting elements.

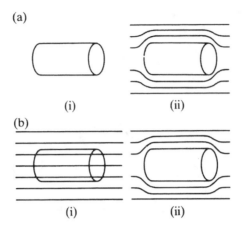

**FIGURE 10.3** (a) (i) Superconductor with no magnetic field. When a field is applied in (ii), the magnetic flux is excluded. (b) (i) Superconducting substance above the critical temperature, $T_c$, in a magnetic field. When the temperature drops below the critical temperature (ii), the magnetic flux is expelled from the interior. Both are called Meissner effects.

It is also found that the critical temperature, $T_c$, changes in the presence of a magnetic field. A typical plot of $T_c$ against increasing magnetic field is depicted in Figure 10.5, where you can see that as the applied field increases, the critical temperature drops.

It follows that a superconducting material can be made non–superconducting by the application of a large enough magnetic field. The minimum value of the field strength required to bring about this change is called the **critical field strength**, $H_c$: its value depending on the material in question and on the temperature. Type I

**FIGURE 10.4** A permanent magnet floating over a superconducting surface. (Courtesy of Darren Peets, UBC Superconductivity Group.)

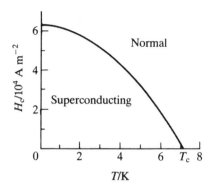

**FIGURE 10.5** The variation of the critical field strength, $H_c$, for lead. Note that $H_c$ is zero when the temperature, $T$, equals the critical temperature, $T_c$.

superconductors, which includes most of the pure metal superconductors, have a single critical field. Type II superconductors, which include alloys and the high $T_c$ superconductors, allow some penetration of the magnetic field into the surface above a critical field, $H_{c1}$, but do not return to their non-superconducting state until a higher field $H_{c2}$ is reached. Type II superconductors tend to have higher critical temperatures than type I superconductors.

Similarly, if the current in the superconductor exceeds a **critical current**, the superconductivity is destroyed. This is known as the **Silsbee effect**. The size of the critical current is dependent on the nature and geometry of the particular sample.

### 10.2.3 JOSEPHSON EFFECTS

In 1962, Josephson predicted that if two superconducting metals were placed next to each other separated only by a thin insulating layer (such as their surface oxide coating) then a current would flow in the absence of any applied voltage. This effect is indeed observed because if the barrier is not too thick then electron pairs can cross the junction from one superconductor to the other without dissociating. This is known as the **d.c. Josephson effect**. He further predicted that the application of a small d.c. electric potential to such a junction would produce a small alternating current — the **a.c. Josephson effect**. These two properties are of great interest to the electronics and computing industries where they can be exploited for fast-switching purposes.

### 10.2.4 THE BCS THEORY OF SUPERCONDUCTIVITY

This section attempts to give a qualitative picture of the ideas involved and to give some familiarity with the terminology.

Physicists worked for many years to find a theory that explained superconductivity. To begin with, it looked as though the lattice played no part in the superconducting mechanism as X-ray studies demonstrated that there was no change in either the symmetry or the spacing of the lattice when superconductivity occurred.

However, in 1950, an **isotope effect** was first observed: for a particular metal, the critical temperature was found to depend on the isotopic mass, $M$, such that

$$T_C \propto \frac{1}{\sqrt{M}} \tag{10.1}$$

The frequency, $\nu$, of vibration of a diatomic molecule is well known to be given by

$$\nu = \frac{1}{2\pi} \sqrt{\frac{k}{\mu}} \tag{10.2}$$

where $\mu$ is the reduced mass of the molecule, and $k$ the force constant of the bond. We can see that a vibration also changes frequency on isotopic substitution such that the frequency, $\nu$, is proportional to $1/\sqrt{mass}$ : this suggested to physicists that superconductivity was in some way related to the vibrational modes of the lattice and not just to the conduction electrons. The vibrational modes of a lattice are quantised, as are the modes of an isolated molecule: the quanta of the lattice vibrations being called **phonons**.

Frohlich suggested that there could be a strong phonon/electron interaction in a superconductor, which leads to an *attractive* force between two electrons that is strong enough to overcome the Coulomb repulsion between them. Very simply, the mechanism works like this: as a conduction electron passes through the lattice, it can disturb some of the positively charged ions from their equilibrium positions, pushing them together and giving a region of increased positive charge density. As these oscillate back and forth, a second electron passing this moving region of increased positive charge density is attracted to it. The net effect is that the two electrons have interacted with one another, using the lattice vibration as an intermediary. Furthermore, the interaction between the electrons is *attractive* because each of the two separate steps involved an attractive Coulomb interaction.

It is the scattering of conduction electrons by the lattice vibrations, phonons, which produces electrical resistance at room temperature. (At low temperatures, it is predominantly the scattering by lattice defects that gives electrical resistance.) Contrary to what we might have expected intuitively, a superconductor will have *high* resistance at room temperature, because it has strong electron/phonon interactions. Indeed, the best room-temperature electronic conductors — silver and copper — do not superconduct at all. *Superconductors do not have low electrical resistance above the superconducting critical temperature, $T_c$.*

In 1957, Bardeen, Cooper, and Schrieffer published their theory of superconductivity, known as the **BCS theory**. It predicts that under certain conditions, the attraction between two conduction electrons due to a succession of phonon interactions can slightly *exceed* the repulsion that they exert directly on one another due to the Coulomb interaction of their like charges. The two electrons are thus weakly bound together forming a so-called **Cooper pair**. It is these Cooper pairs that are responsible for superconductivity. In conventional superconductors, these electrons are paired so that their spin and orbital angular momenta cancel. They are described by a wave function, known as an **order parameter**. In this case the order parameter has symmetry similar to that of the wave function of $s$ electrons and represents a singlet state.

BCS theory demonstrates that several conditions have to be met for a sufficient number of Cooper pairs to be formed and superconductivity to be achieved. It is beyond the scope of this book to go into this in any depth: suffice it to say that the electron–phonon interaction must be strong and that low temperature favours pair formation — thus high temperature superconductors were not predicted by BCS theory. The relatively high critical temperature of $MgB_2$ is thought to be due to the high vibrational frequencies associated with the light B atoms and the strong interaction between the electrons and lattice vibrations. Evidence for the involvement of B atom vibrations comes from the observation that $T_c$ is increased by about 1 K when $^{11}B$ is replaced by $^{10}B$.

Cooper pairs are weakly bound, with typical separations of $10^6$ pm for the two electrons. They are also constantly breaking up and reforming (usually with other partners). Thus, enormous overlap occurs between different pairs, and the pairing is a complicated dynamic process. The ground state of a superconductor therefore is a 'collective' state, describing the ordered motion of large numbers of Cooper pairs. When an external electrical field is applied, the Cooper pairs move through the lattice under its influence. However, they do so in such a way that the ordering of the pairs is maintained: the motion of each pair is locked to the motion of all the others, and none of them can be individually scattered by the lattice. Because the pairs cannot be scattered by the lattice the resistance is zero and the system is a superconductor.

## 10.3 HIGH TEMPERATURE SUPERCONDUCTORS

By 1973, the highest temperature found for the onset of superconductivity was 23.3 K: this was for a compound of niobium and germanium, $Nb_3Ge$, and here it stayed until 1986 when Georg Bednorz and Alex Müller reported their findings. Their Nobel Prize citation states, "Last year, 1986, Bednorz and Müller reported finding superconductivity in an oxide material at a temperature 12°C higher than previously known." The compound that prompted their initial paper has been shown to be $La_{2-x}Ba_xCuO_4$, where $x = 0.2$, with a structure based on that of $K_2NiF_4$, a perovskite-related layer compound: they observed the onset of superconductivity at 35 K. The insight that they brought to this field was to move away from the investigation of metals and their alloys and to study systematically the solid state physics and chemistry of metallic oxides.

The idea was soon born that it might be possible to raise the temperature even further by substitution with different metals. Using this technique, it was Chu's group at Houston, Texas, that finally broke through the liquid-nitrogen temperature barrier with the superconductor that is now known as '1-2-3': This superconductor replaces lanthanum with yttrium and has the formula $YBa_2Cu_3O_{7-x}$. The onset of superconductivity for 1-2-3 occurs at 93 K. The highest critical temperature discovered so far (2003) is 134 K for a doped $HgBa_2Ca_2Cu_3O_8$.

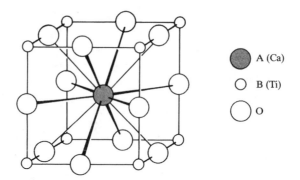

**FIGURE 10.6** The A-type unit cell of the perovskite structure for compounds $ABO_3$, such as $CaTiO_3$.

## 10.3.1 The Crystal Structures of High Temperature Superconductors

All the high $T_c$ superconductors discovered so far, with one exception, contain weakly coupled copper oxide, $CuO_2$, planes. The highest critical temperatures are found for cuprates containing a Group 2 metal (Ca, Ba, Sr) and a heavy metal such as Tl, Bi, or Hg. The structures of all the cuprate superconductors are based on, or related to, the perovskite structure. The one report (in 2000) of a non-cuprate high $T_c$ super-conductor is of surface superconductivity in $Na_xWO_3$. The structure of $Na_xWO_3$ is also based on the perovskite structure.

The perovskite structure is named after the mineral $CaTiO_3$ (see Chapter 1, Section 1.6.3.4): many oxides of general formula $ABO_3$ adopt this structure (also fluorides, $ABF_3$ and sulfides, $ABS_3$). The so-called perovskite A-type unit cell (with the A-type atom in the centre of the cell) is depicted in Figure 10.6. The central A atom (Ca) is coordinated by 8 titaniums at the corners of the cube and by 12 oxygens at the midpoints of the edges. The perovskite structure can be equally well repre-sented by moving the origin of the unit cell to the body-centre: this has the effect of putting Ca (A) atoms at each corner, Ti (B) atoms at the body centre, and an O atom in the centre of each face (Figure 10.7).

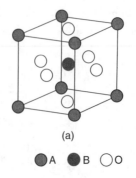

(a)

●A  ●B  ○O

**FIGURE 10.7** B-type cell for the $ABO_3$ perovskite structure.

The crystal structure of the 1-2-3 superconductor, $YBa_2Cu_3O_{7-x}$ is depicted in Figure 10.8. Figure 10.8(a) depicts only the positions of the metal atoms. If we discuss it in terms of the perovskite structure $ABO_3$, where B = Cu, the central section is now an A-type perovskite unit cell and above and below it are also A-type perovskite unit cells with their bottom and top layers missing. This gives copper atoms at the unit cell corners and on the unit cell edges at fractional coordinates $^1/_3$ and $^2/_3$. The atom at the body-centre of the cell (i.e., in the centre of the middle section) is **yttrium**. The atoms in the centres of the top and bottom cubes are **barium**

If, in this structure, all three sections were based exactly on perovskite unit cells, we would expect to find the oxygen atoms in the middle of each cube edge (Figure 10.8(b)): this would give an overall formula of $YBa_2Cu_3O_9$. This formula is improbable because it gives an average oxidation state for the three copper atoms of

$\dfrac{11}{3}$, implying that the unit cell contains both Cu(III) and Cu(IV), which is unlikely as Cu(IV) complexes are extremely rare. The unit cell, in fact, contains only approximately seven oxygen atoms ($YBa_2Cu_3O_{7-x}$): when $x = 0$, the oxygen atoms on the vertical edges of the central cube are not there and two are also missing from both the top and bottom faces (Figure 10.8(c)). A unit cell containing seven oxygen atoms has an average copper oxidation state of 2.33 indicating the presence of Cu(II) and Cu(III) in the unit cell (but no longer Cu(IV)).

In the 1-2-3 structure (when $x = 0$), the yttrium atom is coordinated by 8 oxygens and the barium atoms by 10 oxygens. The oxygen vacancies in the 1-2-3 superconductor create sheets and chains of linked copper and oxygen atoms running through the structure (this is illustrated slightly idealized in the diagrams as, in practice, the copper atoms lie slightly out of the plane of the oxygens): the copper is in fourfold square-planar or fivefold square-pyramidal coordination (Figure 10.8(d)). The superconductivity is found in directions parallel to the copper planes which are created by the bases of the Cu/O pyramids and which are separated by layers of yttrium atoms. These copper/oxygen nets seem to be a common feature of all the new high temperature superconductors. If this superconductor is made more deficient in oxygen, at $YBa_2Cu_3O_{6.5}$ ($x = 0.5$) the superconducting critical temperature, $T_c$, drops to 60 K and at $YBa_2Cu_3O_6$ the superconductivity disappears. The oxygen is not lost at random but goes from specific sites, gradually changing the square-planar coordination of the Cu along the $c$ direction into the twofold linear coordination characteristic of Cu$^+$ and the arrangement of the copper and oxygen atoms in the base of the pyramids is not affected. However, when the formula is $YBa_2Cu_3O_6$ all the square planar units along $c$ have become chains containing Cu$^+$, and the pyramid bases contain only Cu$^{2+}$; the unpaired spins of the Cu$^{2+}$ are aligned antiparallel and the compound is antiferromagnetic. It is not until the oxygen content is increased to $YBa_2Cu_3O_{6.5}$ that the antiferromagnetic properties are destroyed and the compound becomes a superconductor. It is thought that this compound contains copper in all three oxidation states: I, II, and III. $YBa_2Cu_3O_7$ contains Cu(II) and Cu(III), both in the sheets and in the chains. Clearly, the oxidation state of the coppers in the structure (and thus their bonding connections and bond lengths) is extremely important in determining both, whether superconductivity occurs at all, and the temperature below which it occurs ($T_c$).

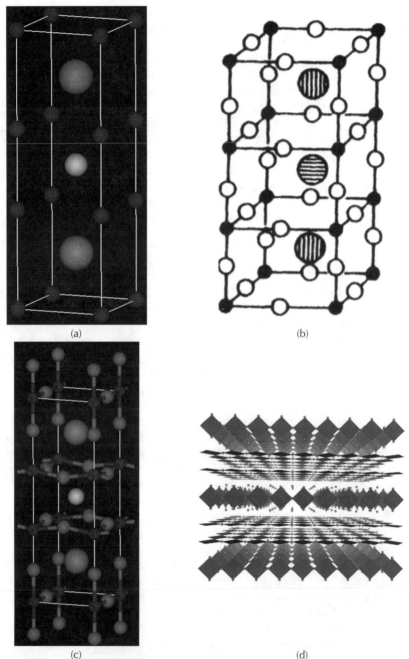

(a)                                                          (b)

(c)                                                          (d)

**FIGURE 10.8** The structure of 1-2-3: (a) the metal positions; (b) an idealized unit cell of the hypothetical $YBa_2Cu_3O_9$, based on three perovskite A-type unit cells; (c) idealized structure of $YBa_2Cu_3O_{7-x}$; and (d) the extended structure of $YBa_2Cu_3O_7$, depicting copper-oxygen planes, with the copper-oxygen diamonds in between. (See colour insert following page 356.) Key: Cu, blue; Ba, green; Y, aqua; O, red.

The Hg-containing cuprate superconductors, of which more than 40 are known and which include the material with the highest known (in 2003) $T_c$ at ambient pressure, are also perovskite-based structures. The structures of these solids fall into one of three classes, 1201 based on $HgBa_2CuO_4$, 1212 based on $HgBa_2CaCu_2O_6$, and 1223 based on $HgBa_2Ca_2Cu_3O_8$. The Hg layers in the parent compounds contain no oxygen and are poor superconductors. In 1993, however, it was found that annealing these compounds in oxygen resulted in oxygen atoms being inserted into the ($^1/_2$, $^1/_2$, 0) position in the Hg layer and $HgBa_2CuO_{4+\delta}$ was superconducting. At pressures above 10 GPa, $HgBa_2Ca_2Cu_3O_{8+\delta}$ has a critical temperature greater than 150 K. By replacing some of the Hg atoms by metal atoms in higher oxidation states, compounds with extra oxygen in the Hg layer were formed which were more stable at normal pressures. The highest $T_c$ at the time of writing was for the 1223 structure with a fifth of the Hg atoms replaced by Tl, $Hg_{0.8}Tl_{0.2}Ba_2Ca_2Cu_3O_{8+0.33}$. The structure of this compound is given in Figure 10.9. The unit cell contains four $CuO_2$ layers with Cu in square pyramidal coordination and with the apices of the pyramids alternately pointing up and down. Ba occupies sites close to the apices of these pyramids and the Ca positions are just above or below the bases. At the top and bottom of the cell, between two layers of Ca atoms are CuO layers with the Cu in square planar coordination. In the centre of the cell is the Hg layer. Overdoping with Tl gives a metallic compound with poor superconducting properties.

In 1-2-3, $La_{2-x}Ba_xCuO_4$ and the mercury cuprates, the average oxidation state of the Cu is greater than 2 and as a result positive holes are formed in the valence bands. The charge is carried by the positive holes and, therefore, such materials are known as **p-type superconductors**. In 1-2-3 and the mercury cuprates, adding oxygen forms positive holes, but other methods are available. In $SrCuO_2F_{2+\delta}$,

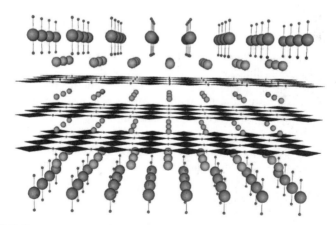

**FIGURE 10.9** Structure of $HgBa_2Ca_2Cu_3O_8$, depicting copper oxygen diamonds between the Ca layers and the copper oxygen layers forming the bases of the pyramids. The apices of the pyramids are in the barium oxygen layers. In the superconductor $Hg_{0.8}Tl_{0.2}Ba_2Ca_2Cu_3O_{8+0.33}$, one fifth of the $Hg^{2+}$ ions are replaced by $Tl^{3+}$ ions and additional oxygen ions are present in the mercury layer. (See colour insert following page 356.) Key: Cu, blue; Ca, green; Ba, aqua; Hg/Tl, grey; O, red.

superconductivity is induced by the insertion of fluorine and in the collapsed oxy-carbonates, such as $TlBa_2Sr_2Cu_2(CO_3)O_7$, superconductivity can be produced by shearing so that a shift appears along one plane every $n$ octahedra (where $n$ is typically 3 to 5), which leaves the $CuO_2$ layers unchanged and mixes the TlO and CO layers. $T_c$ for the oxyhalides can reach 80 K; for the collapsed carbonates, $T_c$ is in the range of 60 to 77 K.

Until 1988, all the high temperature superconductors that had been found were $p$-type, and it was assumed by many that this would be a feature of high temperature superconductors. However, some **$n$-type superconductors** have also been discovered, where the charge carriers are electrons: the first to be found was based on the compound $Nd_2CuO_4$ with small amounts of the three-valent neodymium substituted by four-valent cerium — $Nd_{2-x}Ce_xCuO_{4-y}$ where $x \sim 0.17$ (samarium, europium, or praseodymium can also be substituted for the neodymium). Other similar compounds have since been found based on this structure where the three-valent lanthanide is substituted by, for example, four-valent thorium in $Nd_{2-x}Th_xCuO_{4-y}$. The superconductivity occurs at $T_c \leq 25$ K for these compounds.

It appears clear that these ceramic superconductors all have a common feature — the presence of copper/oxygen layers sandwiched between layers of other elements. The superconductivity takes place in these planes and the other elements present and the spacings between the planes changes the superconducting transition temperature — exactly how is not yet understood

## 10.3.2 THEORY OF HIGH $T_c$ SUPERCONDUCTORS

The consensus is that, in common with the conventional superconductors, high temperature superconductors contain Cooper pairs (see Section 10.2.4). But a difference occurs in the pairs formed. The Cooper pairs in conventional superconductors couple to give a total angular momentum of zero and are described by an order parameter of $s$-wave symmetry. In high temperature superconductors, the Cooper pairs couple to give non-zero angular momentum and are described by an order parameter of $d$-wave symmetry. The pairing state in BCS theory does not need to be caused by electron-phonon interaction, and it is thought that pairing in the $d$ state is due to antiferromagnetic spin fluctuations, or electron-magnon coupling. Such a possibility had been examined by Kohn and Luttinger, who found that a weak collective residual attraction could be generated from the Coulomb repulsion between electrons, but only if the electrons in the Cooper pairs were prevented from close encounters. This could be achieved if the pairs had angular momentum (i.e., if they were described by $p$- or $d$-waves). Their initial estimate of $T_c$ in such systems was very low, but it was hoped that higher temperatures might be found in metals with strong spin fluctuations. As in BCS theory for conventional superconductors, the coupling is a collective property and the pairing is a dynamic process. Spin-mediated coupling is far more effective in bringing about superconductivity in two dimensions than in three and this may explain why high $T_c$ superconductivity is found in quasi-two-dimensional solids such as the cuprates. Experiment suggests that the quasi-two-dimensional organic superconducting polymers (see Chapter 6) are also $d$-wave superconductors.

The presence of $d$-wave symmetry has been demonstrated in an elegant experiment using scanning tunnelling microscopy (STM, see Chapter 2). Collaborators at Berkeley and Tokyo led by J.C. Séamus Davis and Shin-ichi Uchida used STM to measure differential tunnelling conductance on a specimen of bismuth strontium calcium copper oxide with zinc atoms replacing a small number of copper atoms. Differential tunnelling conductance is proportional to the density of states available to electrons tunnelling from the sample into the microscope's tip. This should be zero within a defined energy above the highest occupied levels known as the superconducting symmetry gap. For $d$-wave symmetry, the energy gap is zero in certain directions, resembling the nodes of $d$ orbitals. The Zn atoms broke the superconducting pairs at their positions so that a conductance peak appeared at the positions of the Zn atoms. Emanating from the Zn positions were lines of relatively high conductance in the direction of the nodes expected for the $d$-wave function.

## 10.4 FERROMAGNETIC SUPERCONDUCTORS

As discussed earlier, the original BCS theory predicted that superconductors were perfect diamagnets. It was therefore with surprise that solids were recently discovered that are both ferromagnetic and superconducting.

In 2001, reports were published of three such alloys $UGe_2$, URhGe, and $ZrZn_2$. Perhaps of more interest to chemists is the compound $Sr_2RuO_4$. This has a crystal structure almost identical to $La_2CuO_4$, the parent compound of the high $T_c$ superconductor 1-2-3. However, whereas $La_2CuO_4$ is an antiferromagnetic insulator, $Sr_2RuO_4$ is metallic and below about 1.5 K, superconducting. The ruthenium in $Sr_2RuO_4$ is formally present as Ru(IV) giving an electronic configuration of $4d^4$. The Ru is in the centre of an octahedron of oxygen, and so the $d$ bands are split into $t_{2g}$ and $e_g$. The four $4d$ electrons partly occupy the $t_{2g}$ band. It is the electrons close to the highest occupied levels of this band that are responsible for the superconductivity.

These ferromagnetic superconductors are thought to have Cooper pairs formed by magnetic interactions as in high $T_c$ superconductors, but with their spins aligned giving a triplet state. The wave function of the Cooper pair must be antisymmetric with respect to exchange of electrons. For singlet Cooper pairs, exchange leads to a change in spin and so the order parameter must be even ($s$-, $d$-, etc.). For a triplet pair, both electrons have the same spin and so the order parameter must be odd. It is generally accepted that the ferromagnetic superconductors have $p$-wave order parameters.

## 10.5 USES OF HIGH TEMPERATURE SUPERCONDUCTORS

Low temperature superconducting magnets are currently employed for several important uses. Strong magnets are used to remove impurities from food and raw materials. For instance, the magnetic impurities in china clay discolour the manufactured product if not removed. The latest NMR spectrometers employ superconducting magnets to provide large magnetic fields, so improving sensitivity. They are an important tool in fundamental scientific research, and are used in body-scanners for NMR imaging (magnetic resonance imaging [MRI]) used in medical diagnosis.

Conventional electromagnets (made by passing an electric current through insulated copper wire wound around an iron core) can produce fields of about two tesla. Low temperature superconducting magnets are manufactured from both NbTi and $Nb_3Sn$ wires which can carry much higher currents than copper wire ($\sim$400 000 amp cm$^{-2}$) and so eliminate the need for the amplifying iron core. Such superconducting magnets are both lighter and more powerful ($\sim$10 tesla) than their conventional counterparts, consume very little power, and produce little heat — but they have to be cooled to 4 K to become superconducting.

Josephson junctions are used in **S**uperconducting **Qu**antum **I**nterference **D**evices (**SQUIDs**). These consist of a loop of superconductive wire with either one built-in Josephson junction (RF SQUID) or two (DC SQUID). The device is extremely sensitive to changes in magnetic field, and can measure voltages as small as 10$^{-18}$ volt, currents of 10$^{-18}$ amp, and magnetic fields of 10$^{-14}$ tesla. SQUIDs are being used in medical research to detect small changes in magnetic field in the brain. Geologists are able to employ SQUIDs in prospecting for minerals and oil where deposits can cause small local changes in the earth's magnetic field; they also use the device for research into plate tectonics, collecting magnetic data on rocks of different ages. Physicists use SQUIDs for research into fundamental particles, and it is thought that there may be military applications in detecting the change in the earth's magnetic field caused by submarines. Clearly, SQUIDs with their remarkable sensitivity have a myriad of potential uses. A change to high temperature superconductors in the devices could make them cheaper and more flexible to operate.

Development of magnetically levitating trains — the so-called **maglev** — was begun in the 1970s. The first maglev trains used conventional electromagnets which produce an attractive force between the guideway and the train. Such a system linked Birmingham airport to the railway station in the 1980s. Germany is the only Western country that had solid plans for a maglev railway recently. The Transrapid project would have linked Berlin with Hamburg in 2005. This has now been shelved, but Transrapid is working on exporting its system. German engineers worked on the Shanghai maglev railway in China; this is a 30 km track running from the airport to the financial district in Pudong. The trains reach a speed of 430 kph (270 mph). The Japanese are working on a different maglev train system using low temperature superconductors. In this project, superconductors on the train are repelled by magnetic fields produced by coils of wire on the track, lifting the train 4 to 6 inches above the guideway. In November 2003, a Japanese maglev train running on a test track near Kofu, 68 miles west of Tokyo, reached a speed of 560 kph (347 mph).

Use of the new high temperature superconductors should reduce the refrigeration costs of the project. Unfortunately refrigeration is only a very small percentage of both the capital and running costs and so will probably not make the difference as to whether this becomes a viable mode of transport or not.

Using high temperature superconductors would reduce costs in some applications. The cuprates are brittle ceramic materials not easily formed into wires and the superconductivity in these materials is anisotropic. Progress has been made on overcoming these problems. The most suitable cuprate superconductors for making wires found so far (2003) are the bismuth strontium calcium copper oxides $Bi_2Sr_2Ca_2Cu_3O_{10}$ and $Bi_2Sr_2CaCu_2O_8$. These have been processed into long wires by packing a powdered precursor into a silver tube, heating to form the superconductor and drawing the silver tubes to form a wire. The bismuth-based cuprates were chosen because they are more flexible than other high $T_c$ superconductors due to the weak bonds across the double Bi/O layers which allow the layers to slip. Wires that can carry 100 A per strand at 77 K are available in lengths of 100 m.

Electric motors using high temperature superconductors and magnets containing such materials have been developed. In 1997, a high $T_c$ superconducting magnet was installed in the beamline of a carbon-dating van der Graaf accelerator at the Institute for Geological and Nuclear Sciences in Wellington, New Zealand. In 1998, Oxford Magnet Technology and Siemens Corporate Technology built a prototype MRI body scanner with two coils made from high $T_c$ wires and Hitachi has demonstrated a 23.4 tesla superconducting magnet close to the 23.5 tesla needed for NMR spectrometers operating at 1 GHz.

The absence of electrical resistance could be extremely useful in the distribution of electricity throughout the country, as the copper and aluminum wires now used lose 5 to 8% of the power due to the resistance of these elements. Using superconductors would not eliminate power loss entirely. Although superconductors carry direct current (dc) with no loss of power and without generating heat, a small loss of power occurs when they carry alternating current (ac) due to the production of radio waves. However, this loss is much less than that with copper and aluminium wires. The National Grid is unlikely to be replaced by superconducting cables in the near future but such cables are capable of providing a solution to the problem of providing extra power to cities. Power is run into cities via cables in conduits. Many conduits running power into large cities and towns are full and to provide extra power would mean buying up land and digging new conduits. By replacing the existing cables with high $T_c$ cables, more power can be supplied using the existing conduits. In May 2001, 150 000 residents of Copenhagen received electricity through a 30 m high temperature superconductor cable. In the same year, Pirelli replaced nine copper cables at a substation in Detroit by a 120 m cable consisting of three high temperature superconducting cables. The total mass of this high temperature cable was 70 times less than the copper it replaced. For the first time, power was delivered on a commercial basis to U.S. customers via superconducting wires.

Applications are expected to increase in the future. In 2000, a consortium of European companies estimated that the worldwide market in superconducting products will reach £22 billion by 2020.

## QUESTIONS

1. A party of late-night revellers, returning home, decides to take a short cut across a field. Unfortunately, the night is dark and moonless, and the field is known to contain deep potholes. Someone has the bright idea that if everyone links arms and advances together across the field, then if one

of the individuals does encounter a pothole, she will be lifted clear by dint of collective support! Taking this story as an analogy (not a perfect one, of course) with BCS theory, try to identify as many components as possible with corresponding components of BCS theory.

2. Draw a packing diagram (Chapter 1, Section 1.4.5) of the perovskite A-type cell ($CaTiO_3/ABO_3$), determine the number of $ABO_3$ formula units, and describe the coordination geometry around each type of atom. Repeat this procedure for the perovskite B-type unit cell.

3. Complete the following table, summarising types of superconductivity.

| Material | Symmetry of Order Parameter | Spin State |
|---|---|---|
| Metals and Alloys | | |
| Superconducting Cuprates | | |
| Ferromagnetic Superconductors | | |

4. Calculate the average oxidation state of Cu in the mercury cuprate $Hg_{0.8}Tl_{0.2}Ba_2Ca_2Cu_3O_{8.33}$. Assume Tl is present as $Tl^{3+}$ and Hg as $Hg^{2+}$.

# 11 Nanoscience

## 11.1 INTRODUCTION

We felt that we could not leave this book without a brief look at the latest 'hot' topic. The prefix currently on everyone's lips is *nano-*, as in nanoscience, nanotechnology, nanostructures, nanocrystals, etc. In a now famous lecture given in 1959 entitled 'There's Plenty of Room at the Bottom', Nobel Prize-winning physicist Richard Feynman said, "The principles of physics, as far as I can see, do not speak against the possibility of manoeuvring things atom by atom," and that is the nub of what scientists are now trying to do. The final aim is the ultimate in designer chemistry: to be able to assemble a molecule, film, or solid, atom by atom. Already the dream is beginning to be realized, and some, but by no means all, of this science falls into the remit of solid state chemistry.

In 1905, Einstein, as part of his doctoral thesis, measured the size of a sugar molecule using diffusion techniques and found it to be about 1 nm in diameter. A hydrogen atom is about 0.1 nm in diameter, so for chemists, thinking on the nanometre scale is thinking on the atomic or molecular scale, which is what we are used to doing. Therefore, what is different about nanoscience and nanotechnology from what chemists have always done? Current research has two crucial strands. The first is to investigate and utilise the properties of very small particles, which are found to have very different properties from the bulk solid. Chemists have roles here in synthesizing these particles, investigating their properties, and developing new uses for them; it covers such topics as nanotubes, coatings, new alloys, composites, particles for sunscreens, catalysts, colloids, and quantum dots, to name but a few. The second is to be able to manipulate or manufacture things very precisely on the nanometre scale. Some techniques and processing, such as photolithography for the printing of silicon chips, are so-called **top-down processes** because they are carving out or etching nanometre size structures; these are clearly based in technology, although the processes are often chemical. The building up of structures, self-assembly processes, and the direct manipulation of atoms into nanostructures, so-called **bottom-up processes**, once again often falls into the realms of chemistry.

Nanoparticles are usually considered to have at least one dimension less than 100 nm, although this is not a rigid definition and dimensions of several hundred nanometres can fall into this research. Such particles have a larger surface-area to volume ratio than larger particles, affecting the way they react with each other and with other substances. A 10 nm diameter nanoparticle has about 15% of the atoms on the surface; by comparison, this drops to < 1% for a bulk solid. Because a nanoparticle may only consist of a few atoms, the energy levels associated with extended solids, such as bands in metals, no longer apply and the electronic energy levels are more similar to the quantized levels found in individual atoms, this affect their conductivity, and the way they interact with light and other forms of energy.

Unfortunately because 'nano' has become somewhat of a buzzword and i currently attracting a lot of research funding, the term is being used loosely to cove a huge range of topics and techniques to do with anything that is fairly small, and it can be difficult to separate what is important. The Royal Society has come up with two working definitions:

> **Nanoscience** is the study of phenomena and manipulation of materials at atomic, molecular, and macromolecular scales, where properties differ significantly from those at larger scales.
>
> **Nanotechnology** is the production and application of structures, devices, and systems for controlling shape and size at nanometre scale.

Much of the research centres on the organic/bioorganic area and is involved with using self-assembly methods to build large molecules with particular properties such as molecular machines. We will not discuss this type of work here, but in the following sections will concentrate only on areas that relate in some way to solid-state chemistry. First, we consider the physical and electronic effects of the nano scale, and then we discuss examples which might be useful. To try and bring some order to the discussion, we have grouped the examples into one-dimensional (1-D) two-dimensional (2-D), and three-dimensional (3-D) systems. Finally, we look at atomic force microscopy, a technique covered in Chapter 2, but which can also be used for moving atoms around individually and thus has potential for building atomic and molecular structures from 'the bottom up'.

## 11.2 CONSEQUENCES OF THE NANOSCALE

Properties of nanoscale materials may be very different from those of the bulk material. For instance, small particles may melt at much lower temperatures than the bulk and are often much harder: 6 nm copper grains are five times as hard as bulk copper.

### 11.2.1 NANOPARTICLE MORPHOLOGY

The high surface-to-bulk ratio in nanostructures means that a nanocrystal structure is determined by a balance between bulk terms such as lattice energy, surface energy terms, and terms due to faults (such as dislocations) as all these terms are now significant. This can lead to unusual crystal structures such as thin films of *bcc* copper, compared with the normal bulk structure which is *ccp*.

II-VI semiconductors, such as CdSe and CdS, normally have the wurtzite structure (see Chapter 1) where each element is tetrahedrally coordinated. Under high pressures (2 GPa), these transform to the six-coordinate NaCl (rock salt) structure. However, if pressure is applied to a CdSe nanocrystal of about 4 nm in diameter, it now takes much more pressure, about 6 GPa, to transform it to the rock salt structure. It is thought that this may be a resistance to the exposure of high-index crystal planes

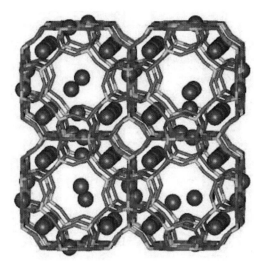

FIGURE 7.8 Framework and cation sites in the Na⁺ form of zeolite A (LTA). (Courtesy of Dr. Robert Bell, Royal Institution of Great Britian, London.)

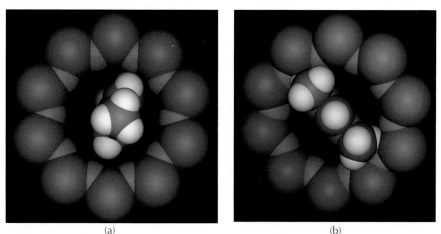

(a)                                                                (b)

FIGURE 7.20 Computer model illustrating how (a) *para*-xylene fits neatly in the pores of ZSM-5, whereas (b) *meta*-xylene is too big to diffuse through.

**FIGURE 7.21(b)** Computer graphic of ethane and methane molecules inside one of the hexagonal pores of MCM-41.

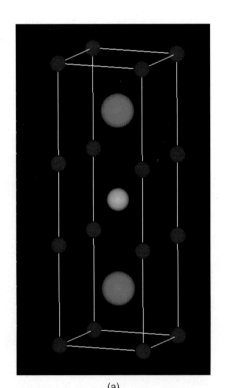

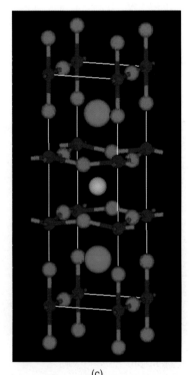

(a)                                              (c)

**FIGURE 10.8** The structure of 1-2-3: (a) the metal positions; (c) idealized structure of $YBa_2Cu_3O_{7-x}$; (d) the extended structure of $YBa_2Cu_3O_7$, depicting copper-oxygen planes, with the copper-oxygen diamonds in between. Key: Cu, blue; Ba, green; Y, aqua; O, red.

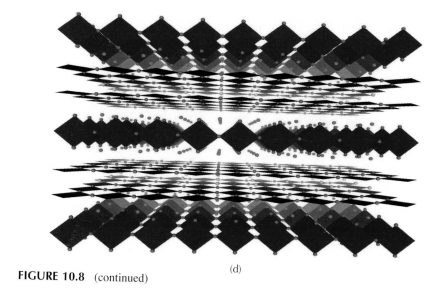

**FIGURE 10.8** (continued)

(d)

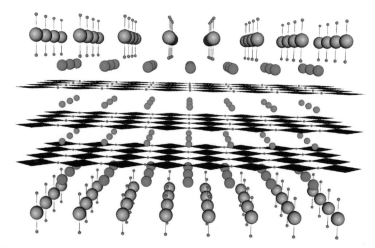

**FIGURE 10.9** Structure of $HgBa_2Ca_2Cu_3O_8$, depicting copper oxygen diamonds between the Ca layers and the copper oxygen layers forming the bases of the pyramids. The apices of the pyramids are in the barium oxygen layers. In the superconductor $Hg_{0.8}Tl_{0.2}Ba_2Ca_2Cu_3O_{8+0.33}$, one fifth of the $Hg^{2+}$ ions are replaced by $Tl^{3+}$ ions and additional oxygen ions are present in the mercury layer. Key: Cu, grey; Ca, red; Ba, blue; Hg/Tl, green; O, aqua.

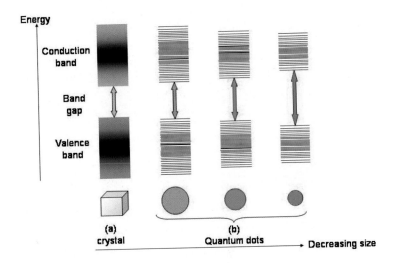

**FIGURE 11.2** The band gap of a semiconductor depends on its size.

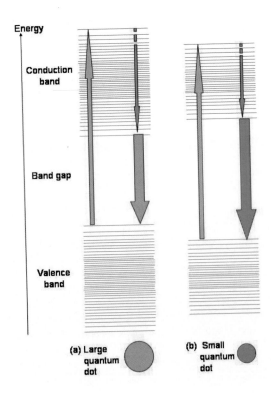

**FIGURE 11.7** The colour of the fluorescence from a nano-sized particle depends on i dimensions.

in the newly formed nanocrystal, which would not have the normal thermodynamically stable morphology.

Varying the conditions of deposition of the film in CVD can alter the morphology of the nanocrystals formed; Figure 11.1(a) and Figure 11.1(b) show nanosized diamond crystals in diamond films grown with *111* (octahedral) and *100* (cubic) faces. Techniques for producing specific morphologies could be very important in the production of catalysts because different crystal faces can catalyse very specific reactions.

Many nanomaterials can be made in different forms. We are familiar with the example of carbon, which we can find as diamond films, carbon black, fullerenes, and multi- and single-walled nanotubes. $MoS_2$ can be made as nanotubes, 'onions' (multi-walled fullerene-type structures), and thin films.

The methods and conditions of nanostructure manufacture are crucial, as we will see that the properties depend critically on the size and shape of particles produced. However, the synthetic techniques used are so particular to each system that they make a subject in their own right and we will not attempt to cover them here.

### 11.2.2  Electronic Structure

In Chapter 4 we discussed how the energy levels of a crystal could be obtained by thinking of a crystal as a very large molecule. For crystals of, for instance, micrometre dimensions, the number of energy levels is so large and the gap between them

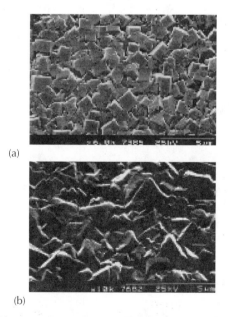

(a)

(b)

**FIGURE 11.1** SEM images of (a) a diamond film grown with methane, hydrogen, and 0.2% $PH_3$, showing (*100*) square facets and (b) a diamond film grown at lower substrate temperature — now the crystals are predominantly (*111*) triangular facetted. (Courtesy of Dr. P.D. May and Professor M.N.R. Ashfold, Bristol University.)

so small that we could treat them as essentially infinite solids with continuous bands of allowed energy. At the nanometre scale, we can still think of the particles as giant molecules but a typical nanoparticle contains $10^2$ to $10^4$ atoms, very large for a molecule but not large enough to make an infinite solid a good approximation. The result is that in nanoparticles, we can still distinguish bands of energy, but the gaps between the bands may differ from those found in larger crystals and, within the bands, the energy levels do not quite form a continuum so that we can observe effects due to the quantised nature of levels within bands. This is illustrated in Figure 11.2, for nano-sized crystals of semiconductor (quantum dots).

The band diagram for a semiconductor is depicted in Figure 11.2(a). The bonding electrons are held in a lower valence band consisting of a continuum of many energy levels, and two electrons can occupy each energy level. The orbital density of states diagrams illustrate that, in general, a lower density of states exists at the top and bottom of the band and a higher density of states exists in the middle (indicated by shading). Above the valence band, at higher energies, a conduction band exists, and separating the two, we see a band gap that can vary in size. If electrons are promoted from the valence band to the empty conduction band by supplying sufficient energy for them to jump the band gap (with heat or light, for instance), then the solid conducts.

As a crystal of a semiconductor becomes smaller, fewer atomic orbitals are available to contribute to the bands. The orbitals are removed from each of the band edges (cf. Chapter 4, Figure 4.6) until, at a point when the crystal is very small — a 'dot' — the bands are no longer a continuum of orbitals, but individual quantised orbital energy levels (Figure 11.2(b)), thus the name quantum dots. At the same

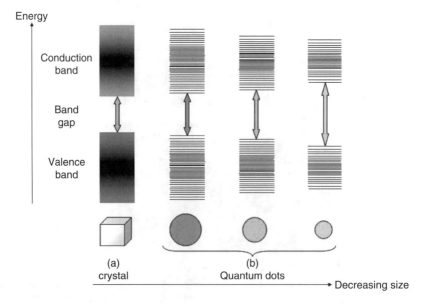

FIGURE 11.2 The band gap of a semiconductor depends on its size. See colour insert following page 356.

time, you can see that this has had the effect of increasing the band gap. As the size of the crystal continues to shrink, so does the number of orbital energy levels decrease and the band gap increases. As the size of the nanoparticles of most semiconductors decreases, the band gap increases. The band gap in CdSe crystals, for example, is approximately 1.8 eV for crystals of diameter 11.5 nm, but approximately 3 eV for crystals of diameter 1.2 nm.

Quantum dots are nanometre scale in three dimensions, but structures that are only nanometre scale in two dimensions (quantum wires) or one dimension (quantum wells or films) also display interesting properties. The quantised nature of the bands in nanostructures can be seen in the density of states. Schematic, theoretical density of states diagrams for bulk material, quantum wells, quantum wires, and quantum dots are pictured in Figure 11.3.

Figure 11.4 presents the density of states for a specific example: a semiconducting nanotube. The density of states for carbon nanotubes is predicted to show sharp peaks (known as van Hove singularities) corresponding to specific energy levels. These can be seen in the figure and have been confirmed by scanning tunnel microscopy (STM) experiments. In Figure 11.4, a gap with zero density about the Fermi energy exists ($E = 0$ on the figure). We associate this band gap with semiconductors. For semiconducting carbon nanotubes, the band gap generally increases with decreasing diameter, but for particular nanotube structures, the band gap becomes zero and the nanotubes are then metallic conductors similar to graphite.

Electrical conductance in solids (other than ionic conductors) depends on the availability of delocalised orbitals close enough together in energy to form bands.

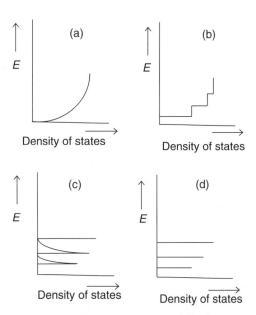

**FIGURE 11.3** Theoretical density of states diagrams for (a) bulk material, (b) quantum well, (c) quantum wire, and (d) quantum dot.

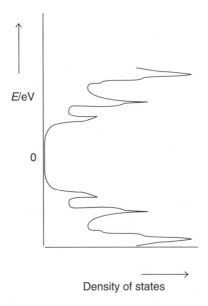

$E$/eV

0

Density of states

**FIGURE 11.4** The density of states for a semiconducting nanotube.

When we have structures on the nanoscale where levels at the tops of occupied bands are not so close in energy as we have seen, then this can affect the conductance. Under certain conditions, electrical conductance through a nanostructure is quantised and increases in a stepwise fashion with increasing voltage. Electrons tunnel into the structure and fill the lowest empty quantised levels until all the levels below the highest filled level providing electrons are full (Figure 11.5). If the thermal energy is insufficient to raise electrons to the next discrete energy level, then no more electrons can tunnel in. The conductance thus drops until the voltage is increased to a value $V$ such that the energy, $e \times V$, is sufficient to raise the energy of the electrons in the adjacent solid so that they can reach the next energy level, when the current increases again. Figure 11.6, for example, depicts the increase of conductance with gate voltage along a quantum wire connecting two GaAs/AlGaAs interfaces in a transistor.

### 11.2.3 OPTICAL PROPERTIES

This section considers two aspects of the interaction of light with nanostructures — differences from the optical properties of bulk solids that arise because of the different energy levels of nanostructures and changes in scattering properties.

The last section revealed that semiconductor band gaps for nanostructures vary with the size of the structure. The wavelength of light emitted when an electron in the conductance band returns to the valence band will therefore also vary. Thus, different colour fluorescence emission can be obtained from different-sized particles of the same substance (e.g., different sized quantum dots of CdSe irradiated with UV emit different colours of light). To produce fluorescence, light of greater photon energy than the band gap is shone onto the nanocrystal. An electron is excited to a

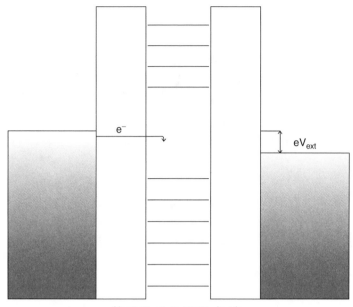

Nanostructure energy levels

**FIGURE 11.5** Electrons fill the lowest empty quantised levels until all the levels below the highest filled level providing electrons are full.

level in the conduction band, from where it reaches the lowest energy level in the conduction band through a series of steps by losing energy as heat. The electron then returns to the valence band, emitting light as it does so (Figure 11.7). Figure 11.7(a) depicts the irradiation of a large quantum dot (smaller band gap). It then decays into the valence band emitting a photon of light, which is the coloured fluorescence seen. If the smaller quantum dot (larger band gap) (Figure 11.7(b))

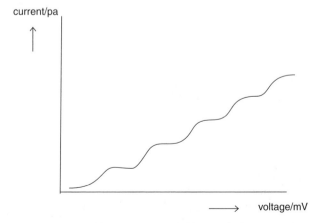

**FIGURE 11.6** The increase of conductance with gate voltage along a quantum wire connecting two GaAs/AlGaAs interfaces in a transistor.

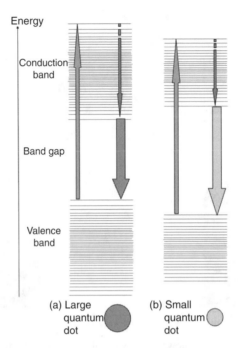

**FIGURE 11.7** The colour of the fluorescence from a nano-sized particle depends on its dimensions. See colour insert following page 356.

undergoes the same process, we can see that the photon emitted as it decays back into the valence band has more energy. From the Einstein equation $E = h\nu$, the higher energy photon will have the higher frequency and thus be closer to the blue end of the spectrum, accounting for the difference in colour. The smaller the nano-structure, the larger the band gap, and hence, the shorter the wavelength of the emitted light. Thus, 5.5 nm diameter particles of CdSe emit orange light whereas 2.3 nm diameter particles emit turquoise light.

In the absorption spectra of nanoparticles of CdSe and other semiconductors, not only can the shift in wavelength be observed, but there are also bands corresponding to absorption to discrete energy levels in the conduction band. For example, 11.5 nm diameter particles of CdSe have an absorption spectrum that shows an almost featureless edge, but particles of diameter 1.2 nm show features resembling molecular absorption bands shifted about 200 nm to shorter wavelengths, as depicted in Figure 11.8.

The colours produced by nanoparticles of gold (colloidal gold) have been used since Roman times where we find them in the glass of the famous Lycurgus cup (British Museum) which appears green in reflected light and red in transmission. Faraday studied them in detail in the mid-19th century and some of his original samples are still held at the Royal Institution in London. In metals, light interacts with surface electrons and is then reflected. The surface electrons in nanoparticles are induced by the light to oscillate at a particular frequency; absorption at this

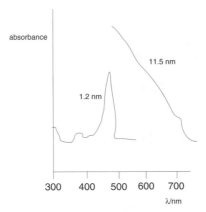

**FIGURE 11.8** The absorption spectrum of 11.5 nm diameter particles of CdSe has an almost featureless edge, but particles of diameter 1.2 nm show features resembling molecular absorption bands shifted about 200 nm to shorter wavelengths.

frequency gives rise to the colour. The oscillation frequency, and hence the colour, depends on the size of the particle. For example, we are familiar with bulk gold which appears yellow in reflected light, but thin films of gold appear blue when light passes through them, and as the particle size is reduced the wavelength of the absorbed light decreases and the film appears first red then, for 3 nm diameter particles, orange. Oscillation frequencies in the visible region are only observed for Ag, Au, Cu, and their alloys, as well as the Group 1 metals.

Particles are known to scatter light as well as absorb it and this produces the white or pale appearance of fine powders. The even smaller nano-sized particles, however, are transparent because the scattering efficiency is reduced. This effect has led to the use of nanoparticles in sunscreens and cosmetics. These will still absorb ultraviolet light but will scatter less visible light.

### 11.2.4 Magnetic Properties

In Chapter 9, the idea of ferromagnetic domains was introduced. Domains typically have dimensions of approximately 10 to 1000 nm. In nanocrystals, therefore we can reach a situation where the domain size and the crystal dimensions are comparable. Such single domain crystals have all the electron spins in the crystal aligned. In larger crystals, the main mechanism for magnetisation and demagnetisation is rotation of the domain walls (see Chapter 9, Section 9.3.1). If the crystal size is reduced, as the single domain region is approached, it becomes harder to demagnetise the crystal by applying a magnetic field. This is because the only possible mechanism is now disruption of spin-spin coupling within a domain. If the particle size is decreased still further, however, the number of spins decreases and the force aligning them becomes weaker. Eventually, this force is too weak to overcome thermal randomisation and in the absence of an applied magnetic field, the spins are randomly oriented. The crystal is then no longer ferromagnetic but superparamagnetic. Figure 11.9 is a plot of the magnetic field needed to demagnetise ferromagnetic particles

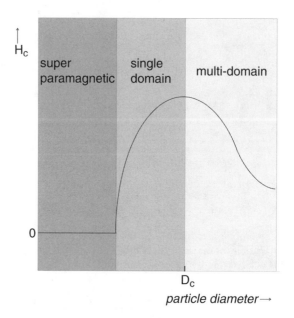

**FIGURE 11.9** Plot of the magnetic field needed to demagnetise ferromagnetic particles (coercivity, $H_C$) as a function of particle size. The particle becomes single domain at radius $D_C$.

(the coercivity, $H_C$) as a function of particle size. The radius $D_C$ is that at which the particle becomes single domain. Values for $D_C$ for the ferromagnetic transition metals and the ferromagnetic $Fe_3O_4$ are given in Table 11.1.

Above a critical temperature, $T_B$, known as the blocking temperature, superparamagnetic particles can have their spins aligned by a magnetic field, behaving in a similar fashion to paramagnetic materials but with a much larger magnetic moment. Below the blocking temperature, there is insufficient kinetic energy to overcome the energy barrier to reorientation of the spins. The blocking temperature varies with the strength of the applied field. In the superparamagnetic state, the particles, like paramagnetic solids, do not show hysteresis.

So far, we have considered only isolated magnetic nanoparticles. When nanosized grains are in close contact, the magnetic effects can differ. This is because spin-exchange coupling can occur between grains. Chapter 9 revealed that thin layers of magnetic material could have the spin state pinned by an adjacent antiferromagnetic

**TABLE 11.1**

**Single domain radii for selected ferromagnetic solids**

| Solid | $D_C$/nm |
|-------|----------|
| Fe | 14 |
| Co | 70 |
| Ni | 55 |
| $Fe_3O_4$ | 128 |

layer. Exchange coupling between grains of hard and soft magnetic materials can produce nanocomposites with high remanence and high coercivity, that is they remain strongly magnetic when the applied magnetic field is reduced to zero and need a large magnetic field to de-magnetise them.

### 11.2.5 MECHANICAL PROPERTIES

Properties of solids such as hardness and plasticity have long been known to vary with grain size. For example, down to $\mu$m-size grains, the resistance to plastic deformation increases with decreasing grain size. However, as the grain size is reduced still further, the resistance levels off or even decreases. The effect has been attributed to a different mechanism for plastic deformation in nano-sized grains. In larger crystals, deformation is mainly governed by the movement of dislocations in the crystal. Such movements are inhibited by grain boundaries. As the grain size is reduced, the ratio of grain boundary to bulk grain increases and so deformation becomes harder. Eventually, as the grain size reaches 5 to 30 nm, movement of dislocations becomes negligible, but deformation through atoms sliding along grain boundaries becomes favourable. This latter mechanism is aided by a large ratio of boundary to bulk grain and so softening increases as the grain size is reduced.

### 11.2.6 MELTING

Nanocrystals show a melting temperature depression with decreasing size. Melting of nanocrystals can be observed in a transmission electron microscope, and gold nanoclusters have been found to melt at 300 K compared with a melting temperature of 1338 K for elemental gold in bulk, and 3 nm CdS crystals melt at about 700 K in a vacuum, compared with 1678 K in the bulk. The surface energy becomes an increasing factor as the size of a crystal becomes smaller; thermodynamics predicts a lowering of the melting temperature if the surface energy of the solid is higher than that of the liquid.

## 11.3 EXAMPLES

To bring some semblance of order to such a huge and diverse area, we have grouped our examples under 1-D, 2-D, and 3-D headings, where 1-D refers to materials with one dimension in the nanometre range and are extended in two dimensions; 2-D is confined to nanometres in two dimensions but extended in one dimension; and 3-D is confined to nanometre dimensions in all three directions.

### 11.3.1 ONE-DIMENSIONAL NANOMATERIALS

#### Nanofilm and Nanolayers

Nanofilms can be used simply as very thin coatings. For example, nanoscale coatings to protect and enhance modern plastic spectacle lenses have been developed, including self-assembling top coatings for nonreflective lenses to protect the antireflective

**FIGURE 11.10** Self-cleaning glass.

layer from dirt, dust, and skin oils as well as super-hard coatings of carbides to protect from scratching.

A spray-on coating containing nanometre-sized particles of $TiO_2$ combined with a photocatalyst can be used on surfaces such as glass to give a dirt repellent surface: the photocatalyst uses UV light to break down organic debris, but the coating also stops rain water from forming droplets, so that it simply runs off collecting dirt as it goes (Figure 11.10). (Similarly, nanoparticles are being used to impregnate fabrics, making them stain resistant.)

The development of electronics using organic compounds (see Chapter 6) has led to nanofilm electronic devices. Such films can be used in flat-screen displays for computer monitors. Two types of displays are used: thin-film transistor liquid-crystal display (TTF-LCD) and organic light-emitting diodes (OLEDs). In TTF-LCD displays, one transistor is connected to each of three LCD elements (red, green, and blue) for each pixel. By making the bank of transistors a film of organic material 20 nm thick instead of a 2 mm thick layer of silicon, thinner, lighter, more flexible screens can be produced. In OLED displays, the colours are produced by an array of light-emitting diodes (LEDs) in which the light-emitting layer is a nanofilm of organic polymer as we described earlier in this book.

The flexibility of thin film organic electronic devices means that they can be printed onto paper and material. An alternative approach is to develop single fibre transistors that could then be woven into cloth. The incorporation of electronics into cloth has inspired some interesting proposals for their use. One suggestion is that military uniforms could be printed with organic films containing electronic circuits that could be used to send commands to soldiers in the field. Another suggestion is for LEDs to be printed onto curtains to provide light when the curtains are closed.

One problem to be overcome is that the organic materials are moisture sensitive and cloth tends to attract moisture.

The electronic and magnetic properties of nanolayers are important in devices formed from electronic materials that are more conventional. We have already discussed quantum well lasers (see Chapter 8) and giant magnetoresistance (GMR) devices used for hard disk read heads (see Chapter 9). Quantum well lasers may be an important component of light-based computers. Other possibilities include magnets with unusual properties (Section 11.2).

## 11.3.2 Two-dimensional nanomaterials

### Nanotubes

In 1991, Sumio Iijima discovered nanotubes. Similar in structure to buckyballs, they consist of sheets of graphite which roll up to form cylinders. Initially multilayered, (multiwalled nanotubes, MWNTs) they can now be made single walled. Single-walled nanotubes, SWNTs (Figure 11.11) have diameters of about 1 nm, and lengths of the order of $10^{-6}$ m. The tubes are capped at each end by half of a fullerene-type structure. They have remarkable properties and are used for a variety of high performance applications. They can adsorb 100 times their volume of hydrogen and, therefore, could be developed as a safe storage medium for hydrogen for fuel cells (see Chapter 5). They have remarkable tensile strength, 100 times greater than steel, but are very light, with half the density of aluminium, and are being used in high performance sports equipment such as tennis rackets. They can also be made as conductors and semiconductors, and one use has been found in car bumpers, where they not only provide strength, but also prevent the buildup of static electricity.

The electronic properties of SWNTs depend on the direction in which the graphite sheet rolls up. Graphite itself is a 2-D metal, exhibiting metallic conductivity within the layers. In the graphite layer depicted in Figure 11.12, notice that the line joining rows of hexagons in the vertical direction is a simple zigzag, but that at right angles to this, so-called 'armchair lines' join the rows. Tubes that roll up along the armchair direction — so-called *armchair* nanotubes — always exhibit metallic levels of conductivity. If graphite sheets roll up along the zigzag lines, with the armchairs along the axis of the tube (*zigzag* nanotubes) or if the sheets roll up along any other

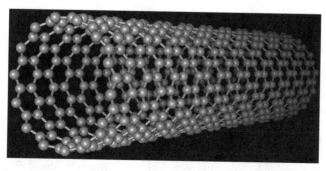

**FIGURE 11.11** Computer simulation of a single-walled carbon nanotube.

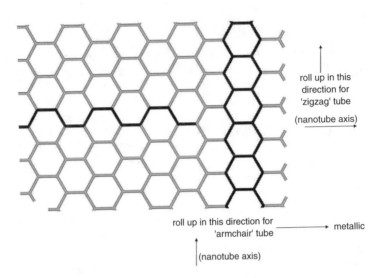

**FIGURE 11.12** If the nanosheets roll up in the direction of the arrow, so that the zigzag lines lie along the axis of the nanotube, then the tubes are usually metallic.

direction except the zigzag or the armchair lines, forming *helical* nanotubes, a band gap is introduced, and the tubes can be semiconducting. It is this property that has allowed the formation of field-effect transistors (FETs), single electron transistors, and diodes from carbon nanotubes. At the time of writing, no preparative method is available for making the type of nanotube needed consistently, they have to be selected by hand from mixtures, but it is hoped that advances will eventually allow the miniaturization of circuitry well beyond that of the silicon chip and ultimately to a molecular computer.

Multiwalled nanotubes have now been synthesized from inorganic compounds; tubes up to 1 μm in length have been formed from $WS_2$ and $MoS_2$.

*Safety Concerns*

Early experiments involving the exposure of mice to nanotubes show that they develop an immune response to them, causing the nanotubes to clump together when inhaled causing lung tissue damage. Both the manufacture and disposal of waste products containing nanotubes will have to be examined carefully.

## Nanowires

Nanowires with lengths of several micrometres and diameters of < 20 nm have been made from various semiconducting materials, for example, silicon, indium phosphide, and gallium nitride. One method involves creating gold nanoparticles using laser ablation and using these as a catalyst with different gaseous starting materials such as a silicon hydride.

Nanowires can also be deposited from a flowing liquid causing them to lie in the same direction and forming arrays. Changing the direction of flow allows an overlay of a second layer, and such arrays can be used for various electronic devices such as transistors and diodes. The thrust of this type of research is aimed at miniaturization, and ultimately with producing a molecular computer.

### 11.3.3 THREE-DIMENSIONAL NANOMATERIALS

#### Nanoparticles

The large surface area of nanoparticles lends increasing dominance to the behaviour of the atoms on the surface of the particle. In catalysis, this is exploited to improve the rate of production in commercial processes, and in the structure of electrodes to improve the performance of batteries and fuel cells (see Chapter 5, Section 5.5.4 and Chapter 7, Section 7.5 and Section 7.6). The interactions between these surface atoms and a surrounding matrix determines the properties of high-performance nanocomposites. Because the dimensions of nanoparticles are less than the wavelength of visible light (400–700 nm), visible light is no longer scattered by them, rendering them transparent; this is a useful property for the manufacture of cosmetics and coatings.

Nanoparticles have been known and used for centuries, think, for instance, of the pigments used to colour stained glass and ceramic glazes, and of the colloidal gold particles used to make 'ruby' glass, known since Roman times. Within the modern era, since the turn of the 20th century, carbon black has been produced for use in printing inks and from the 1940s fumed silica (small particles of silica) has been added to solids and liquids to improve the flow properties. Nanoparticle-sized AgBr is used to coat photographic film (see Chapter 5, Section 5.5).

At the simplest level, nanoparticles of hard substances are useful as polishing powders which are able to give very smooth, defect-free surfaces. Indeed, 50 nm nanoparticles of cobalt tungsten carbide are found to be much harder than the bulk material. Therefore, they can be used to make cutting and drilling tools that will last longer.

#### Carbon Black

Carbon black is made by the vapour-phase incomplete pyrolysis of hydrocarbons to produce a fluffy fine powder. Worldwide, about 7 million tons a year are produced. It is used as a reinforcing agent in rubber products such as tyres (20–300 nm), as a black pigment (< 20 nm) in printing inks, paints, and plastics, in photocopier toner, and in electrodes for batteries and brushes in motors.

#### Fumed Silica

Reacting $SiCl_4$ with an oxy-hydrogen flame makes this very pure form of silica. The resulting $SiO_2$ particles with dimensions 7 to 50 nm, have an amorphous structure. The particles have silanol, $Si-OH$, groups on the surface, and these hold the particles together by hydrogen bonding forming chain-like structures. When added to liquids this 3-D network traps the liquid and increases the viscosity, but when the thickened liquid is subsequently brushed out or sprayed, the liquid and any trapped air is released. When the shear force is removed, the liquid thickens again. This property is known as thixotropy, and is very useful in paints for preventing the settling of pigments, and for improving the flow properties of paints, coatings, and resins. It is also used to improve the flow properties of powdered solids such as pharmaceuticals, cosmetics, cement, inks, and abrasives, helping to prevent caking.

### Quantum Dots

A quantum dot or **nanocrystal** is defined as a crystal of a semiconductor which is a few nanometres in diameter typically containing only $10^2$ to $10^4$ atoms. As discussed in Section 11.2, quantum dots exhibit quantum-size effects in their physical properties, having interesting electronic, magnetic, and optical properties which are a consequence of their size rather than their chemical composition. For instance, a cadmium selenide (CdSe) dot of diameter 2.3 nm gives out turquoise fluorescent light when irradiated with UV light, but a dot of dimension 5.5 nm glows orange (Figure 11.7). Their unusual spectroscopic properties have led to the detection of Si quantum dots in the nebulae of outer space.

Considerable practical difficulties arise in making quantum dots of a regular size and shape, and in preventing the dots, once made, from coalescing into a larger crystal. CdSe quantum dots are currently made by dissolving selenium and dimethylcadmium ($(CH_3)_2Cd$) in hot trioctylphosphine oxide (TOPO), from whence large quantities of CdSe quantum dots precipitate. The size of the dots is controlled by the time they remain in the hot liquid. The shape of the dots is improved by the growth of a thin layer of zinc sulfide, ZnS, on the surface, and they are then coated in TOPO to make them hydrophobic and prevent them from coalescing. Now we must ask how the unusual properties of quantum dots are being exploited.

In Chapter 8, we saw how, in LEDs, a voltage is applied across the $p$–$n$ junction of a semiconductor, causing electrons to move across the junction from the $n$-type side and drop into vacancies in the valence band on the $p$-type side, emitting a photon of light. As we have seen, quantum dots of different sizes emit light of different wavelengths, and if made reliably to a particular size, can be used to make LEDs, but of a very pure colour. This purity of colour is what is needed for use as pixels in colour displays.

The purity of the colour produced by quantum dots is also leading to research into their use in special dyes and polymers which could be used, for instance, to combat counterfeiting, as they will be extremely difficult to reproduce.

The light-emitting properties of quantum dots mean that they can be used as fluorescent probes in biological systems, where they can have many advantages in replacing conventional organic fluorescent probes. They have been used both *in vivo* and *in vitro*; if they can be attached to a biological molecule of interest, such as an antibody or a protein, they can be used to follow the reactions of that molecule. They have the advantages that they are stable, can be excited by broadband excitation, but emit a narrow band of frequencies of high intensity, and are available in many colours; because the colour emitted by each size dot is pure, several different colour dots can be used at the same time to track different processes.

In an area of interest which has already been discussed (Chapter 7, Section 7.8), quantum dots of CdS have been grown inside zeolite cages, thus controlling their size. This gives the prospect of forming new semiconducting materials useful for data storage. A similar idea under investigation is the growth of quantum dots inside diatom skeletons (which are made of silica) as reaction vessels, where the thousands of different species could provide templates for complex arrays of dots.

Research is also looking at the possibility of a quantum dot laser, which would allow rapid light pulses from a minute source to further increase the speed of circuitry and communications. Such lasers should be tunable, as the frequency of light emitted will depend on the size of the quantum dot.

## Explosives

Aluminium particles are made with diameters in the range of 20 to 200 nm, with a protective shell of aluminium oxide about 4 nm thick and, mixed with a suitable oxidising agent, are used for rocket propulsion fuels.

Aluminium powders are also used for the thermite reaction. This is the reaction traditionally used to weld rail tracks together *in situ*: $2Al(s) + Fe_2O_3(s) = 2Fe(l) + Al_2O_3(s)$, which produces so much heat that molten iron is produced, but it can also be used to produce other less easily reduced metals, such as Mo, from their oxides. Using nanometer particles dramatically improves performance, and these reactions are used as priming reactions for munitions and in fireworks.

## Magnetic Nanoparticles

Nanoparticles of iron oxides ($\gamma$-$Fe_2O_3$ or $Fe_3O_4$) are superparamagnetic, having zero remanence, that is, a magnetic field induces magnetism which disappears when the field is removed. This property can be used as a simple switch to separate components of chemical or biological solutions. The particles are coated with a functional group that will attach to the required component in the solution. If a magnetic field is then applied to one side of the container, the particles become magnetised and drift toward the magnet taking the attached component with them.

Iron oxide nanoparticles, acting as contrast agents, are used to enhance magnetic resonance imaging (MRI) scans. Ferromagnetic Fe or Ni nanoparticles can also be used for this purpose, with the advantages of a larger magnetic moment and therefore a larger signal, but are much more difficult to produce.

There are other promising medical uses for these magnetic nanoparticles. If they can be functionalized appropriately, for instance, with an antibody, then they could also be injected into the bloodstream and used to seek out markers of disease or infection such as cancer sites; this would then be detectable by MRI. If a drug molecule could be attached to the surface of a magnetic nanoparticle, then a magnetic field could be used to draw the drug toward the site of the infection, thus targetting the drug more efficiently.

The superparamagnetic properties of iron oxide nanoparticles when suspended in hydrocarbons, are used in **ferrofluids**. These were invented by NASA as a way to control liquid rocket fuels in space, because the flow can be controlled by magnetic fields. With the advent of solid rocket fuels, they are now used mainly as bearing seals in the hard drives of computers and dampers in speakers. A magnetic fluid containing nanoparticles of iron nitride ($Fe_3N$) in kerosene has been produced by reacting iron carbonyl ($Fe(CO)_5$) and ammonia. Similarly, thermal, or sonochemical, decomposition of iron carbonyl in decalin also produces a ferrofluid containing nanoparticles of α-Fe.

Cobalt clusters embedded in silver display giantmagnetoresistance (GMR), with an increase in resistance of up to 20%, and are used for magnetic recording and data storage.

### Medical and Cosmetic Use

*Silver Nanoparticles.* Silver nanoparticles have been found to have very good anti-bacterial action, and are being used to impregnate bandages; they are also used to impregnate socks and added to underarm deodorants because the antibacterial action kills the bacteria responsible for the unpleasant smells.

*Sunscreens.* $TiO_2$ (and ZnO) has been used in sunscreens for many years because of its ability to absorb the ultraviolet (UV) radiation which is harmful to the skin. Nevertheless, $TiO_2$ also has a very high refractive index and scatters light very efficiently — so efficiently, in fact, that it makes a very good white pigment and consequently is used in most white paint because of its high covering power. Its use in the sunscreens and sunblocks with the highest factor numbers has, for many years, made them appear white on the skin and not very attractive. Nanoparticles of $TiO_2$ of about 50 nm in diameter are now being manufactured, and they are transparent because they are too small to scatter the visible light, but they still absorb the harmful shorter wavelength UV radiation and are now being used in sunscreens to get around this problem. Larger nanoparticles of $TiO_2$ are being used to impregnate fabrics to make clothing with a built-in sun protection.

*Safety Concerns.* Many people still feel that there may be safety issues to consider in the use of nanoparticles in sunscreens and cosmetics, as little is known about their metabolism and absorption through the skin. In addition, nanoparticulate $TiO_2$ produces free radicals at the surface in the presence of sunlight and moisture, although it is thought that these would not survive long enough to penetrate through the skin where they could cause damage to live cells. Possible methods of preventing free-radical production include coating or doping the particles.

*Dental Fillings.* Nanoparticles of zirconium oxide ($ZrO_2$) are being used in the new UV-cured dental fillings. They give strength and are transparent to visible light, but are opaque to X-rays. Rare-earth oxides can also be used.

## 11.4 MANIPULATING ATOMS AND MOLECULES

Perhaps one of the most awesome achievements of recent technological advances has been the development of techniques that can deal with single atoms or molecules. Surfaces can be characterised at the atomic level, materials built up atom by atom, and chemical reactions of single molecules studied. One of the drivers for this research has been the production of ever yet smaller electronic circuits. Traditionally, integrated circuits have been assembled using techniques in which a pattern is laid down on the surface of a crystal and the uncovered parts treated to form a differently doped semiconductor, an insulating oxide, or a metallic connector. Currently, the patterns are produced using photolithography, but the spacing of components using this technique is limited by diffraction effects to about 100 nm. To produce integrated circuits in which the components are nanometre scale requires new techniques. It is possible to use variations on the existing approach by, for example, using UV radiation, X-rays, or electron beams in place of visible light. Another approach is to use scanning probe microscopy (see below) to print nanoscale patterns onto a substrate. For example, if a current is passed through the probe tip raising its temperature, it can soften a thermoplastic polymer in the immediate vicinity of the tip producing a small indentation. A completely different approach is to build up the required nano-scale structures atom by atom or molecule by molecule. Because we want to concentrate on chemistry at the single-molecule level, we have chosen to look at methods of producing integrated circuits only when they involve this latter approach.

## 11.4.1 SCANNING TUNNELLING MICROSCOPY

The scanning tunnelling microscope (STM) was invented in the early 1980s by Binnig and Rohrer who were awarded the 1986 Nobel Prize in physics for their work.

As explained in Chapter 2, in STM experiments, a sharp metallic tip formed by etching a wire to a very fine point (as small as 20 nm in diameter) is brought close to a conducting or semiconducting surface. A voltage is applied to the tip, and the tip is scanned across the surface. Electrons from the surface travel to the tip through quantum mechanical tunnelling, producing a current measured in nanoamperes. This current depends on the density of states near the surface and the distance between the tip and the surface. A picture of the surface can be built up by keeping the current constant and recording the changes in distance needed. Chemical information can be obtained by altering the voltage so that electrons from different types of energy levels tunnel across.

As well as presenting a picture of the surface and any molecules on it, STM can also be used to move atoms and molecules across a surface and to make molecules react. If the STM tip is brought closer to the surface than usual, the force between it and atoms/molecules on the surface increases. If this force is attractive, moving the tip across the surface pulls the molecule along after it. A study of 1,4-diiodo-benzene on a copper surface at 20 K, for example, demonstrated that the molecule was made to hop along the surface from copper atom to copper atom. Figure 11.13 clearly illustrates that the 1-4, diiodobenzene has moved relative to the CO molecule on the Cu surface.

Repulsive forces between the tip and surface (produced by reversing the sign of the voltage) can push molecules away from the tip across the surface. An atom or molecule can also become attached to the tip, and can be slid across the surface. The famous picture of xenon atoms forming IBM on a nickel surface was produced by increasing the voltage so that the Xe became attached to the tip, sliding the atom across the surface, and then decreasing the voltage again at the position required.

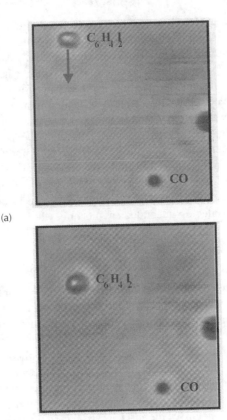

(a)

(b)

**FIGURE 11.13** A 1,4 diiodobenzene molecule on a copper surface at the upper left-hand corner in the STM image (a) is moved to a new position shown in (b). (Reprinted with permission from the *Annual Review of Physical Chemistry*, vol. 54, ©2003 by Annual Reviews www.annualreview.org and by kind permission of the authors, Saw-W. Hla and Karl H. Rieder.)

To perform single molecule chemical reactions, the STM tip is used to channel electrons into a molecule. The energy can both dissociate bonds and provide the activation energy for bond formation.

### 11.4.2 ATOMIC FORCE MICROSCOPY

The invention of STM led to the development of other, similar, techniques with atomic resolution, known collectively as scanning probe microscopy. The most broadly applicable is atomic force microscopy (AFM). AFM, similar to STM relies on a very sharp tip, but in this case, the tip is brought close enough to the surface that the intermolecular forces between tip and surface can be measured. The standard method does not yield any information as to the composition of surface species, but by attaching a particular molecule to the tip, the AFM can be made responsive to certain molecules or groups of molecules and not others. Images of water on a surface, for example, have been obtained using tips treated with a hydrophobic compound. AFM is often used to study biological molecules and some groups have used modified tips to extract particular molecules, for example proteins from cell membranes.

## Writing with AFM Tips

By scanning AFM tips along a surface, it is possible to induce chemical reactions along a line. One example is the application of a voltage to an AFM tip to oxidise strips of silicon with nanometre dimensions. The electric field from an AFM with a nanotube tip will strip hydrogen from the silicon surface, leaving a line of bare silicon which is readily oxidised in air so that the net result is a track of $SiO_2$ on the silicon surface.

## Dip-pen Lithography

In dip-pen lithography, molecules (such as thiols) are placed on the AFM tip and delivered to a substrate surface (such as gold) via a water meniscus. The tip is loaded either by dipping in a solution or by vapour deposition. In damp air, a water meniscus naturally forms between the tip and the sample so that when the tip is close to the substrate the molecules are delivered to the surface via the meniscus, and form a self-assembled monolayer. Surface tension holds the tip at fixed distance from the surface as it moves across it. The size of the deposit is dependent on the tip radius of curvature, the relative humidity (which controls the size of the water droplet) and any diffusion of the molecules across the surface. Currently (2005), structures 5 nm apart can be generated. The technique has the advantage that the same instrument can be used to both build the nanostructure and image it. An array of AFM tips linked together shows great promise. One tip is slightly (0.4 nm) more above the surface than the others and is used for imaging and hence guidance on moving the array across the surface. The other tips all lay down identical structures.

### 11.4.3 NANOSTENCILS

Structures can be built up by directing collimated ion beams in a high or ultrahigh vacuum through openings in a cantilever or membrane.

### 11.4.4 OPTICAL TWEEZERS

The interaction between the electric dipole of light and the polarisability of molecules or dielectric particles produces a force of the order of piconewtons. By using one or more tightly focussed laser beams, this tiny force can be exploited to move molecules or nanoparticles and arrange them into particular structures. Presently, building up a structure in this way is too slow to be technologically useful.

### 11.4.5 MOLECULAR LITHOGRAPHY

An unusual approach to constructing electronic circuits is to use the coding properties of deoxyribonucleic acid (DNA). By modifying the DNA so that a compound only attaches to some sequences in a DNA strand, we can ensure that a following reaction takes place only where that compound is, or only where it is absent. In one example, DNA was reacted with glutyraldehyde, which produced aldehyde groups on the DNA. A protein was then polymerised on a probe DNA molecule. The resulting nucleo-protein filament attached to a specific sequence on the original DNA. Thus, specific lengths of DNA were covered by the protein and others left bare. The DNA was then reacted with silver nitrate. Where aldehyde groups were free (i.e., not covered by the filament), the silver nitrate was reduced to silver. The silver clusters formed catalysed the formation of gold from a solution of $KAuCl_4$, KSCN, and hydroquinone. This produced a gold wire that was a few micrometres long and 50 to 100 nm in width.

## 11.4.6 DATA STORAGE

One use of AFMs currently being investigated by computer firms is for data storage. A method has been put forward that involves a large array of AFM tips and a polymer-coated storage disk. To record data, a current is passed through the tip raising its temperature to 400°C. The polymer in the immediate vicinity of the tip is softened producing a small indentation. To read the data back, the tip is heated to 300°C, which is too low to cause softening during the brief time the tip is in contact but enough to cause a change in resistance when the tip drops into an indentation. This system is expected to increase storage capacity by a factor of more than 20, which is sufficient, for instance, to turn a mobile phone into both a computer and music player.

# Further Reading

To keep up with current developments, the reader is referred to journals such as *Science, Scientific American, Angewandte Chemie, Physics Today, Materials Today, Chemistry World,* and *Physics Web* (http://physicsweb.org).

## BIBLIOGRAPHY

### GENERAL

West, A.R. (1984) *Solid State Chemistry and Its Applications*, John Wiley, New York.

Cox, P.A. (1987) *The Electronic Structure and Chemistry of Solids*, Oxford University Press, Oxford.

Cheetham, A.K. and Day, P. (eds.) (1992) *Solid State Chemistry: Compounds*, Oxford University Press, Oxford.

Bruce, D.W. and O'Hare, D. (eds.) (1992) *Inorganic Materials*, John Wiley and Sons Ltd., Chichester.

Wold, A. and Dwight, K. (1993) *Solid State Chemistry*, Chapman and Hall Inc., New York.

Weller, M.T. (1994) *Inorganic Materials Chemistry*, Oxford University Press, Oxford.

West, A.R. (1995) *Basic Solid State Chemistry*, 2nd edn, John Wiley and Sons Ltd., Chichester.

Rao, C.N.R. and Gopalakrishnan, J. (1997) *New Directions in Solid State Chemistry*, 2nd edn, Cambridge University Press, Cambridge.

Dann, S.E. (2000) *Reactions and Characterization of Solids*, Royal Society of Chemistry, Cambridge.

## CHAPTER 1

### CRYSTALLOGRAPHY AND SYMMETRY

Wells, A.F. (1984) *Structural Inorganic Chemistry*, 5th edn, Oxford University Press, Oxford.

Hyde, B.G. and Andersson, S. (1989) *Inorganic Crystal Structures*, John Wiley, New York.

Hahn, T. (ed.) (1987) *International Tables for Crystallography*, Vol. A, 2nd edn, International Union of Crystallography.

Cotton, F.A. (1971) *Chemical Applications of Group Theory*, 2nd edn, John Wiley, New York.

Fletcher, D.A., McMeeking, R.F., and Parkin, D. (1996) The United Kingdom Chemical Database Service, *J. Chem. Inf. Comput. Sci.*, **36**, 746–9.

### LATTICE ENERGIES AND BORN–HABER CYCLES

Johnson, D.A. (1982) *Some Thermodynamic Aspects of Inorganic Chemistry*, 2nd edn, Cambridge University Press, Cambridge.

## CHAPTER 2

### Physical techniques

Cheetham, A.K. and Day, P. (1987) *Solid State Chemistry: Techniques*, Oxford University Press, Oxford.

Ebsworth, E.A.V., Rankin, D.W.H., and Cradock, S. (1991) *Structural Methods in Inorganic Chemistry*, 2nd edn, Blackwell Scientific Publications, Oxford.

Ladd, M.F.C. and Palmer, R.A. (1993) *Structure Determination by X-ray Crystallography*, 3rd edn, Plenum, New York.

Bish, D.L. and Post, J.E. (ed.) (1989) *Modern Powder Diffraction: Reviews in Mineralogy*, Vol. 20, The Mineralogical Society of America, Washington, D.C.

Niemantsverdriet, J.W. (1995) *Spectroscopy in Catalysis*, VCH, Weinheim.

Catlow, C.R.A. and Greaves, G.N. (1990) *Applications of Synchrotron Radiation*, Blackie, Glasgow.

Young, R.A. (1993) *The Rietveld Method*, International Union of Crystallography, Oxford.

Clegg, W. (1998) *Crystal Structure Determination*, Oxford University Press, Oxford.

Hsu, J.W.P. (2001) True Optical Microscopy, *Mater. Today*, **4**, 26.


## CHAPTER 3

### Synthesis

Rao, C.N.R. (1999) Novel materials, materials design, and synthetic strategies: recent advances and new directions, *J. Mater. Chem.*, **9**, 1–14.

Hagenmuller, P. (ed.) (1997) *Preparative Solid State Chemistry*, Academic Press, New York.

Rao, C.N.R. and Rouxel, J. (eds.) (1997) Synthesis and reactivity of solids, *Curr. Opinion Solid State Mater. Science*, **2**, 129–73.

Rao, C.N.R. and Rouxel, J. (eds.) (1996) Synthesis and reactivity of solids, *Curr. Opinion Solid State Mater. Science*, **1**, 225–94.

Segal, D.H. (1994) *Chemical Synthesis of Advanced Ceramic Materials*, Cambridge University Press, Cambridge.

Mingos, D.M.P (1998) Microwaves in chemical syntheses, *Chem. Society Rev.*, **27**, 213.

Patil, K.C., Aruna, S.T., and Mimani, T. (2002) Combustion synthesis: an update, *Curr. Opinion Solid State Mater. Science*, **6**, 507–12.


## CHAPTER 4

Moore, W.J. (1967) *Seven Solid States*, Chapters II and III, W.A. Benjamin Inc., New York.

McWeeny, R. (1979) *Coulson's Valence*, Oxford University Press, Oxford.

Cox, P.A. (1987) *Electronic Structure and Chemistry of Solids*, Chapters 1, 4, and 7, Oxford University Press, Oxford.

West, A.R. (1988) *Basic Solid State Chemistry*, Chapters 2 and 7, John Wiley, New York.

Duffy, J.A. (1990) *Bonding, Energy Levels and Bands in Inorganic Solids*, Chapters 4 and 7, Longman, London.

Rosenberg, H.M. (1989) *The Solid State*, Chapters 7–10, Oxford University Press, Oxford.

Kittel, C. (1996) *Introduction to Solid State Physics*, 7th edn, John Wiley, New York.

## CHAPTER 5

### NON-STOICHIOMETRY

Tilley, R.J.D. (1987) *Defect Crystal Chemistry*, 2nd edn, Blackie, Glasgow.
Greenwood, N.N. (1968) *Ionic Crystals, Lattice Defects and Non-stoichiometry*, Butterworths.

### BATTERIES

Dell, R.M. and Rand, D.A.J., (2001) *Understanding Batteries*, The Royal Society of Chemistry, Cambridge.

### FUEL CELLS

http://www.fuelcells.org.

## CHAPTER 6

### REVIEWS

Higgins, S.J., Eccleston, W., Sedgi, N., and Raja, M. (May 2003) Plastic electronics, *Education in Chemistry*, 70–3.
Clery, D. (25 March 1994) After years in the dark, electric plastic finally shines, *Science*, **263**, 1700–2.

## CHAPTER 7

Breck, D.W. (1974) *Zeolite Molecular Sieves*, John Wiley, New York.
Barrer, R.M. (1982) *Hydrothermal Chemistry of Zeolites*, Academic Press, New York.
Dyer, A. (1988) *An Introduction to Zeolite Molecular Sieves,* John Wiley, New York.
Catlow, C.R.A. (ed.) (1992) *Modelling of Structure and Reactivity in Zeolites*, Academic Press, London.
Beck, J.S. and Vartuli, J.C. (1996) *Curr. Opinion Solid State Mater. Science*, **1**, 76–87.
Maschmeyer, T. (1998) *Curr. Opinion Solid State Mater. Science*, **3**, 71–8.

## CHAPTER 8

Duffy, J.A. (1990) *Bonding, Energy Levels and Bands in Inorganic Solids*, Longman, London.
Johnson, N.M., Nurmikko, A.V., and DenBaars, S.P. Blue diode lasers, *Phys. Today* (October 2000), 31–6.
Yablonovitch, E. Photonic crystals, *Sci. Am.* (December 2001), 47–55. September.
Goodman, C.H.R. (September 1983) Optical Fibres, *Chemistry in Britian*, 745.

## CHAPTER 9

Awschalom, D.W., Flatté, M.E., and Samarth, N. Spintronics, *Sci. Am.*, (June 2002) 68–73.
Guiner, A. and Julien, R. (1989) *The Solid State from Superconductors to Superalloys*, Oxford University Press, International Union of Crystallography, Oxford.

Newnham, R.E., Trolier-Mckinstry, S. and Giniewicz, T.R. (1993) Piezoelectric, pyroelectric and ferroic crystals, *Journal of Materials Education*, **15**, 189–223.

## CHAPTER 10

### Books

Buchel, W. (2004) *Superconductivity — Fundamentals and Application*, John Wiley-VCH, New York.
Waldram, J.R. (1996) *Superconductivity of Metals and Cuprates*, IOP, London.
Silberglitt, R. et al. (2002) *Strengthening the Grid: Effect of High-Temperature Superconducting Power Technologies on Reliability, Power, Transfer Capacity and Energy Use*, Rand Corporation.
Vanderah, T. (ed.) (1999) *Chemistry of Superconductor Materials*, Noyes Publications, Park Ridge, NJ.

### Reviews

Tallon, J. Industry warms to superconductors, *Phys. World* (March 2000). http://physicsweb.org
Hervieu, M. (1996) Recent developments in the crystal chemistry of high $T_C$ superconductors, *Curr. Opinion Solid State Mater. Sci.*, **1**, 29–36.
Gross Levi, B. Learning about high $T_c$ superconductors from their imperfections, *Phys. Today* (March 2000), 17–8.
Maeno Y., Rice T.M., and Sigrist M.The intriguing superconductivity of strontium ruthenate, *Phys. Today* (January 2001),, 42–7.
Flouquet, J. and Buzdin, A.G. Ferromagnetic Superconductors, *Phys. World* (January 2002). http://physicsweb.org.
Gough C. New metallic superconductor makes an immediate impact, *Phys. World* (April 2001), http://physicsweb.org.

## CHAPTER 11

Nalwa, H.S. (ed.) (2002) *Nanostructured Materials and Nanotechnology*, Academic Press, New York.
Timp, G. (ed.) (1998) *Nanotechnology*, AIP Press, College Park, MD.
*J. Materials Chem.*, (2004) **14** (4). [The whole issue is devoted to new developments in nanomaterials.]
Rao, C.N.R. and Cheetham, A.K. (2001) Science and technology of nanomaterials: current status and future prospects, *J. Mater. Chem.*, **11**, 2887.
Cox, J. (September 2003) A quantum paintbox, *Chem. Britain*, 21–5.

# Answers to Odd-Numbered Questions*

## ANSWERS TO CHAPTER 1

1.  (a) $NF_3$. Figure 1.62(a) shows the threefold ($C_3$) axis and three planes of symmetry.

    (b) $SF_4$. Figure 1.62(b) shows the twofold ($C_2$) axis and two planes of symmetry.

    (c) $ClF_3$. Figure 1.62(c) shows the twofold ($C_2$) axis and two planes of symmetry.

3.  There are four. Unit cell (2) contains two lattice points; (3a) and (3b) each contain three lattice points; (4) contains four lattice points.

5.  The indices are B – $1\ \bar{1}$; C – $01$; D – $21$; E – $2\ \bar{1}$. (By choosing a different line, you may have come up with the answers: $\bar{1}1, 0\ \bar{1}, \ \bar{2}\ \bar{1}$, and $\bar{2}\ 1$. These are equally valid answers.)

7.  Figure 1.63 shows the three planes: (a) $100$ (b) $110$ (c) $111$. The area of the $100$ plane shown, is $a^2$ and contains $(1 + 4 \times {}^1/_4) = 2$ atoms. The $110$ plane contributes an area $a^2 \sqrt{2}$ and contains $(2 \times {}^1/_2 + 4 \times {}^1/_4) = 2$ atoms. The $111$ plane contributes an area of $\dfrac{a^2 \sqrt{3}}{2}$ and contains $(3 \times {}^1/_2 + 3 \times \frac{1}{6}) = 2$ atoms. The relative densities per unit area for these three planes are thus:

    $$100 : 110 : 111 = 2 : 1.414 : 2.31.$$

    Notice that the $111$ planes are the close-packed layers of the structure and so have the densest packing of atoms, as we would expect.

9.  Four. There are $(6 \times {}^1/_2) = 3$ sulfurs at the centres of the faces and $(8 \times \frac{1}{8}) = 1$ at the corners. These four are matched by the four zincs entirely enclosed in the cell.

---

\* Answers to even-numbered questions are found in the Solutions Manual.

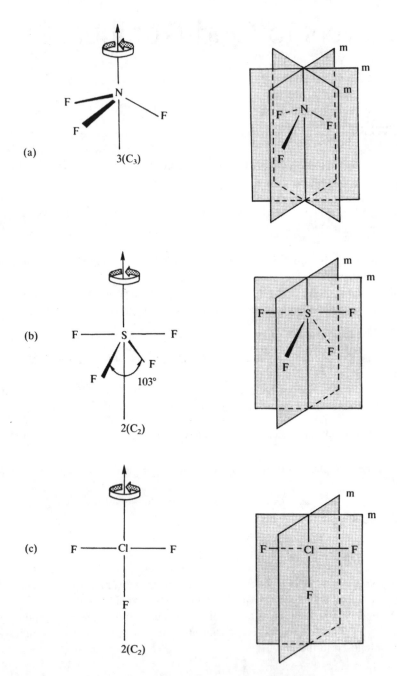

**FIGURE 1.62** (a) threefold ($C_3$) axis in $NF_3$, and three planes of symmetry in $NF_3$, (b) twofold axis ($C_2$) in $SF_4$, and two planes of symmetry in $SF_4$, (c) twofold axis ($C_2$) in $ClF_3$, and two planes of symmetry in $ClF_3$.

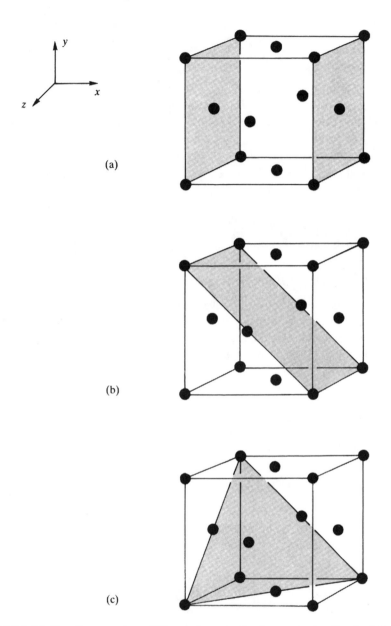

**FIGURE 1.63** The diagrams show different planes in the face-centred unit cell of the *ccp* structure: (a) *100*, (b), *110*, and (c) *111*.

11. The relative molecular mass of NaCl is (22.9898 + 35.453) = 58.4428. Four molecules are in the unit cell, so the relative mass of the unit cell in kg is:

$$\frac{4 \times 58.4428}{6.0220 \times 10^{26}} = 3.8819 \times 10^{-25} \text{ kg}$$

The volume of the unit cell is:

$$(564 \times 10^{-12} \text{ m})^3 = 1.79 \times 10^{-28} \text{ m}^{-3}$$

The density is therefore

$$\frac{3.8819 \times 10^{-25}}{1.79 \times 10^{-28}} = 2.17 \times 10^3 \text{ kg}$$

13. Assuming that anion–anion contact occurs as in Figure 1.46(b), the iodide ion radius is $\dfrac{300}{\sqrt{2}}$ or 212 pm.

15. The *100* and *111* planes are the same as the planes for the cubic-close packed structure in Question 7, Figure 1.63. The only plane that has changed is the *110* (Figure 1.64) which now has two additional atoms. Thus, the relative densities per unit area become:

$$100 : 110 : 111 = 2 : 2.83 : 2.31$$

(Note that the structure is no longer truly close-packed (i.e., the carbon atoms are not touching) because the *110* planes are now more densely packed than the *111*. Remember that the radius of a tetrahedral hole is

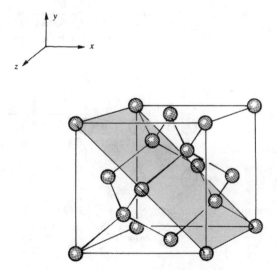

**FIGURE 1.64** The *110* plane of the diamond structure.

only $0.225r$, and the carbon atoms here all have the same radius. Therefore, although we can think in terms of the carbon atoms occupying tetrahedral holes in a close-packed structure, they are actually too big to do so, and so this description this is only useful in terms of the relative geometry, in the same way as we discussed for the fluorite structure.)

17. The Madelung constant, $A = -3.99$.

There are seven ions in the structure in Figure 1.61: six cations and one anion. First, calculate the contribution to the potential energy of interactions of the six cations with the central anion. Each cation is distance $r_0$ from the central ion.

$$E_1 = -\frac{6e^2}{4\pi\varepsilon_0 r_0}$$

Each cation also interacts with a diametrically opposite cation (distance $2r_0$). There are three such interactions, so

$$E_2 = +\frac{3e^2}{4\pi\varepsilon_0 2r_0}$$

Finally, $E_3$ is calculated from interactions between adjacent cations (distance $\sqrt{2}r_0$) of which there are twelve:

$$E_3 = +\frac{12e^2}{4\pi\varepsilon_0\sqrt{2}r_0}$$

$$E = E_1 + E_2 + E_3$$

$$= -\frac{e^2}{4\pi\varepsilon_0 r_0}\left(6 - \frac{3}{2} - \frac{12}{\sqrt{2}}\right)$$

$$= -\frac{Ae^2}{4\pi\varepsilon_0 r_0}$$

So,

$$A = 6 - \frac{3}{2} - \frac{12}{\sqrt{2}}$$

$$A = -3.99$$

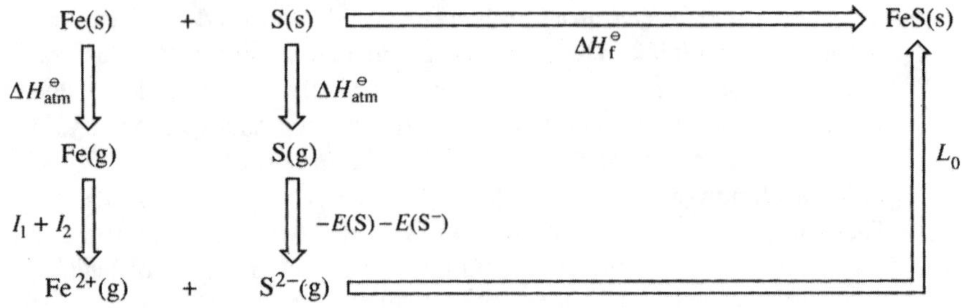

$$Fe(s) \quad + \quad S(s) \xrightarrow{\hspace{6cm}} FeS(s)$$

$$\Delta H_f^{\ominus}$$

$$\Delta H_{atm}^{\ominus} \downarrow \qquad \qquad \downarrow \Delta H_{atm}^{\ominus}$$

$$Fe(g) \qquad \qquad S(g) \qquad \qquad \qquad L_0$$

$$I_1 + I_2 \downarrow \qquad \qquad \downarrow -E(S) - E(S^-)$$

$$Fe^{2+}(g) \quad + \quad S^{2-}(g)$$

**FIGURE 1.65** Born-Haber cycle for the calculation of the electron affinity of sulfur.

19. An appropriate cycle is presented in Figure 1.65. Note that sulfur is a solid in its standard state not a gas.

   We wish to calculate $[E(S) + E(S^-)]$, which we can do if $L(FeS,s)$ can be calculated. Values of all the other terms in the cycle are known. If the lattice energy relationship of equation 1.15 is used, substituting $v = 2$, $Z_+$ $= 2$, $Z_- = 2$, $r_+ = 92$, $r_- = 170$, gives $L = -3\ 295$ kJ mol$^{-1}$. From the cycle in Figure 1.65:

$$E(S) + E(S^-) = -\Delta H_f^{\ominus}(FeS,s) + \Delta H_{atm}^{\ominus}(Fe,s) + (I_1 + I_2) + \Delta H_{atm}^{\ominus}(S,s) + L(FeS,s)$$

$$= (100.0 + 416.3 + 761 + 1\ 561 + 278.8 - 3\ 295)$$

$$= -177.9 \text{ kJ mol}{-}1$$

$$E(S) = 200, \text{ so } E(S^-) = -177.9 - 200 = -377.9 \text{ kJ mol}^{-1}$$

   The value of the electron affinity is the heat given out on the addition of an electron, so this implies that the enthalpy change for the addition of an electron to the $S^-(g)$ anion is endothermic:

$$S^-(g) + e^-(g) = S^{2-}(g); \qquad \Delta H_m^{\ominus} = 377.9 \text{ kJ mol}^{-1}$$

   Not surprisingly, it appears to be energetically unfavourable to add an electron to a negatively charged ion.

21. First, we use equation 1.15 to calculate a value for the lattice energy:

   For $NH_4Cl$, $v = 2$, $Z_+ = 1$, $Z_- = 1$, $r_+ = 151$ pm, $r_- = 167$ pm, giving $L = -679$ kJ mol$^{-1}$.

   From Figure 1.58:

$$P(NH_3,g) = -\Delta H_f^{\ominus}(NH_4Cl,s) + \Delta H_f^{\ominus}(NH_3,g) + \frac{1}{2}D(H\text{-}H) + I(H)$$

$$+ \frac{1}{2}D(Cl\text{-}Cl) - E(Cl) + L(NH_4Cl,s)$$

$$= (314.4 - 46.0 + 218 + 1\,314 + 122 - 349 - 679) \text{ kJ mol}^{-1}$$

$$= 894.4 \text{ kJ mol}^{-1}$$

The addition of a proton to the ammonia molecule is an exothermic process.

$$NH_3(g) + H^+(g) = NH_4^+(g); \Delta H_m^{\ominus} = -894.4 \text{ kJ mol}^{-1}$$

(The experimentally determined value is $871 \pm 15$ kJ mol$^{-1}$, which is quite good agreement!)

## ANSWERS TO CHAPTER 2

1. The spacings for these planes are $d_{100} = a$, $d_{110} = \dfrac{a}{\sqrt{2}}$ and $d_{111} = \dfrac{a}{\sqrt{3}}$
   and so the reflections occur in the order *100*, *110*, and *111*.

3. If the *100* and *110* reflections are absent then the crystal is likely to be face-centred cubic.

5. The $\sin^2\theta$ values do not have the ratio $1 : 2 : 3 : 4 : 5 : 6 : 8......$, or $1 : 2 : 3 : 4 : 5 : 6 : 7 : 8.......$ so the cell is not primitive or body-centred. If it is face-centred then the common factor A will be $0.0746 - 0.0560 = 0.0186$. Dividing the $\sin^2\theta$ values by the common factor gives the $h^2 + k^2 + l^2$ values and these are listed in Table 2.6 rounded to the nearest integer. $h\,k\,l$ only takes values where $h\,k\,l$ are either all even or all odd, so the data fits a face-centred cubic structure.

7. If the mass of the unit cell contents is $M$ and the unit cell volume is $V$ then the density, $\rho$ is given by

$$\rho = \frac{M}{V} = 2.17 \times 10^3 \text{ kg m}^{-3}$$

but

$$V = (563.1 \times 10^{-12})^3 \text{ m}^3$$

**TABLE 2.6**
**Powder diffraction data for NaCl**

| $\theta_{hkl}$ | $\sin^2\theta$ | $h^2 + k^2 + l^2$ | $hkl$ |
|---|---|---|---|
| 13°41′ | 0.0560 | 3 | 111 |
| 15°51′ | 0.0746 | 4 | 200 |
| 22°44′ | 0.1492 | 8 | 220 |
| 26°56′ | 0.2052 | 11 | 311 |
| 28°14′ | 0.2239 | 12 | 222 |
| 33°7′ | 0.2984 | 16 | 400 |
| 36°32′ | 0.3544 | 19 | 331 |
| 37°39′ | 0.3731 | 20 | 420 |
| 42°0′ | 0.4477 | 24 | 422 |
| 45°13′ | 0.5036 | 27 | 511, 333 |
| 50°36′ | 0.5972 | 32 | 440 |
| 53°54′ | 0.6529 | 35 | 531 |
| 55°2′ | 0.6715 | 36 | 600, 442 |
| 59°45′ | 0.7462 | 40 | 620 |

The mass of one mole of

$$NaCl = (22.99 + 35.45) \times 10^{-3} \text{ kg}$$

dividing by the Avogadro constant we get that the mass of one formula unit of

$$NaCl = \frac{(22.99 + 35.45) \times 10^{-3}}{6.022 \times 10^{23}} \text{ kg}$$

and if there are $Z$ formula units in one unit cell, then the mass of the unit cell contents is

$$M = \frac{Z(22.99 + 35.45) \times 10^{-3}}{6.022 \times 10^{23}} \text{ kg}$$

So

$$\rho = 2.17 \times 10^3 = \frac{Z(22.99 + 35.45) \times 10^{-3}}{6.022 \times 10^{23} \times (563.1 \times 10^{-12})^3}$$

and rearranging gives

$$Z = \frac{2.17 \times 10^3 \times 6.022 \times 10^{23} \times (563.1 \times 10^{-12})^3}{(22.99 + 35.45) \times 10^{-3}}$$

$Z = 3.99$, or $Z = 4$ (rounding to the nearest whole number)

9.

$$\rho = \frac{M}{V}$$

$$3.35 \times 10^3 = \frac{Z(40.08 + 15.99) \times 10^{-3}}{6.022 \times 10^{23}} \div (481 \times 10^{-12})^3$$

$$Z = \frac{3.35 \times 10^3 \times (481 \times 10^{-12})^3}{9.311 \times 10^{-26}}$$

$$Z = 4$$

11. From Equation (2.7), we know that $\sin^2\theta_{hkl} = \dfrac{\lambda^2}{4a^2}(h^2 + k^2 + l^2)$.
    A cubic-close packed structure has a face-centred unit cell. The first two reflections observed will therefore be the *111* and the *200*, with $h^2 + k^2 + l^2$ values of 3 and 4, respectively. Thus, $\sin^2\theta_{111} = \dfrac{3\lambda^2}{4a^2}$ and $\sin^2\theta_{200} = $

    $\dfrac{4\lambda^2}{4a^2}$, and $\dfrac{\lambda^2}{4a^2} = \sin^2\theta_{200} - \sin^2\theta_{111} = 0.181 - 0.136 = 0.045$
    $a = 363.5$ pm

    In a close-packed structure where the atoms are considered to be in contact, the radius of an atom, $r$, is $1/4$ of the length of the body-diagonal.

    $$r = \tfrac{1}{4}\sqrt{3}\, a$$

    $$r = (\tfrac{1}{4} \times 1.732 \times 363.5)\ \text{pm}$$

    $$r = 157.4\ \text{pm}$$

13. Both *111* and *222* are observed in P and F lattices, but the *111* is not present for I. The *001* is not observed for F, but would be present in a P unit cell. The Bravais lattice is thus F.

15. The peaks in the spectrum maximize at approximately −88, −93, −99, and −105 ppm. If you mark these values on the chart in Figure 2.26, then you will see that the best correspondence is to the four linkages: $Si(OAl)_3(OSi)$, $Si(OAl)_2(OSi)_2$, $Si(OAl)(OSi)_3$, and $Si(OSi)_4$.

17. The starting sample (a) clearly shows the presence of tetrahedral Al in the framework (peak at 61 ppm). After treatment with $SiCl_4$, (b) the amount of Al in the framework has been reduced considerably, but a very strong peak exists due to $[AlCl_4]^-$ (at l00 ppm) and a peak due to octahedral aluminium at 0 ppm. The first washings (c) remove $Na^+[AlCl_4]^-$ from the sample, and repeated washing (d) removes some of the octahedrally coor-dinated Al.

19. The first exotherm at about 60°C coincides with a sharp weight loss and is due to dehydration of the ferrous sulfate. The second exotherm at 90°C does not coincide with any weight loss and must therefore be due to a phase change (this is the melting temperature). The third exotherm, at about 600°C, again coincides with weight loss and is due to decomposi-tion.

## ANSWERS TO CHAPTER 3

1. This compound could be made from the elements in the correct stoichiometric proportions. The reactants would have to be very well mixed and the reaction vessel would have to be closed to prevent loss of the volatile sulfur (see SmS preparation).

3. Advantages — more homogeneous product with a smaller range of grain sizes; thin layers can be formed for microcapacitors. Disadvantages — the process is more complex and would be expensive to set up on an industrial scale.

5. β-TeI is metastable and contains Te in an unusual oxidation state. A method that employs temperatures of this order and can produce com-pounds in unusual oxidation states is hydrothermal synthesis.

7. The tetrapropylammonium ion acts as a template and directs the formation of the network of channels around the propyl ligands. Heating in air oxidizes the cation and leaves the zeolite framework intact.

9. Because the reaction is exothermic, it is driven to the left by raising the temperature. Thus, the crystals grow at the hotter end of the tube.

## ANSWERS TO CHAPTER 4

1. For the Fermi level in sodium, $E = 4.5 \times 10^{-19}$ J and the mass of an electron is $9.11 \times 10^{-31}$ kg. This gives

$$4.5 \times 10^{-19} = \frac{1}{2} \times 9.11 \times 10^{-31} \times v^2$$

$$v = (2 \times 4.5 \times 10^{-19}/9.11 \times 10^{-31})^{\frac{1}{2}}$$

$$= 9.9 \times 10^5 \text{ m s}^{-1}$$

3.  (a) $N = 10^{-12} \times (2 \times 9.11 \times 10^{-31} \times 4.5 \times 10^{-19})^{3/2}/3\pi^2 \times (1.055 \times 10^{-34})^3$

    $= 2.135 \times 10^{16}$

    (b) $2.135 \times 10^{22}$

    (c) 0.2

Each level can take two electrons and a crystal of $N$ atoms of sodium has $N$ electrons to fill the band. As you can see, the agreement between the number of filled levels predicted by this very simple theory and the number needed to accommodate the available electrons is very good. Note also that this question illustrates how the energy level spacing increases as the electrons are confined to a smaller and smaller volume.

5.  Si, Ge

7.  Carborundum like silicon and germanium has $4N$ valence electrons for a crystal of $N$ atoms. The tetrahedral diamond structure will be favoured because all $4N$ electrons will then be in bonding orbitals and the energy is lower than in the higher coordination structure.

## ANSWERS TO CHAPTER 5

1.  $n_S \approx Ne^{-\Delta H_S/2RT}$ where $\Delta H_S = 200$ kJ mol$^{-1}$ and R = 8.314 J K$^{-1}$ mol$^{-1}$

---

**TABLE 5.10**
**Schottky defect concentration in MX compound at various temperatures**

| Temperature °C | Temperature K | $n_S/N$ | $n_S/\text{mol}^{-1}$ |
|---|---|---|---|
| 27 | 300 | $3.87 \times 10^{-18}$ | $2.33 \times 10^6$ |
| 227 | 500 | $3.57 \times 10^{-11}$ | $2.15 \times 10^{13}$ |
| 427 | 700 | $3.45 \times 10^{-8}$ | $2.08 \times 10^{16}$ |
| 627 | 900 | $1.57 \times 10^{-6}$ | $9.45 \times 10^{17}$ |

---

3.  Increasing the impurity levels does not affect the intrinsic (left-hand side) of the graph. It does, however, increase the value of $\sigma$ (and thus of $\ln\sigma$) in the extrinsic region. As the activation energy, $E_a$, for cation movement stays the same, the slope of the graph is unchanged (see Figure 5.46). The purer the crystal, the lower the transition temperature at which thermally generated defects take over.

5.  The oxide has more highly charged ions ($M^{4+}$ and $O^{2-}$) than a fluoride. This means that the coulombic interactions will be stronger and thus the

energy of defect formation will be higher (see Table 5.1), so reducing the defect concentration at a given temperature.

7. The anion at the body-centre of the unit cell in Figure 5.3(c) is surrounded by six anions at distance $\dfrac{a}{2}$ and by four cations at a distance of $0.43a$.

The interstitial site at the body-centre of the unit cell in Figure 5.3(a) is surrounded by *eight* anions at a distance of $0.43a$ and by *six* cations at a distance of $\dfrac{a}{2}$.

For the normal anion site:

$r = 0.43 \times 537 \times 10^{-12}$ m and $Z = +2$, for interaction with four cations. $r = 0.5 \times 537 \times 10^{-12}$ m and $Z = -1$, for interaction with the six anions.

$$E = - \left( \frac{2.31 \times 10^{-28}\,\text{J m}}{537 \times 10^{-12}\,\text{m}} \right) \left( \frac{4 \times 2}{0.43} + \frac{6 \times (-1)}{0.5} \right)$$

$$= - (4.302 \times 10^{-19}\ \text{J})\ (6.605)$$

$$= - 2.84 \times 10^{-18}\ \text{J}$$

For the interstitial site:

$$E = - (4.302 \times 10^{-19}\ \text{J}) \left( \frac{8 \times (-1)}{0.43} + \frac{6 \times 2}{0.5} \right)$$

$$= - (4.302 \times 10^{-19}\ \text{J})\ (5.395)$$

$$= - 2.32 \times 10^{-18}\ \text{J}$$

The energy of defect formation is the difference in energy between the two sites and so is given by:

$$(-2.32 \times 10^{-18}\ \text{J}) - (-2.84 \times 10^{-18}\ \text{J}) = 5.2 \times 10^{-19}\ \text{J}$$

The experimental value for fluorite is given in Table 5.1 as $4.49 \times 10^{-19}$ J. This calculation gives a very good level of agreement considering that we have ignored the more distant interactions, including internuclear repulsion, lattice vibrations, and lattice relaxation!

9. $I^-$, $S^{2-}$, $Se^{2-}$, and $Te^{2-}$ are all polarizable anions.

11. Unit cell volume is $(428.2\ \text{pm})^3 = 7.8513 \times 10^{-29}$ m³. Mass of contents for *iron vacancies:*

$$[(4 \times 55.86 \times 0.910) + (4 \times 16.00)]/(N_A \times 10^3)\ \text{kg,}$$
$$\text{giving a density of } 5.6534 \times 10^{-3}\ \text{kg m}^{-3}$$

Mass of contents for *oxygen interstitial*:

$$[(4 \times 55.85) + (4 \times 16.00 \times 1/0.910)]/(N_A \times 10^3) \text{ kg}$$
$$\text{giving a density of } 6.2126 \times 10^3 \text{ kg m}^{-3}$$

Comparing these theoretical values with the experimental value, we again see that the evidence supports an iron vacancy model.

13. A crystal containing F-centres contains anion vacancies. We would expect, therefore, that the density would be *lower* than that of the colourless crystal.

15. **Titanium vacancies**: There are 8 at the corners $(8 \times 1/8) = 1$, and $(2 \times 1/2) = 1$ on cell faces.

   **Titanium ions**: cell edges, $(4 \times 1/4) = 1$. Cell faces, $(8 \times 1/2) = 4$ on the top and bottom. There are 5 ions contained within the cell boundary, making 10 in total. The titanium stoichiometry of the unit cell is obviously representative of the whole structure: of the 12 sites, 10 are occupied, and 2 are vacant. This is also true for oxygen.

   **Oxygen vacancies**: cell faces, $(4 \times 1/2) = 2$.

   **Oxide ions**: cell faces, $(8 \times 1/2) = 4$. Cell edges, $(8 \times 1/4) = 2$. There are 4 ions contained within the cell boundary, making 10 in all.

17. The structure has Ti vacancies — every fifth Ti is missing. For every absent $Ti^{2+}$ ion there must be two $Ti^{3+}$ present or one $Ti^{4+}$.

19. Figure 5.50 depicts the shear structure with a unit cell added. Within the boundary, one group of four edge-sharing octahedra and four $[WO_6]$ octahedra are present. The formula is thus $W_4O_{11} + 4WO_3 = W_8O_{23}$.

# ANSWERS TO CHAPTER 6

1. Polyphenylenevinylene has a $\pi$ system delocalised over the benzene ring. It is likely that this is also delocalised over the conjugated double bond and hence to the benzene ring of the next unit. Thus, like polyacetylene, this polymer will have a delocalised $\pi$ system, but it will include the $\pi$ ring orbitals.

3.

$$8(CH)_n + 9\delta nHClO_4 = 8[(CH)^{\delta+}(ClO_4^-)_\delta]_n + \delta nHCl + 4\delta nH_2O \qquad (6.5)$$

5. Fluorine tends to attract electrons to itself, and so $TCNQF_4$ is likely to be a stronger electron acceptor than TCNQ. The conducting properties of solids like TTF-TCNQ arise from partial transfer of electrons from one type of molecule to the other. $TCNQF_4$ is a sufficiently good acceptor to enable transfer of one electron per unit, thus completely filling the conduction band in one stack and completely emptying it in the other stack.

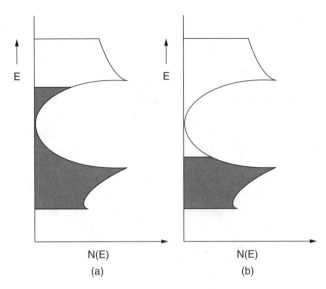

**FIGURE 6.15** Band structure of graphite doped with (a) an electron donor and (b) an electron acceptor.

Therefore, HMTTF-TCNQF$_4$ has no partially full bands and hence is not a metallic conductor.

7. Figure 6.15 illustrates the band structure of graphite intercalated with (a) an electron donor and (b) an electron acceptor.

## ANSWERS TO CHAPTER 7

1. The chart in Figure 2.26 shows that the most likely Si environment for zeolite A is Si(OAl)$_3$(OSi). However, we know that the Si/Al ratio is 1, so this coordination is not possible without systematically breaking Loewenstein's rule. It was spectroscopic work on this structure (and on zeolite ZK-4) that eventually confirmed the Si(OAl)$_4$ structure of zeolite A with strict alternation of Si and Al and led to the extended ranges shown in the chart.

3. This system demonstrates reactant shape-selective catalysis. The branched hydrocarbons are too bulky to pass through the pore openings in the catalyst.

5. Zeolite A (Ca) shows reactant selectivity. The straight-chain *n*-hexane can pass through the windows and undergo reaction but the branched-chain 3-methylpentane is excluded. The selective cracking of straight-chain hydrocarbons in the presence of branched chains is an important industrial process known as **selectoforming**, which improves the octane number of the fuel.

7. As the Si/Al ratio increases, the —OH stretching frequency falls. This indicates a decrease in the covalent bond strength making ionization of $H^+$ easier (i.e., an increase in acid strength).

## ANSWERS TO CHAPTER 8

1. In $Mn^{2+}$, an electron can only go from one $3d$ level to another if it changes its spin. Transitions in which an electron changes its spin are forbidden and so give rise to weak spectral lines.

3. ZnS is a semiconductor with a full valence band and empty conduction band. When the phosphor is illuminated, electrons are promoted to the conduction band. Because the orbitals in this band are delocalised, the energy can easily be transferred to other parts of the crystal, particularly to the dopant atoms.

5. Silicon is an indirect band gap solid with an available non-radiative pathway from the conduction band to the valence band. In photo-voltaic cells, electrons are promoted from the valence band to the conduction band and are then used to do electrical work. The promoted electrons do not return directly to the valence band either by emitting energy or by a non-radiative pathway. In LEDs, the return of the electrons to the valence band by emitting light is important. This return has a low probability because of the indirect band gap and the electrons use the non-radiative pathway instead. Promotion to the conduction band in the solar cell will also be of low probability, but no competing non-radiative route is available.

7. The photonic band gap wavelength increases as the size of the spheres increase, so that it will be larger for the longer wavelength orange-red colours.

## ANSWERS TO CHAPTER 9

1. Because the Mn atoms are further apart, the overlap of the $3d$ orbitals will be less. The $3d$ band will therefore be narrower than in manganese metal. With a narrower band, there is a larger interelectronic repulsion, and a state with a number of unpaired spins comparable to the number of atoms becomes favourable. The alloy is thus ferromagnetic.

3. The $Zn^{2+}$ and half the $Fe^{3+}$ ions are on octahedral sites with spins aligned, and the remaining $Fe^{3+}$ ions are on tetrahedral sites aligned antiparallel. The net moment of the $Fe^{3+}$ ions is zero. As all the electron spins are paired in $Zn^{2+}$ ions, no overall magnetic moment exists, and the compound is antiferromagnetic.

5. The effect of deuterium substitution suggests that the hydrogen atoms are displaced when the ferroelectric phase is formed.

## ANSWERS TO CHAPTER 10

1.  In the analogy, each person represents a Cooper pair and the linking of arms denotes the overlap between the pairs, giving an ordered system. Because of the ordered collective motion, scattering from defects — tripping over potholes — cannot take place! The analogy breaks down in that it allows for overlap only between adjacent pairs rather than over large numbers; in addition, the loss of superconductivity leads to the scattering of individual electrons, not of individual Cooper pairs.

3.

| Material | Symmetry of Order Parameter | Spin State |
| --- | --- | --- |
| Metals and Alloys | $s$-wave | singlet |
| Superconducting Cuprates | $d$-wave | singlet |
| Ferromagnetic Superconductors | $p$-wave | triplet |

# Index